RESTORING
COLORADO RIVER ECOSYSTEMS

A Troubled Sense of Immensity

Robert W. Adler

ISLANDPRESS

Washington • Covelo • London

ISLAND PRESS is a trademark of The Center for Resource Economics.

Library of Congress Cataloging-in-Publication Data

Adler, Robert W., 1955-
  Restoring Colorado River ecosystems : a troubled sense of immensity / Robert W. Adler.
    p. cm.
  Includes bibliographical references.
  ISBN-13: 978-1-59726-056-5 (cloth : alk. paper)
  ISBN-13: 978-1-59726-057-2 (pbk. : alk. paper)
  1. Restoration ecology—Colorado River (Colo.-Mexico) I. Title.
  QH104.5.C6A35 2007
  333.73'153097913—dc22

                                                        2006101562

Printed on recycled, acid-free paper

Manufactured in the United States of America

10 9 8 7 6 5 4 3 2 1

ABOUT ISLAND PRESS

Island Press is the only nonprofit organization in the United States whose principal purpose is the publication of books on environmental issues and natural resource management. We provide solutions-oriented information to professionals, public officials, business and community leaders, and concerned citizens who are shaping responses to environmental problems.

Since 1984, Island Press has been the leading provider of timely and practical books that take a multidisciplinary approach to critical environmental concerns. Our growing list of titles reflects our commitment to bringing the best of an expanding body of literature to the environmental community throughout North America and the world.

Support for Island Press is provided by the Agua Fund, The Geraldine R. Dodge Foundation, Doris Duke Charitable Foundation, The Ford Foundation, The William and Flora Hewlett Foundation, The Joyce Foundation, Kendeda Sustainability Fund of the Tides Foundation, The Forrest & Frances Lattner Foundation, The Henry Luce Foundation, The John D. and Catherine T. MacArthur Foundation, The Marisla Foundation, The Andrew W. Mellon Foundation, Gordon and Betty Moore Foundation, The Curtis and Edith Munson Foundation, Oak Foundation, The Overbrook Foundation, The David and Lucile Packard Foundation, Wallace Global Fund, The Winslow Foundation, and other generous donors.

The opinions expressed in this book are those of the author and do not necessarily reflect the views of these foundations.

RESTORING
COLORADO RIVER ECOSYSTEMS

To Michele, for her unending love and support

Contents

CHAPTER FIVE

Down the Great Unknown: Environmental
Restoration in the Face of Scientific Uncertainty

CHAPTER SIX

Casting of the Lots: Conflicting Methods and
Goals in Environmental Restoration

CHAPTER NINE

Acknowledgments

This book was many years in the writing, during which many people showed extreme patience and support, and many others provided invaluable assistance, perspectives, and information. I want to acknowledge all of their help with deep appreciation, whether I remember to name them or not, while taking full responsibility for the ultimate contents of the book. I apologize in advance to those whom I unintentionally omit from the following list of acknowledgments.

Highest in the patience department comes my family, especially my wife, Michele, my son, Woody, and my daughter, Sierra, who have listened to my ramblings about this book for all too long, and endured my extra hours of work and other inconveniences along the way. As partial compensation, they did get to accompany me on some of my wanderings to wonderful places along the Colorado River. Our dog Habibi listened to some of my discourse about the book during our trail runs in the Wasatch Mountains, but he probably paid less attention. Brian Kamm and other running friends at least had to pretend they were interested.

My editors at Island Press also showed great patience with my procrastination and technical foibles. More importantly, they had tremendous judgment and advice about how to improve the final product. Barbara Dean encouraged and supported this project from the outset, and helped me to reduce such an immensity of information and ideas about such a complex set of issues into a more manageable and I hope readable book. Barbara Youngblood and Erin Johnson helped with the clarity of the writing and the seemingly endless but important technical details. Liz Wilson and her staff expertly shepherded the manuscript through copyediting and production, and Joy Drohan was a very careful and helpful copyeditor, who caught many of my errors.

Many people associated with the University of Utah, S. J. Quinney College of Law, provided research help, support, and guidance. John Bevin and others on the staff at the S. J. Quinney Law Library tracked down dozens of documents through interlibrary loan. Many works were available in the superb rare documents collection of the University of Utah's Marriott Library. My assistants Barbara McFarlane and Sandra Fatt helped in countless ways, with both research and production. Student research assistants included Tracy Bennett-Hecht, Craig Condie, Jeremy Eyre, Heather Green, David McArthur, Andrew Hartsig, Jeff Merchant, Zach Peterson, John Ruple, Paul Sacksteder, Rod Smith, and Kirsten Uchitel. I am also indebted to the many students who participated in my "Protection of water systems" and "Colorado River" seminars, many of whom suggested ideas during class discussion and submitted research papers that helped me with issues addressed in the book.

Many others provided documents, leads, interviews, bibliographies, and other information that helped me to identify and sort through the massive amount of information available about the Colorado River and ongoing restoration efforts. For reasons of confidentiality (in some cases), I will list first the categories of help I received, and then name individuals who provided assistance in one or more of those ways. Some individuals granted interviews or helped me tour key locations and facilities along the Colorado River. Others suggested documents, other key information, and research leads. Others reviewed one or more drafts of this book, or of previous writings from which I drew ideas in writing the book. I was given the privilege of testing some of the ideas that appear herein at lectures and other events sponsored by the University of Utah, S. J. Quinney College of Law; the Rocky Mountain Mineral Law Foundation Natural Resources Law Teachers Institute; the *Virginia Environmental Law Journal* and the University of Virginia College of Law; the Environmental Law Institute; and Chapman University College of Law. Agencies and organizations that provided additional assistance and information include the U.S. Bureau of Reclamation, the U.S. Fish and Wildlife Service, the U.S. Bureau of Land Management, the National Park Service, the Utah Department of Natural Resources, the Arizona Department of Game and Fish, the Colorado River Basin Salinity Control Forum, Environmental Defense, Defenders of Wildlife, Glen Canyon Institute, Glen Canyon Action Network, Sierra Club, Living Rivers, Friends of Lake Powell, Pacific Institute, Upper Colorado River Commission, and Western Area Power Authority.

Included among the many generous people who contributed to this book in one or more of those ways, again with my sincere apologies to those I may have omitted, are: Bert Anderson, Larry Anderson, Craig Anthony Arnold, Jack Barnett, Kathleen Blair, Jeff Brady Baugh, Kay Brothers, Gary Bryant,

Jim Cherry, Michael Cohen, Wayne Cook, Dan Crabtree, Julian DeSantiago, Debbie Felker, Karl Flessa, Lisa Force, Dave Foreman, Suzanne Fouty, Steve Gloss, Barry Gold, Reed Harris, Mike Hawes, Barbara Hjelle, Richard Hobbs, Pam Hyde, Richard Ingebretsen, Rick Johnson, Robert Keiter, Richard Kettenstette, Dennis Kubly, Owen Lammers, Geri Ledbetter, Henry Maddux, Jan Matusek, Chuck McAda, Steve McCall, Jim McMahon, Theodore Melis, LeGrand Nielson, David Orr, Mary Orton, Clayton Palmer, Randy Peterson, Jennifer Pitt, Ginger Reeves, Kirsten Rowell, David Ruitter, Jack Schmidt, Dave Sharrow, Tom Shrader, Ron Simms, Michele Straube, Gary Taylor, Ron Thompson, Shannon Traub, Dave Trueman, Tim Ulrich, Dave Wegner, John Weisheit, and Barry Wirth.

Preface

The Colorado River is a favorite topic of western authors, and for good reason. The wonders and the dangers of the river and its canyons are documented in the travel journals of Fathers Domínguez and Escalante, James Pattie, Lieutenant Joseph C. Ives, and Major John Wesley Powell. Zane Grey used the mystique and majesty of the Colorado as the backdrop to his classic romance *The Rainbow Trail*. Scientists document and explore all aspects of the river's secrets in shelves of books, journal articles, and technical reports. The history of the Colorado, its changes, its battles, and its challenges are chronicled in Wallace Stegner's *Beyond the Hundredth Meridian*, Donald Worster's *Rivers of Empire*, Marc Reisner's *Cadillac Desert*, Philip Fradkin's *A River No More*, and Paul Martin's *A Story That Stands Like a Dam*. Photographer Elliott Porter (*The Place No One Knew*), singer/songwriter Katie Lee (*All My Rivers Are Gone*), and others lamented the demise of some of the river's most treasured features. Edward Abbey lampooned environmental foibles in the region in *The Monkey Wrench Gang* and *Desert Solitaire*. Fradkin commented that "[p]robably no other single natural feature in this country has attracted so many written words, mostly on the technical and legal levels."[1] John McPhee added: "Anyone interested, for whatever reason, in the study of water in the West will in the end concentrate on the Colorado, wildest of rivers, foaming, raging, rushing southward—erratic, headlong, incongruous in the desert." [2]

Unlike most of that vast literature, this book is primarily about neither the past nor the present. It is about the *future* of the Colorado River and its ecosystems and human users, however uncertain those intertwined fates may be. It is less about what we have done *with* and *to* the river over the past century and more about what we might do *for* it in the next. It is less about utility and harm, and more about restoration.

Of course, it is not possible to address the river's future without putting it in the context of the past and the present. We need to know what the river was before we intruded, and what it has become now. We need to know how we have changed this remarkable place, and why. Above all, we need to understand what we have lost, and what we have gained in return. The first four chapters of this book depict the condition of the river through the eyes of some of the key species and other ecosystem resources affected by our past actions. I use these resources largely as focal points to highlight the broader changes to the ecosystem as a whole and why they occurred. No one intended to drive the razorback sucker to the brink of extinction. It is essential to understand why it does cling to that precipice, however, before restoration is possible. For these purposes, to some degree this book traces the paths of its many superb antecedents, with due respect and hopefully proper attribution to the many great authors and other experts on whom I rely. I discuss the past, however, only to the extent necessary to move on to the future.

Chapters 5–9 address our current efforts to restore the river and its adjacent habitats. Hundreds of smart, dedicated, hard-working, and well-intentioned people labor to heal the river, its species, and their habitats. Those actions affect the lives and livelihoods of a large number of people and groups, whose voices and values may differ greatly. These many individuals and organizations do their jobs under difficult circumstances. Some have been heroic in their efforts to achieve what has seemed impossible in the past. Others strive to balance, as if on the edge of the river's sharp cliffs, dozens of competing and conflicting interests and concerns. Despite all of those good intentions and endeavors, it is my job to raise some of the difficult questions about their work. Criticism is not my underlying intent, but critical analysis is. I hope to do so with respect and without any suggestion of fault by specific individuals. My purpose is to help inform and perhaps to inspire and improve the future of those efforts.

My main aspiration, however, is not to describe and evaluate the many details of restoration program structure and implementation. Nor is it my goal to explain or evaluate in great detail the scientific complexities inherent in river restoration, and as a student of law and policy rather than science, I am not qualified to do so. Details help to organize and to focus the inquiry, but I want to ask broader questions about ecosystem restoration choices and goals, and how we might achieve them. What do we mean by "restoration"? For a river that has changed dramatically over the course of geological and ecological history, to what previous state do we seek to restore the system, and for what purposes? What do we now reap from the river (water, power, recreation) that we are willing to give back in order to recover what has been lost?

Or are there other ways to attain those benefits that allow us to have it both ways? Who should make those hard choices, and how? Despite a veneer of laws, policies, and directives designed to guide those decisions, in most cases the hardest questions remain unanswered. In addition to restoring the river, I will urge that we need to restore the open process by which those difficult, value-laden choices are made.

Once we break this impasse and decide what we want to do, we are left with equally daunting questions of how. Do we tinker with how we operate what is now an artificially plumbed river, or do we take more dramatic steps to dismantle some of the hydraulic hardware that dots the system from its headwaters to its end in the Sea of Cortéz? To some, the "granddaddy" of all Colorado River restoration efforts would be to decommission Glen Canyon Dam, to set the now-shackled river free, at least through the Grand Canyon. The idea is difficult, bitterly controversial, and a lightning rod for attention. But it is impossible to separate what happens at (or to) the dam from other restoration efforts both upstream and down. They are all closely connected, as are the various resources associated with the river itself.

At the same time, pointed ideas often stimulate the most useful debates, whether or not they are actually adopted. Forcing people to answer the most difficult questions often generates the most useful results. Because some of the value-laden questions necessary to restore ecosystems are so perplexing, and because they force us to make choices, we defer or avoid them or pretend falsely that we can have it all. The more important part, I believe, is to ask those hard questions.

Although this book is mainly about restoration of Colorado River ecosystems, it is also about uncertainty, conflict, competing values, and the nature, pace, and implications of environmental change. It is about our place in the natural environment, and whether there are limits to that presence we ought to respect. It is about our responsibility to the ecosystems we live in and use. At bottom, it is about environmental restoration more generally—what I will argue is the third major strategy of the modern environmental movement, and perhaps the most difficult of all to design and accomplish. Similar programs are under way to restore watersheds and ecosystems all over the country. Those efforts struggle with the same core issues as we do for the Colorado.

That broader inquiry suggests some even more fundamental lessons about the degree to which we can hope to succeed in restoring seriously degraded large ecosystems. Around the country, there are dozens of examples of successful restoration at a much smaller scale. Urban streams have been freed from their concrete prisons. Once straightened rivers-turned-canals meander again through their original channels and floodplains. Formerly denuded riverbanks have been replanted with native riparian species. Native fish species have been reintroduced to their indigenous habitats. Smaller projects may involve difficult technical and other problems, but they are within our conceptual and political reach.

Replicating those successes in places such as the Colorado River, the Columbia River, and the Everglades is another matter. Restoring such vast ecosystems requires us to think and act on a much larger scale. We must deal with much larger sets of scientific uncertainties and many more variables. Much more is demanded of the army of scientists who must predict, experiment, and rethink their hypotheses amid incessant political jockeying and interference. Significantly more fundamental choices are needed about conflicting values and trade-offs. These projects demand that we consider ecosystems, issues, and solutions on a larger scale and with more open minds than we have in past.

That realization explains the subtitle of this book. I am professionally trained and accustomed to dealing with complex information on a massive scale. I have litigated cases involving hundreds of thousands of pages of documents. Yet I was still overwhelmed by the immensity of the information needed to obtain even a good general understanding of the intricate, intertwined web of issues involved in restoring the Colorado. Relevant information is generated by disciplines ranging from my own field of law to a dozen or so branches of science to economics, politics, anthropology, history, archaeology, and sociology. Throw in a dose of philosophy and even some religion and you begin to approach the full set of ideas invoked to debate the issues involved. Most disturbing of all, but not surprisingly, even if one can absorb and assimilate that jumble of information, it all seems to point in different directions. Potential solutions to some problems might create or exacerbate others. Some theories suggest one set of actions, while equally plausible hypotheses advise the opposite.

No wonder I decided on the book's subtitle the moment I stumbled on its source quote at the visitors center at the North Rim of the Grand Canyon. To pioneering 19th century geologist Clarence Dutton, the canyon's scale posed almost dizzying problems for a human brain that was trained to see and to understand the world from an entirely different perspective: "Dimensions mean nothing to the senses, and all that we are conscious of in this respect is *a troubled sense of immensity.*"[3]

Yet some action is needed, and decisions must be made in the face of conflicting, incomplete, or even nonexistent information. Choices are essential. We are at a multidirectional crossroad in the river's future, facing a set of decisions that transcends technical judgments and legal choices to an honest confrontation with the cacophony of competing voices and values about what the river's future *should* be. The most that I can hope to accomplish in these pages, and if I succeed that will be quite enough, is to convince readers that there are no easy answers, no simple solutions, and certainly no perfect ones. Success will lie in our ability to work *with* uncertainty rather than trying either to ignore it or to overcome it. We need to make some very difficult choices, because we cannot have it all.

CHAPTER ONE

Introduction: Retaking Old Ground

With the power of modern engineering, we increased dramatically our ability to change the natural world. Especially in the European tradition, and in some interpretations of Judeo-Christian liturgy, some believe it to be our very *obligation* to mold nature to our benefit.[1] Those of us who are wealthy enough to enjoy modern conveniences protect ourselves from the elements in climate-controlled buildings. We do not feel the effects of famine and drought because of mass-produced and chemically preserved food delivered by truck and train, and stored water piped through indoor plumbing. We blast through mountains to route our highways, fill in swamps to grow our food and to build our cities, and even change the course of mighty rivers and the destiny of colossal canyons.

Thus, it was with supreme confidence that late 19th century engineer Robert Brewster Stanton promoted his proposed railroad along the Colorado River through the length of the Grand Canyon: "That the proposed road is feasible and practicable, and at a reasonable amount of cost, is beyond question." He was probably correct, given that railroads had been built in even more challenging locations by that time, although the railroad through the Grand Canyon was never built.[2] But other, even more profound engineered changes were not too far distant.

The massive Glen Canyon Dam seems almost more permanent than the sandstone cliffs into which it is anchored. Engineers designed the dam with massive blocks of concrete, cured with refrigerated coils to ensure its strength. The natural canyon consists largely of more fragile rock, sandstone whose very aesthetic fame is caused by erosion. The dam, however, is only as strong as its host canyon. As the walls erode, so goes the dam. Regardless of whether it lasts another decade, another century, or another millennium, it cannot last forever. After all, the hydrological power it restrains helped to cut the Grand Canyon out of the Colorado Plateau in the space of just several million years, a "flicker of a nod" in geological time.[3]

More long-lasting than the dam itself may be the changes it has brought to the Colorado River, especially combined with the effects of the dozens of other dams, water diversions, dikes, levees, and other artificial changes to the river, its channel, its banks, its tributaries, and its biota. We molded (or more accurately, remolded) the river to serve the needs of human communities, but with significant impacts to the river and its ecosystems. Now, many scientists, public officials, and citizens hope to reshape nature in a different way, to *restore* portions of the river and its ecosystems. Environmental restoration on the scale of the Colorado River basin, or even significant portions of the basin, is an immensely challenging endeavor. In addition to an almost overwhelming array of technical difficulties, it is fraught with perplexing questions about the appropriate goals of restoration and the extent to which environmental restoration must be balanced against environmental changes designed to sustain the human economy.

This book explores those questions and challenges. It evaluates the relationships among the laws, policies, and institutions governing use and management of the Colorado River for human benefit, and those designed to protect and restore the river and its environment. It examines and critiques the challenging interactions among law, science, economics, and politics within which restoration efforts must operate, often in the face of tremendous scientific uncertainty. Ultimately, this book proposes that the concept of "restoration" must include more than efforts to rehabilitate individual patches of habitat or specific features of the river. Restoration should include changes in how we use and manage the resources of the river, or ways to replace those resources, to strike an appropriate balance between human and environmental needs (to the extent that those goals can be separated), for the Colorado River and elsewhere.

This chapter introduces the major challenges inherent in an undertaking as complex as restoring the ecosystems of the Colorado River. First, however, although these issues will be explored in more detail in chapters 2–4, we need some background on how we have changed the river, and why.

Taming the Colorado—How We Changed the River to Meet Human Needs

The Colorado River plummets from its headwaters more than 14,000 feet above sea level through the magnificent canyons of Utah and Arizona (figure 1.1). It continues through the deserts of Arizona, California, Baja California, and Sonora to its languid end in the Colorado River delta and the Sea of Cortéz (Gulf of California). Wild, volatile, and unpredictable, the natural river varied dramatically from its headwaters to the delta, from year to year, and from season to season. The river once raged through its canyons every spring, sending hundreds of thousands of

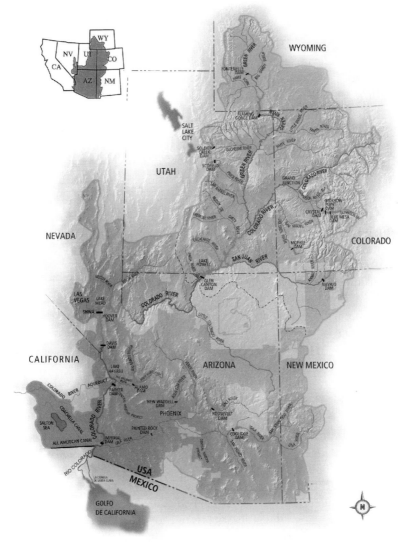

Figure 1.1. Colorado River basin. *Courtesy of the U.S. Bureau of Reclamation*

cubic feet per second of snowmelt from the Rockies to the sea, only to subside to a comparative trickle during the fall and winter. Each summer, after the snowmelt receded, sand and gravel bars emerged and formed habitats used by native fish. Those unique conditions supported a system with one of the world's largest percentages of endemic fish, species found nowhere else on the planet. In its lower reaches, the river often spilled out over a broad floodplain, nourishing thousands of acres of wetlands that hosted lush vegetation and a bounty of waterfowl and other wildlife. The river also carved the sublime, rugged, and remote canyons for which the Colorado Plateau earned its scenic fame.[4]

But while the wild Colorado was bountiful and beautiful, it was hardly friendly to human users. Hohokam Indians built dams and irrigation canals similar in concept, although not in size and impact, to those built by European settlers centuries later.[5] The Hohokam structures were relatively easy to build, but also easy for the river to wash away during a spring surge. Some early southwestern explorers and traders traveled the waters of the lower Colorado in sternwheeler steamboats, but at times they ran their craft in reverse so the wheel could plow rather than ply through the river's frequently shifting sandbars. Upriver, brave (or foolish) trappers and explorers plummeted through roaring whitewater canyons in flimsy boats and canoes. Farmers and ranchers found plenty of water during the spring runoff, but only a silt-laden trickle during summer and fall when they needed it most for crops and pastures.

That once volatile Colorado is now relatively tamed by what has been called a massive system of plumbing, designed to provide more reliable water supplies to farmers and urban users, to control flooding, and to generate hydroelectric power. Dams include Glen Canyon, Hoover, Davis, Parker, Imperial, Laguna, and Morelos on the main stem down to the Mexican border; Fontanelle and Flaming Gorge on the Green; the Aspinell Project on the Gunnison; Navajo on the San Juan; Coolidge on the Gila; and Saguaro, Canyon, Apache, and Theodore Roosevelt on the Salt. These monuments—and hundreds of smaller dams and diversions throughout the system—leveled the river's natural volatility and transformed it into a step-series of placid lakes joined by remnants of flowing river. By storing the region's spring runoff behind dams, communities could divert for human use waters that once flowed to the sea. Farms and communities now could use the river's liquid gold when and where it was needed, not when and where it happened to flow. Storing water behind dams also protected communities from the annual floods that once discouraged extensive settlement along the river's broad floodplains.

The Colorado River's artificial plumbing system contributes immensely to the human economy and lifestyle in the arid Southwest.[6] It provides water and electricity for the economies of seven U.S. states and two in Mexico. Water stored behind dams within the basin and diverted to users both in and out of the watershed now serves population centers with more than 30 million people, and irrigates more than 3 million acres of land. That same water drives turbines that generate 11.5 billion kilowatt-hours of hydroelectric power a year, enough to supply about 13 million households. (Average annual household consumption in the United States is 908 kilowatt-hours.[7]) Hydroelectric plants are particularly suited to generate power during periods of peak demand, such as hot summer days in the Southwest when air conditioners run almost constantly. This is power that comes without the need to dig for coal or drill for oil, without the smoke that fouls human lungs, and without the greenhouse gases that transform the global climate. Hydroelectric power is *sustainable* in that it will continue so long as the river flows.

If those benefits were not enough, the massive reservoirs behind the dams built to fuel the region's economic growth are also playgrounds for millions of visitors who come to boat, camp, fish, and enjoy the splendor of the Colorado's canyons, and who pour buckets of money into local and regional economies. Land once subject to flooding is now protected behind dikes and levees because the dams hold back and release water gradually, except during exceptionally wet periods when storage capacity is insufficient. Colorado River reservoirs can store an impressive 60 million acre-feet (maf) of water, four times the average annual flow of the river. (An *acre-foot*, intuitively enough, is the volume of water necessary to cover an acre of land to a depth of one foot [approximately 326,000 gallons]). The vast majority of that capacity is in Lake Mead and Lake Powell, the two largest reservoirs in the United States, with a combined maximum storage of more than 50 maf.

But those benefits come with a serious price. From its headwaters in the mountains of Wyoming and Colorado to its delta in the Sea of Cortéz—where the river flows at a tiny fraction of its former majesty—human change has taken its toll on the biological and aesthetic resources of the Colorado River. Hundreds of miles of formerly flowing river now lie below artificial reservoirs, and the dams fragment the river both physically and biologically. With a few exceptions in the headwaters, upstream of any dams, even those stretches that remain "river" have been changed significantly. In most places, spring flood flows are significantly smaller and flows from the dams are much higher and more uniform during other parts of the year. Especially below the Hoover Dam, hundreds of miles of river channel are imprisoned within levees or constrained by artificially "armored" rock banks. Water released from the dams is typically much colder and more uniform in temperature than in the natural river, and is starved of the sediment and nutrients that used to flow downstream. This loss of sediment changed the former patterns of bars and eddies that provide habitat for native fish. It also robbed sand from beaches formerly used by river runners, other recreationists, and wildlife. Water quality has deteriorated due to direct releases of sewage and industrial waste, and polluted runoff from farms, roads, mines, mining wastes, and other sources adds salts and other contaminants.

On top of all that, we have introduced dozens of nonnative ("exotic" or "alien") species, some intentionally and others by accident, which can prey on or outcompete native species. Once-thriving populations of endemic fish are now gone from parts of their former range, replaced and outcompeted by artificially introduced trout and other species. The U.S. Fish and Wildlife Service has listed four Colorado River fish species—the humpback chub, bonytail chub, razorback sucker, and Colorado pikeminnow—as endangered.[8] (Ichthyologists changed the popular name of the last to pikeminnow from Colorado squawfish, which was an offensive slur to Native Americans. The new name will be used in this book, except in quotations from earlier sources.) Riparian communities once dominated by cottonwood, willow, and other native plants are now overrun by tamarisk. (This plant is often known as salt cedar, although this name is technically correct for only one species of tamarisk. For convenience I will refer to all species collectively via the singular "tamarisk.") For hundreds of miles, levees and other structures separate the river from its natural floodplain, and reduced flows have eliminated the river's natural spring overflows. Birds and other species that once thrived in those habitats, like the southwestern willow flycatcher and the Yuma clapper rail, are endangered due to the loss of riparian habitat. Many of the canyons that once contributed to the majesty and mystery of the region are drowned below hundreds of feet of water and silt. Among the affected portions of the ecosystem is the Grand Canyon, that towering symbol of American conservation and undisputed international treasure.[9]

The Colorado River is just one example of what we have done to almost all rivers in the United States since European settlement. And no wonder, given the degree to which waterways fueled economic activity. Virtually from the beginning of recorded history, people have settled at the water's edge. Waterfront areas were among the first to be developed, because proximity to water was useful for navigation, irrigation, industry, defense, and household use. Flat, fertile floodplains were easy to build on and productive to farm. From before the earliest days of the American republic, waterways were critical for transportation and trade, settlement, fishing, and defense.[10]

This use and development of U.S. rivers and other waterways produced some of the most dramatic ecological change in human history. Most people know that we have dumped immense amounts of chemical pollution into our waters, but do not consider the many other ways we have changed the face and the shape of the aquatic environment. Experts are virtually unanimous that the biggest problem facing aquatic ecosystems is not pollution, but the destruction and alteration of aquatic habitats. The banks of most rivers and streams no longer support their natural riparian plants, with a loss of as much as ninety-five percent of natural vegetation in some areas. Floodplain development destroyed about half of the natural woody riparian habitat in the contiguous states. We filled or otherwise destroyed more than half of the wetlands in those states, and although the rate of wetlands destruction has slowed, we continue to lose about 60,000 acres a year, despite a stated national goal of "no net loss" of wetlands. More than six hundred thousand stream miles are inundated by reservoirs, and diversions from those structures seriously alter natural stream flows and habitats. Many more waters are levied, diked, armored, rip-rapped, and channelized, largely to aid or protect development at the water's edge. We justify this massive hydraulic machine by arguing that economic progress is needed if not inevitable, change is essential to foster that growth, and we must accept a highly modified environment as the price of our modern lifestyle.[11]

The growing environmental restoration movement, however, openly defies such concepts of futility in the name of utility. Although restoration as a strategy for environmental protection is not new, it is becoming more prominent and more ambitious. In fact, it can be viewed as the third major strategy of the modern environmental era (beginning in the 1960s).[12] Let's take a moment first to explore briefly the way in which restoration fits within the overall suite of environmental management strategies, issues to which we will return in more detail in chapters 5–9.

Restoration and the Environmental Movement

In the second half of the 20th century, environmental management involved two main strategies. First, we tried to mitigate the increasingly severe environmental damage caused by our accelerating industrial economy and our thirst for more and bigger things. Asthmatics choked on smoky air; rivers caught fire; and solitude dwindled as millions of Americans sought the great outdoors in their cars and trailers. Mitigation is somewhat like a paramedic treating an accident victim. The immediate task is to stop the bleeding and to minimize the resulting harm. We built fish ladders, for example, to reduce the effects of dams on migratory fish, although often with questionable success. The dams were designed to accommodate water development with as little harm to fish as possible. We set treatment requirements for polluted effluent from factory pipes and public sewage systems, but continued to release somewhat cleaner residues into the nation's rivers. With this treatment strategy, we aimed to reduce impacts while allowing the growth of American industry, population, and quality of life.[13]

Mitigation, of course, costs money, resources that could be used to feed and clothe and provide other economic goods and services. The costs of environmental controls stimulated debate about whether pollution controls and other forms of mitigation provided society with more benefit than they cost. Economists sought "optimal" levels of pollution. How much we should spend on this environmental emergency room became the subject of cost-benefit analysis, cost-effectiveness determinations, and the like. Others asked whether and how we could place a price on such amenities as clean air, clean water, healthy ecosystems, and the spiritual, emotional, and intrinsic values of wild places in their natural state.[14]

Gradually, we learned that it might be cheaper and more effective to *prevent* environmental harm by providing the same or similar goods and services in ways that cause less damage to the environment, thereby avoiding debates over the value of environmental mitigation. This second-phase strategy is like the efforts of an epidemiologist to prevent accidents and causes of disease in the first place, rather than treating patients once they become injured or ill. We use laws such as the National Environmental Policy Act and the Endangered Species Act, discussed extensively in later chapters, to predict adverse impacts in advance and to consider feasible but less harmful alternatives. That information can be used to decide whether a dam or a highway should be built, or whether more efficient water use or mass transit could serve the same ends with less environmental harm. Congress designed antipollution laws such as the Superfund statute, the Resource Conservation and Recovery Act, and the Pollution Prevention Act to encourage or require industry to produce the same or similar goods or services with less environmental harm.[15]

Prevention strategies work when innovation generates economic strategies that produce the same goods and services with less environmental harm, and at the same or lower costs. Otherwise, we continue to face difficult choices. Will we pay more for white paper manufactured without dioxin-causing chlorine bleach, or for cars that carry the same load, at the same speed, but that emit less pollution and use less gas? Even perfect prevention, moreover, cannot undo our past mistakes, and new mistakes are inevitable. A serious legacy of environmental damage remains, and is likely to endure for the foreseeable future.

The third, most recent modern environmental strategy, then, is to take affirmative steps to restore the health of ecosystems that have been altered or damaged by our past actions. Restoration is the holistic medicine of environmental policy. Holistic medicine might help a patient to recover and to prevent further illness through a combination of treatment, exercise, stress relief, diet, and other changes in lifestyle. It requires us to look at the whole patient rather than individual symptoms or body parts. In some cases, it requires the patient to choose between good health and cheeseburgers. To that extent, holistic medicine combines elements of prevention as part of a broader strategy of restoring and maintaining a patient's health.[16]

Similarly, environmental restoration requires us to look at all parts of the ecosystem's anatomy and physiology, particularly for large-scale restoration efforts like those in the Colorado River basin. It requires us to make hard choices about the value of a healthy environment compared to material wealth, such as the choice between water for off-stream economic use and the value of a free-flowing river. Scientists and others involved in Colorado River restoration efforts have grappled with those choices for many years and noted the impossibility of optimizing all river uses and values simultaneously.[17] Often, however, politicians and the public want perfect, "win-win" solutions. In restoration we seek to redress the cumulative effects of human actions on ecosystems rather than focusing only on specific environmental media (e.g., air, water, land, wildlife) or on particular human activities (e.g., steel or power production, farming, hunting). As I will argue in chapter 9, environmental restoration in the Colorado River and elsewhere should incorporate elements of prevention as part of a broadly defined strategy to restore and maintain the health of the river and its ecosystems.

Some environmental philosophers question whether restoration is appropriate. They argue that restored ecosystems are human artifacts passed off as real "nature," that restoration illustrates hubris about human ability to dominate and mold the natural world, and that the practice will lead to more environmental ills because of the false view that future harm can be reversed through restoration. One philosopher asserted that captive breeding programs for razorback sucker in the Colorado River region, in a futile effort to save a doomed species, are morally suspect and analogous to the impairment of a person's right to die with dignity. As other philosophers and practicing restoration ecologists counter, however, these arguments ignore the practical reality that tremendous harm has already occurred, and more is likely despite our best mitigation and prevention strategies.[18] The Colorado River is a good example of this. More than a century of human use has left a serious legacy of harm, including species perched on the brink of extinction, effects that are not likely to be reversed absent affirmative restoration.

That conclusion, however, raises perplexing issues of what we mean by a "healthier" ecosystem. Restoration requires difficult judgments about what we want to restore, to what condition, and to what purpose. In its seminal report on restoration of aquatic ecosystems, a National Research Council committee defined "restoration" as "the reestablishment of predisturbance aquatic functions and related physical, chemical, and biological characteristics." But what do we mean by "predisturbance"? Over the course of even recent geological history the Colorado River has hardly been a static place. About a million years ago, a lava dam rose to 2,330 feet within the Grand Canyon—over three times higher than the Glen Canyon Dam—and may have backed up the river for 179 miles upstream of Lees Ferry until the river eroded away the dam in about a quarter of a million years. In the face of massive changes wrought by natural forces, as well as significant artificial modifications, to what condition should we "restore" the river now? The Society for Ecological Restoration International defined restoration as "the process of assisting the recovery of an ecosystem that has been degraded, damaged, or destroyed."[19] This, of course, still requires judgments about what we mean by "recovery."

Scientists have developed a variety of tools to characterize "natural" (or predisturbance) ecosystems, many of which we explore later in the book. But because ecosystems constantly change, due to both natural and human forces, there is no single, fixed target condition for purposes of restoration. As just one example discussed in more detail later on, humans introduced wild horses and burros to the American West, species now protected by federal law but not native to North America at the time of European arrival. Does restoration of western range lands require removal of wild horses and burros? A more ecologically significant and plausible restoration target may be to identify natural processes (such as annual inundation of floodplains) and natural ranges of variation within which species and ecosystems fluctuate.[20]

Restoration ecologist Eric Higgs described this quandary as "the dark waters of defining restoration," and noted that "[t]here is no *original* condition for an ecosystem in any meaningful sense; one cannot fix a specific point in time." Ultimately, Higgs argued that defining restoration goals is as much a function of values as of science. Below Glen Canyon Dam, for example, full restoration of the ecosystem competes with environmental amenities such as a superb trout fishery that have thrived under the new conditions. Defining values, then, requires attention not just to scientific but also to political processes. Although scientists continue to make progress in defining ways to identify restoration goals and means to achieve them, in the Colorado River basin and elsewhere our legal and political institutions have not always followed so quickly in helping us to make the choices necessary to adopt and implement restoration efforts with clearly defined purposes.[21]

Even when we identify restoration goals, uncertainty can plague our efforts to achieve them. Often the reasons for ecosystem impairment are difficult to identify, especially in complex systems with multiple sources of harm. We may not know which of those causes to redress, in which order, and to which degree. A large degree of real-world experimentation may be necessary to test which restoration strategies are effective, usually under variable conditions that can cloud experimental results with additional uncertainty, due to both natural and artificial changes. Did fish return to a particular stretch of river this year because of increased flows from the upstream dam (the restoration action), or was it because of an unusually large supply of food (environmental variability)? Did fish continue to decline because the restoration action (more water) was ill-conceived, or because of polluted runoff from a new construction project?

These issues regarding the goals and methods of environmental restoration are explored in the chapters ahead, through the lens of Colorado River restoration efforts. The issues become even more complex as the scale of ecosystem restoration increases from a local wetland to a small river within a single state or community to a huge watershed like that of the Colorado River, which crosses both state and international boundaries. Professionals are now beginning to grapple with the problem of restoration on a large, ecosystem scale, rather than at a smaller, project-specific level. From the Columbia River watershed to the Florida Everglades, from the Chesapeake Bay to the San Francisco Bay delta, we are trying to reverse long-term ecosystem declines. We are doing so not just by reducing or eliminating new sources of harm, but by rejuvenating some of the physical, hydrological, and biological features and processes that support those systems.

Choosing Restoration Strategies: Restoring Places or Restoring Processes

In her austere account of the Southern California desert east of the Sierra Nevada mountains, *The Land of Little Rain*, Mary Austin wrote of her neighbor Naboth's field, "not greatly esteemed of the town, not being put to the plough nor affording firewood, but breeding all manner of wild seeds that go down in the irrigating ditches to come up as weeds in the gardens and grass plots." This community at the turn of the 20th century, bent on "reclaiming" the desert for productive uses, viewed Naboth as irresponsible, if not guilty of the heinous sin of waste. All he really did, however, was nothing, allowing nature to take back her own: "It is interesting to watch this *retaking of old ground* by the wild plants, banished by human use. Since Naboth drew his fence about the field and restricted it to a few wild-eyed steers, halting between the hills and the shambles, many old habitués of the field have come back to their haunts. The willow and brown birch, long ago cut off by the Indians for wattles, have come back to the streamside, slender and virginal in their spring greenness, and leaving long stretches of the brown water open to the sky." Austin described a basic clash of values about the human relationship to land. Naboth was "not greatly esteemed of the town" because he didn't put his land to economically productive use, and worse yet, sent his weeds into his neighbors' manicured gardens. But Austin lauded him for allowing this "retaking of old ground."[22]

I saw a similar phenomenon along the lower Colorado River, just north of Yuma, Arizona. My real purpose was to visit the Mittrey Lake/Pratt Lease restoration project sponsored by the U.S. Bureau of Reclamation. On the way to the restoration site I passed a sign to the Betty's Kitchen Wildlife and Interpretive Area operated by the Bureau of Land Management (BLM). It looked interesting, but I continued diligently to my primary destination. The Pratt Lease was planted in spring 1999 and December 2000, with long, straight rows of willows and cottonwoods, one of several efforts to restore native vegetation to riparian areas along the lower Colorado River.[23] Some trees appeared to be up to twenty-five feet high already, in fairly dense stands. I walked through, confirmed that this was growing into a thicket and not just a tree farm, and saw a few juncos and one northern harrier. But the restoration area was quite small, and soon I emerged on a much larger field of irrigated cauliflower. As addressed in chapter 7, this acre by acre strategy of replanting native vegetation is slow and expensive, and success depends on individual site conditions and other factors.

On my return, I stopped to visit the Betty's Kitchen area, which seemed far more "natural" than the Pratt Lease site, at least to my untrained eye. The site was empty, with no sign of other recent use. A sign explained: "If you were here prior to the Colorado River flood of 1983, you would have been in the middle of a residential area, complete with a café named Betty's Kitchen. The flood of 1983 greatly changed many things along the Colorado River. It damaged, beyond repair, the residential homes located here, leaving behind an area for nature to reclaim. We now have the opportunity to observe how nature rebuilds its own kitchen. We invite you to see what's on the menu."

So I did, by walking the short trail. Nature "reclaimed" the site so well in seventeen years that there was essentially no evidence of where Betty Davidson's small eatery stood before the flood. If this were private land, undoubtedly the owners would have rebuilt the homes and the restaurant, perhaps courtesy of subsidized national flood insurance, only to stand once again in the path of a surging river sometime in the hydrologically near future. But both the residences and the café sat on leased public land, and the BLM allowed this floodplain to remain undeveloped. Interpretive signs identified seep willow, cottonwood (a few of which were planted), arrowweed, honey mesquite, and quailbush. A few trees were ringed with protective fences. For the most part, nature alone seemed to have done quite well. The natural revegetation apparently was fueled by seeds lying in wait in the river sediment and riparian soils, or floating down from upstream. I began to wonder whether natural floods and land use reforms, rather than artificial planting and irrigation, ultimately will "restore" more habitat than the planned, expensive, agricultural-style restoration I saw at the Pratt Lease. The Society for Ecological Restoration International refers to this strategy as returning ecosystems to their historic trajectories. Although the manner in which the system will respond cannot be forecast precisely, it will move within a generally predictable range of directions.[24]

This concept of restoring broad ecological processes rather than just individual places, however, does not imply that ecosystem restoration can rely on passive strategies alone in most cases (as was true for the Betty's Kitchen area). Letting spring floods create natural conditions for native plants to reseed requires us to send more water downstream from the river's many dams, and then to allow this water to flood the banks. These efforts may take long periods of time, and lead to uncertain results. "Natural" reseeding, for example, might favor alien species over native species, depending on what lies upstream or upwind, and on germination conditions relative to different seed characteristics.

Restoring ecosystems impaired by dams could involve conservative strategies such as modification of reservoir releases to more closely match natural flow patterns and temperatures. It could also entail more comprehensive approaches such as dam removal and long-term efforts to restore the natural hydrology and ecology of the system. At least 500 dams, mostly small ones, have been removed in the United States in recent years for various reasons, some to restore the natural flow and connectivity of rivers and streams. To put this in perspective, however, about 80,000 large dams and an estimated 2 million small ones remain.[25] Along the Colorado River, and most pointedly for Glen Canyon Dam, perhaps the most fundamental question is whether the strategy of modifying dam operations will succeed in achieving large-scale restoration, or whether more ambitious approaches are needed.

Choosing Restoration Methods: Dam Modification or Dam Removal

Major efforts to restore key components and values of the Colorado River ecosystem are either proposed or under way, by a laundry list of government agencies and private entities, for the upper river, the Grand Canyon reach, and the lower river to the Mexican border. Another long-standing effort focuses on reducing excess salinity in the river as a whole. These programs use a range of tools, such as restocking native fish, eradicating alien species, releasing artificial "floods" designed to simulate natural flow patterns, replanting riparian vegetation, and restoring natural meanders and backwaters in the river channel. None of those efforts come free, and few are without controversy. All bring costs in dollars, impacts on water, power, and recreation, and economic and social dislocation. The existing programs, however, share one fundamental characteristic: they seek to restore natural ecosystem characteristics and functions while preserving the basic interests of those who use the river for economic benefits.

The reactions of traditional water and power interests to most proposed restoration programs range from severe to moderate opposition to tacit support. For specific users of the river's economic bounty, the level of support or opposition depends on whose acre-foot or kilowatt-hour or visitor-day is curtailed in return for often unpredictable environmental benefits. As a result, most ongoing restoration efforts seek to have it both ways—to keep in place the delicate legal and political structure that doles out the Colorado's water and power values while minimizing the long-term environmental impacts of that system. These efforts tinker with the Colorado's existing plumbing system rather than pulling the plug entirely.

The most far-reaching of Colorado River restoration proposals, on the other hand, would do just that: pull the plug, at least on the Glen Canyon Dam. The Glen Canyon Institute is a small but dedicated group (along with others, such as Living Rivers) devoted to restoring Glen Canyon. For auction at the Institute's 2001 annual banquet was a picture of an icon of the modern environmental movement, the late David Brower, impishly "pulling the plug" on what the group refers to as "Reservoir Powell" to highlight its artificial status. A former board member and later executive director of the Sierra Club, and in John McPhee's characterization the "Archdruid," Brower was one of the handful of people most responsible for stopping dams in places such as Dinosaur National Monument and Grand Canyon National Park. In the process, as part of a political compromise, Brower consented to construction of Glen Canyon Dam. Later, he called this deal the worst mistake of his long career, and "the biggest sin I ever committed." During the last years of his life, Brower proposed to drain the reservoir but not to destroy the dam altogether: "The dam itself would be left as a tourist attraction, like the Pyramids, with passersby wondering how humanity ever built it, and why."[26]

Of course, Brower was not the first to suggest this "radical surgery" approach to restoring the Colorado. Edward Abbey facetiously suggested an even more radical solution before the dam was completed: "some unsung hero with a rucksack full of dynamite will descend into the bowels of the dam; there he will hide his high explosives where they'll do the most good, attach blasting caps to the official dam wiring system in such a way that when the time comes for the grand opening ceremony, when the President and the Secretary of the Interior and the governors of the Four-Corner states are all in full regalia assembled, the button which the president pushes will ignite the loveliest explosion ever seen by man, reducing the great dam to a heap of rubble in the path of the river." Abbey later cited as a literary and ideological forebear Henry David Thoreau, whose only disadvantage in the quest for radical supremacy was in living before the age of explosives: "I for one am with thee, and who knows what may avail a crowbar against Billerica dam." In the Pacific Northwest in the early 20th century, citizens upset with the effect of new dams on salmon runs did more than fantasize. They floated a boat filled with dynamite into the Grant Pass Dam and damaged it beyond repair.[27]

According to proponents, decommissioning Glen Canyon Dam would provide significant environmental benefits without wreaking havoc on the

economy and welfare of the Southwest, and would do so by restoring natural processes rather than by relying on expensive, uncertain efforts to
restore individual habitats.[28] The more natural flow and temperature of
water and sediment and nutrients would help to restore the natural
ecosystem in Grand Canyon and the native species that live there. Because
Lake Mead and other downstream reservoirs do not have enough storage
space to hold the water formerly contained behind Glen Canyon Dam,
proponents argue that increased flows would also help to restore downstream biological resources throughout the lower river and all the way to
its delta. Most of all, they claim, over time it would return Glen Canyon
to its former splendor, like unveiling a long lost Da Vinci portrait, a priceless maze of river and twisting side canyons, infinitely varied in texture
and hue, filled with light and spirit and sound.

Although restoration projects of the "tinkering" variety prompt
moderate reactions, the proposal to drain Lake Powell provokes responses that range from ridicule to near hysteria, with elements of religious fervor and patriotism to spice up the debate. At a 1997 hearing in
the U.S. House of Representatives, California Congressman John
Doolittle warned: "Elimination of this foundational piece in the interlocking water puzzle would throw the entire Colorado River system into
chaos." Utah Congressman Chris Cannon: "Mr. Chairman, I have seen
environmental proposals in my district that can only be described as
dumb, some monumentally dumb. But now, Mr. Chairman, we have
dumb and dumber." Arizona Congressman John Shadegg: "Time moves
in one direction, and that is how God intended it. In this life, each of us
is called to look forward, not backward. . . . Let us not forget as we consider this issue that man is one of God's creations and that man's creations
often honor his God." Colorado Senator Ben Nighthorse Campbell
referred to the proposal as "a certifiable nut idea," "just plain silly," and
"absolutely ridiculous to contemplate." Of course, in the 1950s, some
thought the idea of building Glen Canyon Dam so ludicrous that it
could never really happen.[29]

Equally vehement reactions came from representatives of those who
rely on the resources and economic benefits provided by Lake Powell.
According to Rita Pearson, who at the time was director of the Arizona
Department of Water Resources: "Life as we know it here in the West
would be impossible without the Lake Powell Reservoir." Joseph
Hunter, executive director of the Colorado River Energy Distribution
Association, thought the proposal "seems more than a bit absurd."
Larry E. Tarp, then chairman of a group called Friends of Lake Powell,
offered perhaps the most dramatic response: "[T]he people involved in

daily life, commerce, and the free enterprise system surrounding the area will oppose until their deaths any person or persons that attempt to disrupt our personal rights, freedoms, and opportunities for existence around Lake Powell." Apparently, not just the fate of a dam but the future of American liberty is at stake in the battle over Glen Canyon. Yet there is even room for disagreement over whose side Lady Liberty would join. Environmental philosophers from Thoreau to Aldo Leopold to Gary Snyder linked the protection of wild places to the preservation of personal freedom and collective political liberty.[30]

So rather than accepting the futility of the idea, or bowing to the ridicule it has provoked, proponents are even more vehement in the righteousness of their cause. David Brower: "Once again Grand Canyon would make its own sounds and, if you listened carefully, you would hear it sighing with relief." Scientist Dave Wegner, formerly of the U.S. Bureau of Reclamation and now one of the chief proponents of Glen Canyon Dam decommissioning: "We are people who believe in the resources. We are people who believe in the fish. We are people who speak for the birds." The passionate and eloquent Katie Lee, actress turned folk singer turned songwriter turned river rat and environmental activist, writing of Glen Canyon, said: "It was unlike any other place on earth and should be treated as such—as a haven, a refuge, an ivory tower, a sanctuary—the heart of the Colorado Plateau should be allowed to beat again."[31]

Who is right? That depends on answers to two very different, but closely related, sets of questions. Is the proposal to drain Lake Powell unnecessary because other restoration methods will achieve equivalent results without disrupting the economies and lifestyles of the American Southwest? A multidecade set of experiments is under way to try to address that question, and while more is learned about the efficacy of those strategies every year, clear answers remain elusive. Meanwhile, millions of people rely on the river for water, power, recreation, and other resources. If current restoration efforts fail, are we willing to decommission the dam and forgo its economic benefits to restore the river? This question cannot be answered by science alone. It requires an inquiry into the intersection of law, science, and public policy.

Preserving Economic Institutions or Meeting New Restoration Goals

On July 1, 1999, Interior Secretary Bruce Babbitt, the governor of Maine, and thousands of observers watched from the shores of the Kennebec River as workers breached the Edwards Dam, a structure that had stood for 163 years. The goal was to restore at least one of the long-lost runs of Atlantic salmon that once thrived throughout Maine's large coastal river system.[32]

Two years to the day earlier, on July 1, 1997, the sun finally set on the British Empire as troops from the People's Republic of China marched into Hong Kong. The world watched along with Hong Kong's 6.3 million residents as a peaceful modern invasion transferred sovereignty from the colonialists to the communists. The transition occurred only after intensive negotiations in which China agreed to keep Hong Kong's capitalist economy in place, to guarantee a large degree of political autonomy for at least another half century, and to maintain private property rights and the political freedoms to which Hong Kong citizens were accustomed.[33]

What do these two historic but unrelated events have in common, and what have they to do with the Colorado River? Both represent examples of how societies deal with changes in formerly settled expectations. Both also reflect how difficult it can be for existing legal, political, and social systems to anticipate the long-term impacts of current actions and policies.

In the late 19th century, Imperial China signed treaties granting Great Britain ninety percent of what is now known as Hong Kong, under a ninety-nine year lease scheduled to end in 1997. At the time, neither party had any idea of how the world would change over the next century, nor did they contemplate the implications of that change for this tiny, remote corner of China. Britain undoubtedly believed it would remain a world power with the military and political might to keep Hong Kong after the initial treaty term expired. Neither party could predict that this small but convenient trading post for British merchant vessels would become a global economic powerhouse, with $20 billion in direct U.S. investments alone. Neither side had any idea that the nascent ideas of Karl Marx would create an international political and economic schism, and that under the terms of the treaty Hong Kong would be transferred from one side of the new geopolitical fence to the other. So the British and their Chinese partners in Hong Kong built a whole society premised on the assumption of stability, guaranteed in part by international law and in part by military hardware.

Likewise, when Edwards Dam was built in 1836, no one knew that bountiful Atlantic salmon populations would plummet throughout the eastern seaboard. No one knew very much about the environmental impacts of a single dam, much less the huge systems of dams that would be built on America's rivers. No one knew that the ideas of John Muir and Aldo Leopold and Rachel Carson would change American values and environmental politics so profoundly, or that laws like the National Environmental Policy Act and the Endangered Species Act would follow those value shifts. We built Edwards and other dams on the premise that the U.S. legal system would provide stability, and that investments made in reliance on the dams would be protected.

After all, one major role of law is to establish or codify consistent rules and expectations on which people can rely to guide their behavior. If government makes a major societal decision or a business makes a large capital investment, the law provides some degree of certainty that the rules will not change without good reason. But our legal and political system also has to reflect and respond to change. Society cannot hold blindly to the goal of stability in the face of new and better information and understanding, or widespread changes in values. Deciding when and how such change should occur is one of the very purposes of government. Democratic governance is designed to ensure that those decisions reflect a full range of interests and values in society.

Efforts to restore the Colorado River ecosystem, from the smallest riparian revegetation project to the dramatic proposal to drain Lake Powell, exemplify this tension between stability and change. Early in the 20th century, commissioners from the seven Colorado River basin states and Secretary of Commerce Herbert Hoover labored through months of difficult negotiations to hammer out the Colorado River Compact, the constitution—or to some, the Bible—of Colorado River law. In the ensuing decades, other components of what is known as the "Law of the River" were set in place, some through careful negotiations and others through hard-fought litigation or political battles. That compilation of compacts, treaties, laws, regulations, policies, contracts, and court decisions forms the platform of legal certainty on which economic development is built from the Colorado Rockies to Southern California, and from irrigation projects in Wyoming to urban growth in Phoenix.[34]

Disputes over who may use Colorado River water are far from resolved, especially when drought makes that resource increasingly scarce. Legal wrangling continues over how California will live within its allocated river share while still accommodating ongoing urban growth. Some of the tribal rights to the Colorado have yet to be settled four decades after a U.S. Supreme Court decree established them in principle, as highlighted in a recent lawsuit brought by the Navajo Nation.[35] Still, while continuing to squabble among themselves, members of the traditional water establishment assert that any major changes in the Law of the River—and the water development hardware that supports it—would destabilize if not devastate the region.

The Colorado River Compact is a relatively terse but monumentally significant seven-state agreement that allocates water usage among the "upper basin" and "lower basin" states. The term "basin" reflects a curious mixture between a geopolitical and hydrological divide. The artificially established dividing line between the upper and lower basins is near Lees Ferry, which sits at the head of Marble Canyon, approximately sixteen miles below Glen Canyon Dam. The upper basin states include Wyoming, Colorado, Utah, New Mexico, and a small part of Arizona. Lower basin states include California, Nevada, Arizona, and a small corner of Utah. (To complicate matters further, for some purposes the compact assigns legal rights and obligations to the "upper division" states of Wyoming, Colorado, Utah, and New Mexico, and to the "lower division" states of California, Nevada, and Arizona.) The upper and lower "basins" are not separate watersheds or river basins. Lees Ferry is, however, roughly halfway between the river's origins and its terminus. Moreover, because it is relatively accessible yet sits within deep canyons with few nearby water users, it was a logical place to measure water levels at the bottom of the upper and top of the lower basins.

Water is divided within the upper basin by another interstate agreement, the Upper Colorado River Basin Compact, and within the lower basin by federal statute and a U.S. Supreme Court decision and decree. The United States also agreed to convey annually a minimum of 1.5 maf of water to Mexico in a treaty signed in 1944. Details are added by a host of other documents, all of which collectively dictate the amount, timing, and location of water use and delivery allowed or required all the way up and down the river.[36]

Although this book is not primarily about the Law of the River, and other notable works have been written on that subject, a basic understanding is important to appreciate how ecological restoration goals relate to settled legal and economic expectations, and the political forces they serve. One of the purposes of the compact was to provide some degree of *certainty* (although considerable uncertainty remains in the manner in which the compact is interpreted and implemented, especially in times of drought). By the second decade of the 20th century, the lower basin states, especially California, were far ahead in the race to secure Colorado River water rights under the prior appropriation doctrine ("first in time, first in right"), a major tenet of water law in the West. Under this principle, those who divert water for a legally recognized "beneficial use" can obtain a right to that amount of water, for that use, with a "priority date" based on the time of first diversion and use. In times of shortage, those with earlier priority dates ("senior" right holders) can withdraw all of their water before latecomers ("junior" right holders) receive any.[37]

A major impetus for compact negotiations was a 1922 decision of the U.S. Supreme Court that prior appropriation applied to interstate water disputes.[38] Early water diversion and use in California, especially in the rapidly growing agricultural mecca in the Imperial Valley, ensured that California would have Colorado River water rights senior to other basin states, which were much slower to develop in population, economy, and water use. California, however, lacked financial and other resources needed to build dams and canals to store and transport yet more Colorado River water as demand increased, or to protect its low-lying farmlands when the river overflowed. Congress viewed an agreement among the basin states as a prerequisite to investing in dams, canals, and other water hardware in the basin.

Boiled down to its essence, in the Colorado River Compact California and the rest of the lower basin agreed to reserve a large chunk of the river's flow for the upper basin states. This paved the way for federal financing and construction of Hoover Dam, a major new canal to the Imperial Valley (dubbed the "All-American Canal" because it replaced a politically tenuous existing canal through Mexico), and other major structures along the lower river. California secured more liquid water, as well as flood control and significant amounts of hydroelectric power to fuel the growth of urban Southern California. The upper basin states, in return, received guaranteed *future* rights to water whenever development did occur (thus creating a major exception to prior appropriation law). Just how much water the upper basin received in the compact remains the subject of some dispute. According to the most basic reading of the compact, each basin has the right to the beneficial use of 7.5 maf of water per year, with the lower basin entitled to use another million acre-feet in "surplus" years, when more water than the minimum is available in the lower basin. In return for its deferred development rights, the upper basin agreed to deliver enough water to fulfill the lower basin's entitlements, plus half of the U.S. treaty obligation to Mexico.

To ensure that the lower basin states receive this entitlement of 7.5 maf of water per year, in the key operative provision of the compact, the upper basin agreed to deliver 75 maf of water every ten years as a "rolling average." This means that each year, the total previous ten-year flow at Lees Ferry must equal at least 75 maf. In practice, more is required to include the Mexican treaty requirements, and under the Bureau of Reclamation's long-term operating criteria for Glen Canyon and Hoover dams, a minimum of 8.23 maf of water must be delivered to Lees Ferry, if available.[39] Understanding what this requirement means in terms of the certainty of water rights and use in the upper basin, and how the upper basin responded to its major compact obligation, requires a better understanding of the inherent *uncertainty* in the river's annual flow. Annual virgin river flows (those occurring absent human changes to the river and watershed) vary dramatically, according to historical records, from a low of 4.4 maf to a high of more than 24 maf per year. In addition to this annual variation in the river's flow, compact negotiators in 1922 based their allocations on serious overestimates of average river flows, an issue to which we return in more detail later.

During years in which river flows are high, the upper basin can easily satisfy its delivery obligation. In fact, because the upper basin currently uses less than its maximum compact allotments, more than enough water would flow downstream with or without Glen Canyon Dam. But what happens during dry years, when flows are not adequate to meet both existing upper basin uses and the delivery obligations to the lower basin and Mexico? An arrangement in which the upper basin states had to deliver half of all *available* water to the lower basin would have divided the risk (or burden) of drought between the two basins. By imposing on the upper basin a firm requirement to deliver the entire lower basin share, during protracted droughts the compact can deny the upper basin its full share of water. This placed the risk of drought squarely on the upper basin states, a key factor that will affect restoration efforts throughout the basin unless enough water is stored in reservoirs during wet years so that the upper basin can meet its delivery obligations and still have enough water for its own use during the dry ones.

That, in a nutshell, was the main purpose for Glen Canyon Dam, which sits much too far downriver to serve most upper basin water use needs. The dam and reservoir also provide hydropower, recreation, and other benefits, but the main purpose was to serve as a hedge against drought for the upper basin, and to provide the lower basin with more physical as opposed to legal certainty that its share of water would be delivered. Other dams and reservoirs, upriver and in the tributaries, provide additional storage and serve the upper basin's actual water needs.

Other more detailed and technical components of the Law of the River also affect the manner in which restoration efforts can or will be implemented. For example, Congress directed in 1956 that, although compact and other water law requirements must be met first, the hydropower plants associated with the dams must be operated "so as to produce the greatest practicable amount of power that can be sold at firm power and energy rates." A decade later, in 1968, Congress prescribed that water levels in Lakes Mead and Powell be maintained, "as nearly as practicable," at equal levels, and codified a set of basic water release priorities from the reservoirs, based generally on the requirements of the compact.[40] Restoration efforts either need to accommodate these provisions of Colorado River law, or the law will have to change to accommodate restoration needs.

In this legal and institutional context, Glen Canyon Dam does not stand alone. It is part of a complex system of physical structures (dams, diversions, canals, power plants, and the like) which, along with the Law of the River, divide the waters of the Colorado among its many competing economic users and try to keep peace within the basin. Decommissioning Glen Canyon would cause a cascade of effects that would affect the entire system of water distribution developed over the past century. Because water is fundamental to all economic growth in the arid West, the basin states will resist any restoration efforts that jeopardize the fundamental "deal" struck in the Colorado River Compact. The lower basin states will protest changes in either the hardware or the delivery obligations that ensure they will continue to receive their share of Colorado River water. The upper basin states, which still have not developed and used their full compact shares, will reject programs that jeopardize their future water rights. All of the states—and the various interest groups that enjoy existing or future Colorado River water rights—will cling to whatever degree of certainty is afforded by the compact.

Environmental advocates, by contrast, urge social and political change where existing policies result in more harm than benefits to shared environmental, economic, and social values. If the only value at stake were natural ecosystems, the case for change might be clear. Of course, it is not, and the value choices are far from simple. Is the water stored or the power generated by the dam more valuable or more important than the native ecosystem it destroyed? As a matter of personal values, has the Colorado River *improved* as a result of the dam? Below Glen Canyon and other major dams we replaced native fish, such as humpback chub and razorback sucker, that held little interest to sportfishers, with some of the premier trout fisheries in the world, supported by the clear, cold tailwaters released from the reservoir. In places, the trout feed populations of formerly endangered bald eagles, whose numbers have increased dramatically since the dam was built. The new, more consistent flow regime from the dam also stimulated the formation of riparian wetlands, which are increasingly rare elsewhere in the

Southwest, along with enhanced populations of birds and other wildlife that rely on those new habitats.[41] Was the pristine, rarely visited wilderness known as Glen Canyon more important than the mecca for houseboats and jet skis now called Lake Powell, with its millions of visitors each year?

Draining Lake Powell would affect far more than the natural ecosystem in ways reminiscent of the political, social, and economic drama that is still unfolding in Hong Kong. Whether or not environmentalists are correct that the Glen Canyon Dam causes environmental harm that exceeds its economic and social benefits to the country as a whole, the dam has supported major recreational and other economies in the region. Page, Arizona, whose population, infrastructure, and business community grew based on the assumed longevity of Glen Canyon Dam and the reservoir it creates, could become a ghost town if the reservoir is drained. Among the secondary economic beneficiaries is the adjacent Navajo Nation. The dam and the tourism bring badly needed jobs and income to the Nation, which otherwise faces significant unemployment. Farther downriver, resort communities and luxury homes are built virtually to the river's edge in reliance on the flood control provided by Glen Canyon, Hoover, and other dams. What rights do businesses and communities hold on the strength of those expectations? Or are they no different from the steel industry of the Midwest, which bowed to more efficient foreign competition, or the owners of livery stables at the turn of the 20th century, whose businesses necessarily were replaced by gas stations? Change is a given. We can sit back and watch it happen, or we can shape it in more desirable ways.

As it turns out, the doomsday predictions that preceded the 1997 Hong Kong transition did not materialize. Day-to-day life in Hong Kong did not change significantly. The economy remained strong, and the transition did not rock international investors. Hong Kong was among the quickest to rebound from the Asian market crisis in 1997. At least 200,000 residents who left fearing the worst later returned. Would the many communities and economies that now rely on the stability provided by the Law of the River and the region's extensive plumbing system similarly recover if more dramatic changes are in store? Or would they be replaced by different but equally valuable centers of commerce? Chapter 9 of this book will evaluate long-term economic strategies that might be used to meet the same basic purposes as existing dams, while facilitating ecological restoration efforts.

At its most basic level, the fate of Colorado River restoration efforts provides a compelling example of the conflicts among competing value systems that are characteristic of major environmental disputes. The proposal to drain Lake Powell presents the starkest example, but the same is true for other restoration efforts, all of which affect to some degree the status quo in Colorado River use. Those who value the economic uses of water, power, and recreation provided by the dams and reservoirs, the pumps and canals, and the turbines and transmission lines are pitted against those who value the ecological, aesthetic, and recreational values of the natural system that the dam replaced. To even begin to unravel this complex set of conflicting interests, we need to understand in more detail the manner in which the natural system of the river changed, and with what effects. That is the subject of chapters 2–4.

CHAPTER TWO

The Living Artery: Disruptions to the River's Linear Connections

The Dewey Bridge crosses the Colorado River near the small town of Cisco, Utah, about thirty road miles northeast of Moab. Built in 1915–16 to hold six horses, three wagons, and 9,000 pounds of freight, this was the longest suspension bridge in Utah and the second largest west of the Mississippi. The bridge provided the first direct link between isolated communities in southeastern Utah and badly needed markets and supplies in Colorado. In his classic account of Mormon Country, Wallace Stegner described the Colorado as a "barrier rather than a highway," with only a few possible places to cross. "Along the whole eastern and southern sides of the Mormon Country the Green and the Colorado are a Chinese Wall, and in all its thousand miles there are only three gates."[1] When the river still posed a formidable barrier to travel and trade, the Dewey Bridge facilitated the flow of people, goods, and information.

The old Dewey Bridge—now open only to pedestrians—was restored in 2000 and is listed on the National Register of Historic Places. Although we rehabilitated the bridge to preserve its historic value, the human connections it formed were never broken. The material and data that support human needs in the modern world now flow across a new bridge, along with other roads, rails, air routes, and a growing collection of wires and satellites, airwaves and microwaves. Restoring an old bridge, even to its original condition, is a relatively easy undertaking. Efforts to restore broken ecological connections are not so simple, especially when those natural linkages are disrupted rather than formed by modern technology.

Construction of dams and other human engineering of the Colorado River and its tributaries significantly interrupted a series of physical and ecological connections that bind the headwaters to the river delta more than 1,000 miles downstream. Those broken links in the linear river chain and its hierarchy of watersheds constitute one of several major ecological transformations we have brought to the Colorado River system, and is the main subject of this chapter. We will look first at disruptions in the movement of the unique species of fish that once dominated the Colorado River, and then evaluate a series of less obvious changes caused by dams in the flow of materials and energy throughout the Colorado River system.

Blocking the Movement of Fish

Before we turned the Colorado into perhaps the most regulated river on the planet, it held a much more noble distinction. In 1959, Dr. Robert Rush Miller, one of the preeminent biologists studying the fish of the Southwest, identified the Colorado as having the largest percentage of endemic fish—species that live nowhere else—of any river system in North America. Sixteen fish species evolved exclusively in the Colorado River system, an estimated seventy percent of the species present. During millennia of ecological isolation, a unique aquatic fauna adapted to the harsh conditions of a watershed characterized by large fluctuations in flow. The fish survived raging spring torrents and languid summer lows. They tolerated the massive concentration of sediment borne by the river each year, and the sudden onslaught of summer flash floods. They withstood variations in annual flow ranging from 6 million acre-feet (maf) in drought years to 24 maf during years of heavy snowfall and accompanying spring runoff. "Those that did survive are often described as ugly or bizarre, but all are remarkably adapted, with slender, hydrodynamic bodies, dorsal humps, and large, fanlike fins for swimming in swift currents; thick, leathery skin and few scales to resist the abrasive sediment; and a highly developed lateral line and sensory system to navigate and locate food in dark waters."[2]

Ironically, the largest of the species that evolved in the predam Colorado River was a minnow. But do not be deceived by its name and think of those puny fish you may have caught in a net to use as bait. The Colorado pikeminnow is the largest minnow native to North America, and was the top predator in the predam ecosystem. Under natural conditions, pikeminnows reportedly grew to almost six feet and eighty pounds, huge by the standards of any freshwater fish species.

Of course, as with all fish caught by anglers, we expect at least modest exaggerations.

A U.S. Fish and Wildlife Service (FWS) official, Fred Quartarone, collected some of those "fish stories." One former resident of Green River, Wyoming, claimed: "They was as big as a junior high school kid, 90 pounds." He reported watching his grandfather catching "squawfish" on setlines and dragging them out of the river with his truck. Pikeminnows were prized by sport anglers, and would strike such diverse bait as "frogs, swallows, rabbits, mice, liver, chicken parts, grub worms, earthworms, hellgrammites, sculpins, and parts of fish," as well as cottontail rabbit heads, bacon, and artificial lures. Gene Bittler, of Maybell, Colorado, said "it was one of the most thrilling fish I ever caught if you want to know the truth." Max Stewart, of Vernal, Utah, remembers as the "height of his angling career," at just eight years old, the twenty-five-pound pikeminnow he literally dragged out of the river, falling twice along the way. Stories like that stimulated a promotional advertisement in the August 18, 1938, edition of the *Vernal Express*, which boasted that "white fish . . . are worthy game for those deep-sea fishermen of the President Roosevelt brand, who think in terms of swordfish and tuna." Although pikeminnow may not resemble salmon to a fisheries biologist, early settlers knew them as "white salmon," "Colorado River salmon," or "silver salmon," among other popular names.[3]

The native fish of the Colorado now have a reputation as "trash fish," unworthy of competition with trout and certainly not fit for the table. However, some old-timers disagreed, according to Quartarone's collected tales. Ted Cook, of Green River, Wyoming: "[I]t wasn't a trash fish you know. It was pretty good eating. . . . It was a pretty good fish." A former rancher from Browns Park, Colorado, added: "If you don't know what an ol' whitefish is, then you don't know much about fish. They're just like halibut." One resident recalled that "it was a big treat when we'd catch a whitefish," and another said that, after canning, it was "almost like eating salmon." Lyndon Granat of Palisade, Colorado, said of pressure-cooked Colorado River fish: "Damn, they made salmon taste bad. The bones and everything else."

Colorado pikeminnows were an important source of food during the Great Depression, but their culinary value has a much longer lineage. The giant fish were eaten by American Indians and by early settlers. Robert Brewster Stanton described "Colorado River Salmon," presumably meaning pikeminnow, at a Christmas feast at Lees Ferry during his second (late 1889) railroad survey. Ellsworth Kolb, early river runner, photographer, and fixture in the Grand Canyon tourism industry, also described "Colorado River Salmon" in his travel journal, saying they were "not gamey, but afford a lot of meat with a very satisfying flavour." Reportedly, there was a modest commercial fishery in the area of Grand Junction, Colorado, and one cannery in the lower river distributed pikeminnows across the country as "white salmon."[4]

As important as this remarkable species may have been to aboriginal settlers or more recent arrivals, it was considerably more important to the aquatic ecosystem of the Colorado River. One expert warned: "Loss of the Colorado squawfish . . . would signal a final collapse of the most endemic riverine fish community in North America . . . and perhaps foretell the doom of many of the large migratory fishes of the world as well." As suggested by the large variety of baits one can use to land them, pikeminnow are voracious omnivores, the top predator and the largest fish in the native Colorado River ecosystem. Although they are mainly piscivores that feed on other fish of various sizes and ages, adult pikeminnows also eat any other animals that fit in their mouths. One resident of Craig, Colorado, Chuck Mack, recalled an incident in the early 1950s in Lodore Canyon on the Green River, in which hundreds of baby swallows were leaving their nests. Many fell into the water and drowned. Mack reported that "[e]very big squawfish in the Green River must have migrated to the canyon to feast on the swallows." Every fish that he and his companions caught and gutted had "a stomach plumb full of baby swallows!"[5]

Species that sit atop the food chain in any system often are important indicators of the health of the system as a whole. A significant decline in the population of the top predator might be a symptom of problems elsewhere. Predator populations may decline due to changes in food supplies, habitats, or other factors important to the survival of those species. Those same changes also affect other species. Alternatively, declines in the top predators can affect other species more directly. Predation may no longer control those populations, and less fit individuals are no longer weeded from the gene pool.[6]

In the Colorado River, severe declines in populations of pikeminnow accompanied significant increases in introduced species such as carp and northern pike, which now fill the top predator niche in the ecosystem. Habitat losses and changes that harm pikeminnows likely affect other species as well. In spring, for example, pikeminnows prefer the same floodplain habitats as other fish, which serve as their prey. The same kind of competition and predation from nonnative fish may impair other native species as they do pikeminnows. This has been a pattern in the decline of southwestern aquatic ecosystems: "At first, individual species disappear. Now, whole subfaunas are collapsing." Just two years after his seminal paper on the unique nature of Colorado River fish, Dr. Miller published a second paper warning of the impending peril to those communities. As of 1955, all seven of the native species that had inhabited the Colorado and Gila rivers near Yuma a century earlier (1846–54) were no longer present. By 1943–44, none of the thirteen native fish that lived in the Salt River near Tempe as reported in 1890 remained. Both pikeminnow and bonytail already were nearly extirpated from the lower river, and Professor Miller predicted—we now know correctly— that ongoing dam construction would result in their complete elimination from that reach. "Perhaps nowhere else in North America," he wrote, "has the upset of natural conditions been more strikingly reflected by biotic change than in the arid Southwest."[7]

Although pikeminnows and other native fish in the Colorado have declined for multiple reasons, clearly one of the most significant causes is habitat loss and alteration, including disruption of migrations due to dams and other obstructions. By contrast to their concrete and earthen cousins in the Columbia River basin, most of the structures impeding the flow of the Colorado and its tributaries have no fish passages. Fish ladders have been required in the Northwest since early in the 20th century, but are much more recent arrivals on the Colorado.[8]

In May 2002, on the Gunnison River just upriver from its confluence with the Colorado in Grand Junction, I toured a fish ladder that was completed in 1996 at the Redlands Diversion Dam. This was the first artificial fish passage installed in the Colorado River system, and it took at least ten years of planning to do so. A feasibility study on the Redlands fish passage was prepared in 1986, but the project initially was placed on hold due to lack of funding and uncertainty about its efficacy. The Redlands ladder is designed to reestablish connections between downstream and upstream habitats for endangered fish, connections that were broken by dams, diversions, and other manmade structures. But we do not know whether these new, artificial fish passages will work any better on the Gunnison than they have for salmon in the Columbia River basin, where their efficacy is hotly debated.[9]

The spring of 2002 followed what turned out to be just one in a series of dry winters, symptoms of one of the driest cycles in the basin in recent human history.[10] In the Colorado River system, the vast majority of annual flows come from spring snowmelt rather than summer rain. Even in early May, the snow line was high up in the Rockies, with the annual snowpack a small fraction of average conditions. At the ski area at the top of Monarch Pass, above 11,000 feet, I saw only a few isolated patches of snow. Smaller reservoirs such as Taylor Park at the headwaters of the Gunnison were already low, and were predicted to be half full by the end of the year, with little prospect for help from that spring's sparse runoff.

Chuck McAda, a fisheries biologist and program manager at the FWS, explained the connection between water flows and fish flows. As of then, the Colorado pikeminnow was the only endangered fish species documented to have used the ladder. "These fish are not salmon," McAda noted. Lacking the strong homing characteristics of Pacific salmon species, apparently they need more water, or "attraction flows," to move upstream. So although the FWS holds limited water rights from the series of dams operated by the U.S. Bureau of Reclamation upstream in the Gunnison watershed, he was not yet sure whether adequate water would be available that year to meet the needs of all water users. Recall from chapter 1 that in the West, in times of shortage, water goes first to those with "senior" water rights. Assuming that McAda was given some water for the fish, he planned to save it for the critical period beginning in early July, when pikeminnow and other fish might be moving upstream.

According to McAda, however, who has worked on endangered fish recovery efforts in the Colorado basin since 1986, the ultimate problem was not the drought. From an evolutionary perspective, these species evolved in an arid, variable climate to deal with extended periods of low water flows. Pikeminnows can live for more than fifty years, a very long life span for fish. This allows a population to survive a protracted drought in which little or no successful reproduction occurs. When procreation fails for several years in a row, overall population levels still can be maintained due to the "storage effect" of successful breeding during other years.[11]

Before the dams, when pikeminnow occurred in several large congregations around the basin, if the entire population in one reach of the river was eliminated (known as "extirpation"), fish could reinvade from unaffected populations elsewhere in the system. Fish from the predam San Juan River population, for example, could migrate upstream from the San Juan–Colorado confluence and repopulate habitats in the Green River or upper Colorado River (Grand River) reaches. Genetic similarities among pikeminnow populations in these three regions suggest that such cross-fertilization occurred in the past. Studies in which individual fish are marked and recaptured still confirm pikeminnow migration between the Green River and upper Colorado River populations. Now, at least some of those connections have been severed. It is not clear, for example, whether pikeminnows successfully navigate the reservoir created by Glen Canyon Dam (Lake Powell), which inundates the confluence of the Colorado and the San Juan rivers.[12]

Why, though, does the presence of dams necessarily hurt populations of this large, predaceous fish, so long as there is adequate habitat in between the blockages? Hundreds of river miles separate the two dams at Flaming Gorge and Glen Canyon, and pikeminnow populations remain in several stretches of river in between. More than 300 river miles through Marble Canyon and Grand Canyon also separate Glen Canyon from Hoover Dam. Yet pikeminnow are gone from historic habitats in that stretch. In the rest of the lower Colorado River, pikeminnow have been scarce since Hoover Dam was completed in the 1930s, and absent since the 1970s.[13]

One big hint is provided by the congregations of pikeminnows seen directly below obstructions immediately before the spawning season. Pikeminnow migrate long distances within their habitats, with reported round trips almost 600 miles. According to one expert, this is a logical adaptation for a species that lives in a highly variable system. Fish can seek out optimal spawning sites under different conditions in any given year. And pikeminnow use different aquatic habitats during different life stages. Adults live mainly in deep pools and eddies. As water temperatures begin to rise, fish seek gravel and cobble substrates, which have been cleansed by the spring runoff, on which to lay their eggs. Absent barriers, the fish migrate considerable distances to find ideal spawning sites. Apparently, the fish also show considerable fidelity to particular sites, although it is not clear whether they are driven by sophisticated "homing" behavior like salmon, or whether they are simply drawn upstream by spring flows and temperature cues until they locate an appropriate site. After spawning, pikeminnow larvae drift into calm and warm backwater habitats, where abundant food sustains juvenile fish to adulthood. As they mature, subadults migrate back upstream, sometimes to their parents' original habitat, and sometimes mixing with other populations. This provides a source of genetic connection between populations.[14]

Thus, pikeminnow life cycles and population dynamics rely on and create several kinds of connections within the Colorado River ecosystem. Individual adults require upstream and downstream habitat connections to locate appropriate places to spawn. Larvae and juveniles need connections between the main channel and backwater habitats as they mature. (Lateral connections are discussed more in the next chapter.) The fish then either connect back to their home populations or provide connections between different populations, allowing genetic mixing and a source of replenishment if a local population is depleted by unfavorable conditions or other factors. Such mixing also promotes redistribution when a local population thrives and exceeds resource limits within its habitat. As discussed in more detail in chapter 5, it is not just the flow of fish, but of their genes, that has maintained healthy populations that are resilient to changing environmental conditions. Gene flow within and among discrete populations is critical to maintain genetic diversity, and hence adaptability to environmental variation, in the species.[15]

All of those connections are jeopardized by the obvious fact that dams and other human structures pose physical barriers to fish migration. Other broken connections within the river system are equally important to the river's ecology. The most fundamental is the degree to which we changed the timing and volume of water flows.

Changing the Flow of Water

According to one key statistic, the Colorado River is the most artificially "developed" river system on the continent. With an active capacity of almost 62 maf, the river's massive system of reservoirs stores four times the average annual flow of the river, a higher ratio than anywhere else in the country.[16] Likewise, a higher percentage of the river's flow is put to human off-stream use than any other U.S. river of comparable size, guided by an equally imposing network of diversions, canals, and irrigation works. To accomplish this hydrological feat, the dams, diversions, and other engineered changes to the Colorado extend virtually the length of the river. Those modifications had significant impacts on the river's ecology.

The Grand Ditch sits high in the Colorado Rockies, on the west side of Rocky Mountain National Park. Diversion of water from the Colorado River to the east side of the Rockies started here in 1890. Due to clever hydraulic engineering, the project moves water across the Continental Divide with no pumping. Water is taken from the western side of the watershed at a point higher in elevation than Poudre Pass on the Continental Divide, just six or seven miles from where the river officially "begins." The ditch flows downhill at a steady one percent grade until it reaches the pass, from which it flows down into the Cache la Poudre River toward Fort Collins. Marc Reisner wrote in *Cadillac Desert* that in the West, water flows uphill to money.[17] In this case, it seems, water flows downhill to money. The ditch now diverts roughly half of the water in the Colorado River at that early point in its journey, doubling the flow of the Cache la Poudre River and robbing water from wetlands adjacent to the Colorado's headwaters.

Slightly farther south, more Colorado River water is transported eastward out of the drainage through the Colorado–Big Thompson Project, where it was used initially for irrigation water, later for urban growth, and later still for snowmaking by Front Range ski resorts. Water from Grand Lake is diverted *underneath* the mountains through eleven miles of pipes that exit on the east side of Rocky Mountain National Park, again taking water from the riparian wetlands lining the Colorado headwaters, and from the river itself: "The Colorado River in Kawuneeche Valley is severely depleted by upstream diversions. As a result it has become prematurely senescent, carving lazy oxbows where springtime meltwater floods once would have kept its channel scoured."[18]

Nearly 1,300 river miles from the Grand Ditch sits the Morelos Dam in Mexico, the last point from which large volumes of water are diverted from the river for artificial use. In between are seven large dams on the main stems of the Green and Colorado rivers alone, with more on the major tributaries. At least 122 dams dot the basin, forty-six of which "significantly impede" the river and all that it carries, upstream and down.[19] Every drop of water in the system is put to human use an estimated seventeen times. A full one-third of the river's average flow is diverted out of the basin to the Colorado Front Range, the Salt Lake City region, Phoenix, and Southern California. More than 1 maf of water per year evaporates under the desert sun from reservoir surfaces.[20]

During the wettest years in the basin over the past several millennia, as much as 24–25 maf of water passed through the delta into the Sea of Cortéz. Now, what little remains consists of salty, polluted return flows from thousands of acres of irrigated agriculture on both sides of the border. As reported memorably in Phillip Fradkin's *A River No More*, in many years the river runs bone dry through many of its last miles in Mexico: "Suddenly, the Colorado disappeared. The river sank into the sand and became a braided, dry riverbed until, like magic, it popped up again near the intersection of some unlined irrigation canals. . . . The languid river was now the turgid product of pesticide and saline-laced return flows from the agricultural fields below Mexicali." This was true for the entire fifty years from completion of Hoover Dam through the early 1980s, until the reservoirs were full and high flows caused a brief respite. Although annual flow to the delta averaged almost 16.7 maf from 1896 to 1921, between 1984 and 1999 it dropped to 4.2 maf. Even under the highest flow conditions, as during the wet years of the early to mid-1980s, no more than 10–12 maf have reached the delta since Glen Canyon Dam was built.

Likewise, the Gila River in Arizona, once a wide, shallow, perennial stream and one of the river's major tributaries, runs dry long before its intersection with the Colorado at Yuma.[21]

This reduction in the flow of water through the Colorado River system caused significant changes in the aquatic ecosystem and how it functions. Heavy spring floods served as spawning cues for native fish. Those signals may now be gone. Reduced flows during the spring runoff also steal water from adjacent backwater habitats, wetlands, and riparian ecosystems (addressed in more detail in the next chapter). Changing water levels from fluctuating reservoir releases can also strand eggs and fish larvae. In the Colorado River estuary, some marine species need adequate freshwater flows from the river to keep salinity concentrations low for breeding and egg-laying. At least two aquatic species are now listed as endangered in the delta: a once plentiful and commercially important fish called the totoaba (*Totoaba macdonaldi*) and the vaquita (*Phocoena sinus*), a small dolphin. Likewise, reduced flows of freshwater and accompanying nutrients may have harmed delta shrimp populations.[22]

But it was more than the raw volume of flowing water that characterized the natural Colorado, both hydrologically and ecologically. The *timing* of water flows was also critical, as were the many properties that make "water" as diverse a substance as exists on the planet. By regulating the flow of the river so that water is available for human use when and where it is needed, we changed dramatically the when and the where of the Colorado. Water that once surged through the canyons during the spring snowmelt now is stored in huge reservoirs, and is released when needed for irrigation, power generation, and other uses. As a result, we transformed hundreds of miles of once-flowing river to artificial lakes, with significantly altered physical and chemical conditions and vastly different assemblages of fish and other aquatic life. More water now flows during some parts of the year than in the natural system, and less in others. The river's "hydrograph," a plot of river flow over time, is much flatter, or more evenly distributed, than before the dams, when maximum yearly floods at Phantom Ranch in Grand Canyon averaged 80,000–85,000 cubic feet per second (cfs), but minimum flows averaged just 4,000 cfs. Now, water released from Glen Canyon Dam flows through the canyon far more evenly throughout the year.[23]

This smoothing out of the river's flow changed many of the river's physical features and energy characteristics, which are determined largely by the highest flows that occur only during several days of the year.[24] Those changes, in turn, affected the river's ecology. Natural spring floods swept gravel beds clean in spawning areas for native fish.

When artificial changes reduced those flood peaks, native fish species that had evolved and adapted to those turbulent conditions no longer enjoyed as much natural advantage over introduced species.

The dams also changed daily flow patterns. The Glen Canyon power plant was originally operated to match the diurnal fluctuation in demand across the southwestern regional power grid. Hydroelectric plants can be "ramped up" to full power far more quickly than coal-fired or other thermal power plants, providing more flexibility to meet rapid rises in demand (known as peak demand), for example to provide power for air conditioning during hot southwestern summer afternoons. The Navajo Generating Station in Page, a coal-fired plant right next to Lake Powell, takes twenty-four hours to reach full capacity. Glen Canyon Dam can do so in minutes. The resulting diurnal fluctuation in dam releases "markedly affected the riparian zone and fish and invertebrate productivity in the various tributaries," contributing to the decline of a number of native fish. River runners are also affected by these changes in the river's timing. After the dam operation began, rafters and kayakers had to learn to ride the daily "tides" in river flows. As described by river historian David Lavender: "Boaters well down the river often awoke in the morning to find their rafts high and dry on the sand, or floating on the ends of their ropes, though they had been securely grounded the evening before." Glen Canyon Dam operation has since been modified to address both ecological and recreational concerns, as discussed in chapter 6.[25]

Dams and water withdrawals also changed water temperature and quality. Below the biggest dams—Flaming Gorge, Glen Canyon, and Hoover in particular—river temperatures are now much cooler during the summer and warmer in the winter than under natural conditions. Again, a natural regime characterized by significant annual variation has been smoothed out, with extreme temperatures lopped off of both ends of the annual spectrum. Before the dam, the waters below Glen Canyon ranged from lows of forty degrees Fahrenheit in the winter to sixty to seventy by early spring and seventy-five to eighty-five during the summer. Early explorers reported ice in parts of both Glen and Grand canyons. Water temperatures now run approximately forty-eight degrees Fahrenheit year-round, although ironically, temperatures exceeded sixty degrees in 2005, when drought lowered Lake Powell enough that warmer surface waters were released through the dam. Studies conducted shortly after the construction of Flaming Gorge Dam linked cold-water releases to the disappearance of all four species of endangered "big river fish" from more than 62 miles of river below the dam.

According to one theory, reproduction in native fish is triggered by warmer spring water temperatures, which no longer occur. Cold water can also shock larval fish, reducing their chance of survival.[26]

Likewise, the river's chemical properties changed due to the retention and mixing effects of the reservoirs. The dams now trap sediment and nutrients, which are now carried downstream in much smaller amounts. One might think that this moderating effect of the dams would be "good" for the Colorado, because harsh, extreme conditions often make life difficult. But the native species that inhabited the natural Colorado *evolved* under those conditions of extreme seasonal flow changes, and were adapted to the temperatures, turbidity levels, and other water conditions that prevailed for hundreds of thousands of years. To those species, change was not a good thing, and rather sudden change was devastating. One of the most dramatic of these changes has been in the flow of sediment.

Stemming the Movement of Sediment

On the main stem of the Colorado River, about twelve miles west of Glenwood Springs, Colorado, is a small town called Silt. According to the Silt Historical Society, railroad workers named the town for the nature of the local soils. Track layers needed to be careful here because the soil consisted of silt washed down from gulches throughout the region. They called the section marker Silt, and the name stuck.[27]

More than 1,000 miles downstream, on the Colorado River delta, are miles and miles of mudflats. In December 2001, I provided what help I could to University of Arizona Professor Karl Flessa and his students, who research environmental change in the delta. Parts of the area resemble the stark beauty of the Colorado Plateau. As I lugged scientific gear across the mudflats, I realized that, in one sense, we *were* hiking on the Colorado Plateau. These were the sediments of canyon country, carried down the Colorado by millennia of spring snowmelts and summer flash floods. Like the Colorado Plateau, the mudflats are immense. It seems like you could walk on them forever, unless you are swallowed by the mud. Unlike the others, who were more experienced with appropriate footwear, I wore low-top scuba booties. Mine sometimes pulled right off into a half foot of mud, forcing me to balance on the other leg as I retrieved the boot and reshod myself.

Members of U.S. Army Lieutenant Joseph Christmas Ives' 1857–58 expedition felt the same way as they explored the delta to investigate the navigability of the lower Colorado: "Sinking at every step half way to our waists in the soft, gluey mud, we waded slowly and laboriously to the higher ground." To Ives, the river appeared to be ten to twelve miles wide. But it was impossible to discern where the water ended due to the massive complex of low, flat banks and bars, in places "an unbroken sheet of soft and tenacious mud, into which one sinks deeply at every step."[28]

Chances are, if you walk far enough along the delta and if your foot sinks in deep enough, a grain or two of the silt now clinging to your boot traveled all the way from what is now Silt, Colorado. Before the dams, the Colorado River was a natural sluice box through which millions of tons a year of sediment flowed. Sediment loads in the river, measured at Phantom Ranch in the Grand Canyon in 1927, averaged 374 metric tons a day, with a record of an amazing 27,164 metric tons in one day. And although estimates vary, the river carried as much as 195 million tons of sediment per year through the Grand Canyon before major dams upstream began to act as traps. More sediment was added as the river ran south, through erosive soil and rock. Every rainfall and every snowmelt washes particles of dirt into gulches, arroyos, and streams that eventually reach the Colorado River. The sediment varies widely in size, shape, color, and composition; it is part silt, part clay, part sand, part cobbles, and during the heaviest of downpours, even boulders.[29]

Some sense of just how much material was transported by spring and summer floods can be gleaned from the work of Godfrey Sykes, who in the early 20th century laid the groundwork for later research about the delta and much of the lower river. In the summer flood of 1922, the river flowed at 115,000 cubic feet per second at Yuma and brought 6 maf of water to the delta during the month of June alone. Sykes estimated that, during this period, the river carried 27,500 *acre-feet* of silt to portions of the delta, depositing material six to eight feet deep over some 300 square miles of the region. Methods for estimating sediment loads were even less certain than they are now, but precision is not necessary with numbers like that. Nor was this an isolated incident. Typically, as much as seventy percent of the river's natural sediment load was carried to the delta, which at one time covered 3,000 square miles. All told, the river deposited an average of 100,000 acre-feet of sediment a year on the delta region. "Too thick to drink, to thin to plow," bemoaned early settlers.[30]

In fact, sediment more than water explains the river's name. El Río Colorado. The red river. Spanish explorer Juan de Oñate named the river for the muddy red hues bestowed by the sediment carried from canyons in Utah, Colorado, and New Mexico. Most of the sediment comes from side canyons in lower elevations, especially during short but intense summer thunderstorms. The mountain headwaters contribute more than three-quarters of the water, but only thirty-one percent of the sediment at Grand Canyon. Tributaries draining semiarid canyon country, such as the San Rafael, Paria, Chaco, and Little Colorado rivers, contribute just fifteen percent of the water but about sixty-nine percent of the sediment. Sykes reported that the source of floodwaters in the lower river could be identified by the colors of sediment carried from different tributaries. Before the dams, the river was usually reddish-brown, but flash floods from the Little Colorado ran "brighter and more vivid red," while those from the San Juan and Gila were darker, "almost black." Perhaps this gives a clue as to river conditions when Francisco de Ulloa, a captain under Cortéz, embayed at the head of what he called the "Vermillion Sea" in 1539 (later the Sea of Cortéz).[31]

To Europeans accustomed to rivers that were clear, or blue, or perhaps green, the Colorado's unusual cast seemed terribly unnatural. Novelist Zane Grey wrote: "Even in the dark it seemed to wear the hue of blood." Although John Wesley Powell wrote fondly of the "deep, cold, emerald waters" that give birth to the Colorado in the Rocky Mountains, he bemoaned that "the waters that were so clear above empty as turbid floods into the Gulf of California." Powell dismissed one sediment-laden tributary as "but a muddy creek," and members of his expedition reserved even worse disdain for the Dirty Devil River, which they so named because "the water is exceedingly muddy and has an unpleasant odor."[32]

Despite these ecological prejudices formed in Europe and the eastern United States, sediment was a key component and defining characteristic of the Colorado River ecosystem. More importantly, the constant *movement* and placement of that sediment was a critical, ongoing process in shaping the habitats in which the big river fish and other species evolved. When the summer storms subside, sediment deposits in the river channel where it is "stored" during the low flow periods of late summer, fall, and winter. When the river rages again in the spring, the current propels this stored sediment downstream, or laterally into eddies, where it forms beaches and backwater habitats.

Because of this constant movement of sediment, the river and its delta always changed, yet in another sense stayed the same. The river washed sediment downstream, but it was replaced by new materials during each flood, in an ever-shifting dance of demolition and creation, molding and remolding. Lieutenant Ives reported that the river channel in the delta often shifted from bank to bank over night: "The shifting of the channel, the banks, the islands, and the bars is so continual and so rapid that a detailed description, derived from the experiences of one trip, would be found incorrect, not only during the subsequent year, but perhaps in the course of a week, or even a day." John Steinbeck, whose biological expedition to the Gulf is recorded in *The Log from the Sea of Cortez* (although he never made it all the way north to the delta), poked fun at the authors of the *Coast Pilot* navigational guidebook: "This is a good and careful description by men whose main drive is toward accuracy, and they must be driven fanatic as man and tide and wave undermine their work." But despite this constant change, the same basic conditions remained in slightly different shapes. In portions of the Green River, the Hayden map of 1877 showed general patterns of sandbars almost identical to what was observed by scientists Luna Leopold and Gordon Wolman in 1953.[33]

Although frequent change may have posed problems for navigators, it was the essence of the natural ecosystem. "*Change* is the handmaiden Nature requires to do her miracles with," wrote Mark Twain after his journeys in the American West.[34] Spring flushing of sediment serves at least two important ecological functions for pikeminnow and other native fish. Pikeminnow lay their eggs in gravels and cobbles at preferred spawning locations. Summer flows wash these "birthing rooms" clean, allowing fertilized eggs to incubate in the spaces between the cobble. In turn, the sediment cleansed from those channel bottoms settles at other locations, along the river's margins and at natural bends and traps, forming complexes of sandbars and eddies downstream. Once spawned, juvenile fish are swept into backwaters behind emergent sandbars, where they spend the first year of their lives relatively protected from predators. These backwater habitats also provide abundant sources of food for growing fry. Thus protected and nourished behind their shifting sedimentary havens, enough young fish survive and grow to migrate safely into the main river channels to replenish the population.

Likewise, the annual downstream transport of sediment nourished the riparian ecosystem. Alluvial sediment in the floodplains of the Colorado formed a moist, nutrient-rich substrate for riparian communities dominated in the native system by cottonwoods and willows. Those riverside thickets, clinging to narrow margins within cliff-bound canyons and stretching miles wide within broader river valleys, provided important habitat for birds, reptiles, mammals, and other species that otherwise have little refuge in the arid Southwest. When floods were powerful enough, the plants themselves could be washed away. But with each new load of material the natural river carried seeds from upriver plants, more water to allow those seeds to germinate, and nutrients and organic matter to help them mature.

These sediment-fed riparian zones once supported rich and expansive ecosystems. Ives reported fringes of cottonwoods and willows or
thickets of high reeds in the lower river near the delta, and broader
groves of willows and mesquite from one to seven miles wide farther
upstream. The Mohave Valley and Cottonwood Valley, according to
Lieutenant Ives, boasted "graceful clusters of stately cottonwoods in full
and brilliant leaf." Aldo Leopold, who traveled through the Colorado
River delta by canoe in 1922, wrote of a "verdant wall of mesquite and
willow," with egrets at each bend and fleets of cormorants, avocets, wil-
lets, yellow legs, mallards, wigeons, and teal. He also reported bobcat,
raccoons, coyotes, burro, deer, the elusive great jaguar, and a diversity of
small game that was "too abundant to hunt" and "of incredible fatness."
The reason for this diverse and abundant fauna, according to Leopold,
was the riparian vegetation fed by the annual spring floods and the rich
sediment it carried: "The origin of all this opulence was not too far to
seek. Every mesquite and every tornillo was loaded with pods."[35]

Indeed, the delta once formed a huge complex of nearly 2 million
acres of riparian, freshwater, brackish, and tidal wetlands, with an estimated 200–400 species of vascular plants. Those highly productive
habitats, in turn, supported the "legendary richness" of the Gulf of
California below. Take away the floods of freshwater, take away the
annual supply of nourishing sediment from far upstream in the basin,
and you break the critical connection that fed the delta and its ecosystem for millennia. This wetland system has now shrunk to some 150,000
acres, and far more sediment now erodes into the gulf than is delivered
from the river above.[36]

Similar riparian ecosystems along the length of the river and its
major tributaries were also nourished by this constant flow of material.
In its 1950 report evaluating proposals for more dams in the basin, the
National Park Service noted that the plant and wildlife habitats that
once existed along the banks of the Colorado "were almost unbelievable." In her beautiful natural history of the Green River, *Run, River,
Run*, Ann Zwinger lamented the absence of riparian vegetation along
the banks of Fontanelle Reservoir. The Hayden expedition, she noted,
described this stretch as having "broad river bottoms richly carpeted
with good grass and big cottonwood groves. Now there is no shade
whatsoever." John Wesley Powell named Glen Canyon for the lush
riparian groves that lined the banks where the river curved and trapped
layers of sediment on which plants could take root. Katie Lee wrote of
vegetative riches from "old huge cottonwoods, to stream bank willows,
to monkey flowers and gilia and maidenhair fern, [to] grass and moss."[37]

Humans, too, found habitat in the sediments of the Colorado, amidst the otherwise harsh aridity of the region. Like ancient Egyptians along the Nile, American Indian tribes that settled along the lower Colorado River, including the Cocopa, the Yuma, and the Mohave, farmed flood-plains nourished annually with water and sheets of nutrient-rich alluvial sediment. The ancient Anasazi planted crops in the floodplains along the river, in places such as the Unkar Creek delta and along the alluvial plains in Glen Canyon. These sites were occupied for some 350 years (A.D. 850–1200), when, according to the most recent theories, the Anasazi left the region due to drought and moved farther south. European trap-pers, explorers, and later recreational boaters also used sandbars as they navigated the river. While traversing the rapids of Lodore, John Wesley Powell's party camped in "a little patch of flood plain," a frequent situ-ation throughout his two expeditions. Today, recreational boaters use sand beaches along the river's margins as the only viable campsites in Grand Canyon and other stretches of the river.[38]

Just as the dams on the Colorado severed the connections between fish and their native spawning grounds, and between the gene pools of related populations, they trap the flow of sediment that once helped to form and reform the aquatic and riparian ecosystems of the Colorado River, its delta, and its major tributaries. There is no longer much chance that a given particle of silt from the Town of Silt will make its way to the mudflats of the Colorado River delta, or pro-vide a substrate in which a willow seed might germinate along the banks of the lower river.

Laguna Dam, built from 1907 to 1909, was the first dam constructed by the U.S. Reclamation Service (predecessor to the Bureau of Reclama-tion) on the lower river. Laguna was designed as a diversion dam for one of the early irrigation canals on the lower river, but also as a settling basin for the sediment flowing downstream. Apparently, it worked too well. It filled with sediment within weeks of completion. Imperial Dam, built just upstream from Laguna thirty years later, filled with so much sediment that it contracted from a capacity of 85,000 acre-feet to just 1,000, despite the fact that much of the river's sediment by then was fill-ing in parts of Lake Mead. When Hoover Dam was completed in 1935, the flow of sediment past Yuma dropped to 13 million tons a year, a small fraction of the former load. Lake Mead, in turn, began to fill at the rate of 137,000 acre-feet per year. The slow filling of Lake Mead was one reason for the construction of Glen Canyon Dam three decades hence. Much of the sediment that once flowed down to Lake Mead now remains in Lake Powell. And so on up the river.[39]

Indeed, dam proponents viewed "sediment control" as one of the beneficial purposes of water development projects in the basin, to prevent sediment from clogging irrigation canals and other facilities. According to President Truman's 1950 National Water Commission Report, the two primary water *pollutants* in the Colorado were mineral salts and *silt*. In its comprehensive planning document for the Colorado, tellingly entitled "A Natural Menace Becomes a National Resource," the Bureau of Reclamation argued that "[c]ontrolling the silt load of the Colorado River and its tributaries to prevent damage is an important phase of water conservation for beneficial use." Water developers viewed the downstream movement of sediment as a problem to be controlled, rather than as an important component of the natural ecosystem. Although Godfrey Sykes was fascinated by the ever-shifting channels of the river through the delta, he supported upstream dams because the resulting sediment control would cause the river to form a single, clear channel to the sea.[40] We needed to train the river to behave, like an unruly puppy.

From the perspective of reclamation engineering, the fact that dams trap sediment is a problem because it reduces reservoir capacity. From an ecological view, the string of artificial sediment traps all along the river has filled some areas and robbed others of material, causing major changes in habitats for pikeminnow and other native fish. Hundreds of miles of river bottom, including areas used as spawning grounds when flushed of sediment by the annual spring flows, are now buried under layers of accumulated materials. Downstream of each major dam, the once muddy river is replaced with clear tailwater releases. That relatively clear water is "hungry," meaning that it has a huge physical-chemical capacity to scour sediment from the stretches of river immediately below the dam. The river now erodes sediment from the bottoms of those reaches, which become "armored" with rocks that are too large to be moved by steady flows from the dams, and incised below their former beds. After completion of Hoover Dam, clear water scoured sediment from the river below at the rate of 36 million tons per year, and incised the bed as much as sixteen feet deep. Those conditions of incision and sediment depletion remain until sufficient sediment enters from downstream tributaries to restore the balance.[41]

The enormous power of the untamed river also shaped the river's channel and flow characteristics by flushing downstream the immense debris flows that wash from steep tributaries into narrow stretches of the canyons during short but intense summer storms. Vivid descriptions are provided by some of those fortunate enough to have witnessed, but lucky enough to have survived, the sheer force and the unbelievable volume of water and earth carried into the main river channel during debris flow storms. Railroad engineer Robert Brewster Stanton witnessed one such storm as his party began its escape from "Point Retreat" in the Grand Canyon after three members of his first expedition drowned:

> As the rain increased, I heard some rock tumbling down behind us, and, looking up, I saw one of the grandest and most exciting scenes of the crumbling and falling of what we so falsely call the everlasting hills. . . . In a few moments, it seemed as though the slopes on both sides of the whole Canyon, as far as we could see, were moving down upon us, first with a rumbling noise, then an awful roar. As the larger blocks of rock plunged ahead of the streams, they crashed against other blocks, lodged on the slopes, and bursting with an explosion like dynamite, broke into pieces, while the fragments flew into the air in every direction, hundreds of feet above our heads, and as the whole conglomerate mass of water, mud, and flying rocks, came down the slopes nearer to where we were, it looked as if nothing could prevent us from being buried in an avalanche of rock and mud. It was a scene of the wildest fury of the elements!

River guides caught in a similar recent debris flow in Diamond Creek described a flood "knocking over trees like they were matchsticks," and carrying the expedition's supply trucks all the way down to the main channel, where they washed 1,000 feet downriver. Such flows can carry boulders that weigh tens or even hundreds of tons downslope to the river below.[42]

Debris flows that wash into the main river at tributary junctions can constrict channel width considerably. Because the same volume of water must travel an equal distance within a narrower conduit, water velocity and turbulence increase, and the resulting rapids become more difficult to navigate. One of the most difficult rapids in the Grand Canyon, at Crystal Creek, was relatively mild until a debris flow choked the channel in 1966. Before Glen Canyon Dam was erected, the river roared through the canyons each spring, reworking channel constrictions caused by periodic debris flows and washing cobbles and even large boulders downriver. Now, the balance between these two forces is disrupted. Debris flows continue to occur periodically, but the river's ameliorating force has waned.[43] Some analysts fear that if this trend continues, portions of the Colorado in the Grand Canyon run by thousands of tourists every year will no longer be navigable.

Water, fish, sediment, even rock. All once flowed or traveled with relative freedom down (and in the case of fish, also up) a river unimpeded by artificial obstacles. But the river carried much more than water and sediment. It also carried nutrients, organic matter, and other sources of food and energy. Disruptions to the movement of those materials caused serious ecological impacts as well.

Changing the Flow of Food and Energy

Like many people, I love to sit or walk on the banks of rivers and streams and watch the current flow. Not being a very successful fisherman, I can cast a fly for hours just to watch it float downstream, unmolested. Undoubtedly, the fish prefer it that way. When I think about it, though, it is not the water I watch with such calm fascination. It is whatever the water carries. Sticks, leaves, insects, thin mats of soil, or dislodged plants. Much of what the river bears, of course, cannot be seen by the naked eye. This continuous aquatic pipeline, one of food and energy and raw materials, forms yet another series of ecological connections.

As in many fields of natural science, we used to view rivers as a series of component parts. Hydrologists evaluated annual and seasonal patterns of water flow. Fish biologists studied the distribution and life histories of particular species.[44] Water chemists looked at nutrient cycles or sources and fates of pollutants. Within those fields, scientists studied connections between upstream and downstream reaches. Fluvial geomorphologists, for example, suggested that stream channel morphology and sediment loads existed in a state of dynamic equilibrium. The channel constantly adjusted to flows of material from upstream, and those changes in shape in turn influenced patterns of sediment flow.

In the 1970s and 1980s, researchers began to make connections among different components of river systems. They began to posit relationships between biological communities and physical conditions extending from a river system's headwaters to its mouth. According to the "river continuum concept" first proposed by a group of aquatic biologists in an article published in 1980, to understand river systems better we must focus on gradients of physical components and processes within entire river networks. The constant flow, storage, and use of ecosystem "parts," such as energy, food, and nutrients, all are "regulated" by the physical flow patterns of water, sediment, and other materials. Translated into plain English, and at some risk of oversimplification, the same kinds of connections and processes that govern the natural flow of water and sediment also explain the constant flow and cycling of food, energy, nutrients, and other materials that feed and mold river ecosystems. The pioneers of this concept proposed that the two systems—physical and biological—are logically connected. This is one of those ideas that seems obvious *after* someone else proposes it, but scientists needed to jump out of their discrete disciplines to do so.[45]

In one major tenet of the river continuum concept, different river reaches are characterized by how they produce and consume food. "Autotrophic" (meaning essentially self-feeding) communities produce a large percentage of the food they consume through primary productivity—conversion of light energy to food via photosynthesis. "Heterotrophic" communities receive much of their food from external (or "allochthonous") sources. These might include leaf pads or pieces of wood and their associated insects drifting downstream, or terrestrial insects or other foods that fall or are washed from land to water, like those baby swallows devoured by Colorado pike-minnows. The relative degree of autotrophy or heterotrophy in any given river reach can be measured by the ratio of primary production to consumption, which can be measured by the ratio of photosynthesis, in which food energy is produced from sunlight by green plants, to respiration.[46]

Often the headwater tributaries are relatively shaded, allowing less light penetration and therefore less photosynthesis. But these narrower parts of a stream usually have abundant external sources of insects, organic matter, and nutrients from the close canopies of riparian vegetation. As that material drifts downstream and the river widens, allowing more light penetration, more food is produced by aquatic plants, including algae and rooted vascular plants. This shift is fueled by the downstream transport of soluble organic matter and larger particles of floating or drifting detritus from upstream. Whether food and energy are exported or created internally, any excess material moves downstream and is used by communities below. Nature rarely wastes food and energy.[47]

All of this has important implications for river systems like the Colorado, whose natural connections have been severed by a string of dams and other artificial obstructions. The implications of these severed connections were recognized by aquatic ecologists Jack Stanford and James Ward: "The river continuum, however, is profoundly interrupted when dams are employed by man to impound or divert river flow." As an example, Stanford and Ward cited the Gunnison River, one of the major tributaries to the upper main stem of the Colorado. In its natural state, the macroinvertebrate community was dominated by the lumbering willow fly (*Pteronarcys california*). Willow fly nymphs were present in such numbers that, according to these biologists, one could fish for an entire day from the specimens collected from beneath a few rocks. Willow fly nymphs also would float downstream on leaf packs, but "willow flies, leaf packs, and much of the river are gone today," replaced by an entirely different macroinvertebrate community in the wake of the dams that disrupted the river's flow, causing what Stanford and Ward refer to as "profound changes."[48]

Based on this recognition of the impact of dams, Stanford and Ward proposed the "serial discontinuity concept." The discontinuity theory effectively replaces the continuum model of free-flowing rivers to explain more accurately the structure and function of regulated (dammed) rivers. In free-flowing rivers, the natural flow of materials is in a state of dynamic equilibrium. Natural aquatic ecosystems are not, of course, free from barriers. Plants and animals retain energy and material, as do physical structures such as beaver dams or debris dams. However, those ecosystem components can be disrupted by floods or other factors that allow more material to flow downstream. In a "stable" ecosystem those competing factors of retention and release are in relative balance. The patterns of flow change (hence are dynamic), but vary with conditions and reflect balance over time (thus equilibrium).[49]

Dams and other artificial structures transform those systems into an alternating series of flowing rivers and impounded lakes. Artificial barriers interrupt the flow of water, sediment, energy, living organisms, and detritus. Stanford and Ward suggested that this causes upstream, downstream, or sometimes neutral shifts in stream characteristics such as temperature, species abundance, community diversity, or ecosystem measures such as photosynthesis/respiration ratios. These changes are more or less significant depending on the river and the degree of modification.

The Colorado is the epitome of a regulated river that can be described according to the serial discontinuity concept. The roughly 300-mile stretch of the Colorado River from Glen Canyon to Lake Mead exemplifies the profound changes a dam can cause in the aquatic ecosystem by interrupting a river's flow of energy and materials. Biologists believe there was an "instant change" in the Grand Canyon aquatic ecosystem when Glen Canyon Dam began operation, causing "a fundamental difference in the very nature of the riverine ecosystem, a change no less profound than the sudden illumination of a darkened room."[50]

Before the dam, the river in this reach was generally warmer and far more turbid than it is now. The sediment-laden waters protected juvenile fish from predators, but reduced light penetration through the water column, resulting in relatively low levels of primary productivity. Conditions were not highly favorable to aquatic plants, and biologists described the river as an "aquatic desert." Production of benthic (bottom-dwelling) invertebrates in this portion of the Colorado was low even relative to other rivers in the desert Southwest. Unlike many river systems in other parts of the country, even this far downstream the river was heterotrophic; it did not produce most of its own food. Instead, fish and other species in the Colorado River food web relied on drifting materials from upriver. Large wood flows carried nutrients, organic matter, and large numbers of aquatic and terrestrial invertebrates. Those stowaways on flotillas of detritus became unwitting sources of food for downriver biological communities. Early river runners boasted of the giant bonfires they could build on the beaches of Glen Canyon with the massive amounts of driftwood in the predam system. Undoubtedly, they did not realize that they were robbing downstream biota of their food. The much larger disruption, however, came when the dam virtually eliminated this natural downstream flow of matter and energy.[51]

When Glen Canyon Dam began operation, the turbid, relatively warm water that once flowed past Lees Ferry became uniformly clear and cold. Sunlight now reaches much farther down into the water column, providing the energy needed for higher rates of photosynthesis. The system generates more food. At the same time, the once silty bed of this stretch of river is now scoured of sediment, and provides a favorable substrate for aquatic plants. This section of the river, relatively unproductive before, is now covered by dense beds of large, filamentous algae called *Cladophora*, growing in "veritable underwater forests, covering the stream bottom from bank to bank." Billions of tiny but highly productive diatoms attach to the *Cladophora*, providing nutrition and shelter for small crustaceans, insects, and other sources of food for fish. As a result, the Glen Canyon tailwaters have the second highest reported levels of gross primary productivity in North America.[52]

This new aquatic bounty, along with similar changes in the riparian zone, helps to support the blue-ribbon trout fishery for which the Lees Ferry reach is now famous. But while the river is now far more "productive" than it was before the dam, the predam ecosystem has largely been displaced. The last humpback chub was reported from the dam's tailwaters in 1971. By blocking the natural flow of materials from upriver, the dam changed the basic nature of the aquatic ecosystem from a heterotrophic one that relied primarily on food drifting downstream to an autotrophic one that relies most heavily on instream photosynthesis. Whether those changes are "good" or "bad" is a value judgment that depends as much on human as on ecological considerations (and is addressed in later chapters).[53]

Lake Powell, however, serves as a sink for nutrients, especially phosphorous. In the natural Colorado, phosphorous constantly flowed downstream, bound to particles of sediment. Now, more than eighty percent of the river's nutrients remain behind the dam along with the silt. This blockage affected the next artificial ecosystem downstream in Lake Mead, which boasted one of the most thriving artificial fisheries in the Southwest before Glen Canyon Dam was built. Lake Mead supported a highly productive community of phytoplankton, large populations of thread-fin shad that ate the plankton, and a booming population of trophy-sized striped bass that fed on the shad. The phytoplankton, then shad, and then bass declined precipitously in the 1970s. Researchers traced this decline to the trapping of phosphorus in Lake Powell. What little phosphorus escaped Lake Powell was used by the new community of algae immediately below Glen Canyon Dam. Of course, the retained phosphorous now fuels phytoplankton growth in Lake Powell, causing a process known as eutrophication. As phytoplankton die and sink toward the bottom, their decay causes oxygen depletion and related adverse impacts.[54]

The severe reduction in the transport of nutrients now held within Lake Powell, Lake Mead, and other reservoirs also affects the river downstream. Nutrient levels below Parker Dam remain unnaturally low, except where replaced by agricultural runoff. Apparently, this change in nutrient levels caused a parallel shift in phytoplankton communities to "nonedible" forms characterized by biologists as a "trophic 'dead end'" for a stable food chain.[55]

Artificial obstacles and flow modifications also blocked the natural flow of organic materials throughout the river corridor. Godfrey Sykes reported that powerful spring floods along the lower river uprooted dense thickets of willows, poplars, or cottonwoods and swept them down the channel. By 1921, after the first dams had been built north of Yuma, areas formerly covered with driftwood were largely bare. Only larger logs, timber pilings, and railroad ties remained. A decade later,

Sykes reported, once "widespread and conspicuous" accumulations of driftwood were now virtually gone. "The marvelous amount of this drift material when the river was carrying its several burdens freely into tidewater was one of the most striking features of the lower delta." Driftwood once accumulated "in great fields and windrows" up to two miles back from the shore. These massive accumulations of detritus fed biological material to the delta and formed low ridges or crests that served as natural levees that contained the lateral spread of spring floodwaters. The dams trapped this material upstream, causing significant changes in the morphology of the lower river and delta.[56]

Although the river's ecosystems can be harmed by the reduction of some constituents, other materials cause harm when their concentrations increase beyond natural ranges of occurrence or in places they do not belong. Dams and other river and watershed modifications in the Colorado River basin also led to those kinds of impacts.

Concentrating Pollutants

Travelers along U.S. 191 just west of Moab drive past a huge pile of uranium mill tailings in the Colorado River floodplain. Rainwater and groundwater leach chemicals from the pile into the river, with potential impacts to fish and other aquatic life. The U.S. Department of Energy recently decided to truck those wastes to an upland landfill.

Since the mid- to late 19th century, parts of the Colorado Plateau have been mined extensively for gold, silver, copper, and more recently, radioactive materials such as radium, vanadium, and uranium. By mining, milling, and processing those materials, we create huge piles of rock and low-grade ore called tailings, from which pollutants erode into the river and its tributaries. Several large mines, hundreds of smaller ones, more than twenty processing mills, and innumerable tailings piles dot the basin. In 1979, a tailings pond burst near Church Rock, New Mexico, on the Puerco River, a tributary to the Little Colorado River. Almost 100 million gallons of liquid and 1,100 tons of sediment overflowed more than twenty-five miles downstream.[57]

Damming the river concentrates chemicals that used to be flushed and dispersed by the free-flowing river. Glen Canyon Dam disrupted the movement of mercury and other toxic and radioactive pollutants through the system, causing them to be concentrated in the sediment and water column in Lake Powell and further bioconcentrated in fish and probably other species (such as birds that eat the fish). Bass with tissue concentrations of more than 700 parts per billion (ppb) of mercury have been reported in Lake Powell, in excess of the U.S. Environmental Protection Agency's (EPA) recommended limits for human consumption. In the turbid upper portions of the reservoir, mercury concentrations in the water were measured at six ppb, levels that inhibit phytoplankton primary productivity. We do not know whether that problem is worse during prolonged dry spells, because no one collects the necessary data. The natural runoff into the Colorado River system also is high in selenium. Relatively high selenium concentrations have been found in Lake Powell fish, at levels that may inhibit egg production in pikeminnows and other native fish. Similar concentrations of selenium and other pollutants at Kesterson National Wildlife Refuge and elsewhere caused significant bird mortality and other adverse effects. We know little about the magnitude and effects of these pollutants in the sediments, waters, and biota of Lake Powell and other reservoirs in the Colorado River basin, primarily because they have not been studied extensively.[58]

Other toxins also accumulate farther downstream. Lake Mead is now experiencing pollution from pesticides, perchlorate, and pharmaceuticals, and plans are emerging to relocate the sewage outlet from Las Vegas to alleviate the pollution of Las Vegas Bay, an arm of Lake Mead. Selenium, other heavy metals, persistent pesticides and herbicides, and other toxic pollutants are found along the lower Gila and Colorado rivers, sometimes at levels that may cause reproductive failure in fish and other adverse effects. Return flows and wastewater from the Mexicali valley can carry 70,000 tons of fertilizer and 110,000 gallons of insecticide annually. Selenium in the Sea of Cortéz has been measured at levels up to fourteen times those recommended by the EPA in the United States.[59]

Water development and land use changes in the basin also made the river more salty. The Colorado River system is naturally more saline than most other U.S. rivers. Throughout much of its geologic prehistory, large portions of the watershed were covered by saline inland seas. Marine sediments deposited over millions of years left soils and subsoils with high concentrations of soluble salts. Due to natural erosion into the river and its tributaries, as well as salt leaching by subsurface flows and discharges from natural saline springs and seeps, the Colorado River historically carried salt loads of between 200 and 1,000 milligrams per liter of total dissolved solids, depending on flow conditions and other factors. Salinity was higher during annual low-flow periods, when less water was available to dilute the salts.[60]

Human-induced changes exacerbated this naturally high salinity. The EPA estimated in 1971 that human development more than doubled the average mass of salts that reach the river. Inefficient irrigation of arid lands overlying saline formations causes excess water (water not used by crops, evaporated from the soil, or transpired by the plants) to seep through saline soils and groundwater into the Colorado River or its tributaries. Land disturbance in the watershed from grazing, road construction, and development increases erosion of saline soils, especially during summer storms and flash floods. Smaller but more concentrated discharges of salt come from oil, gas, and mining operations and other industrial facilities. Water use and diversion, as well as evaporation from reservoir surfaces, concentrates salt in the river. The EPA estimated in its 1971 study that thirty-seven percent of salinity concentration in the river was caused by irrigation, and another twelve percent by reservoir evaporation. This doubling of the river's salinity causes significant adverse effects to agricultural, municipal, and industrial water users, including reduced crop yields, removal of lands from farming, and damage to household, municipal, and industrial equipment. The Bureau of Reclamation estimated losses of between $311 and 831 million per year in 1986 dollars. In the lower river, irrigation return flows pumped from saline, oversaturated soils at times caused U.S. water deliveries to Mexico to be extremely saline, endangering Mexican agriculture and prompting international protests and negotiations.[61]

The Living Artery

In his classic memoir of the Southwest, *Desert Solitaire*, Edward Abbey referred to the Colorado as "the living artery of the entire area."[62] Although Abbey used this metaphor to contrast the river and its narrow but lush riparian zone with the arid bleakness of the land it traverses, the analogy is even more appropriate than he suggested. In its natural state, the Colorado served as a healthy artery that transported the energy and materials needed for a fully functioning continuum of aquatic and associated terrestrial ecosystems—water, sediment, energy, nutrients, and other sources of food, and not least of all, the fish and their genes.

When human arteries clog, blocking the flow of blood, oxygen, nutrients, and waste products, the health of the whole body is threatened. The same is true in the Colorado River and in other river systems that have been fragmented by dams and other obstructions. Relatively few rivers in North America, or even significant stretches of river, remain free-flowing. Dr. Arthur Benke of the University of Alabama calculated that in the entire continental United States, all but a single river segment longer than 621 miles was substantially altered in some way. Of more than 100 river segments longer than 124 miles, only forty-two remain free-flowing. Similarly, a 1982 analysis by the EPA and the FWS found that in eighty-one percent of the nation's waters, including more than half of all perennial waters, fish communities were adversely affected by various structural changes.[63]

The same is true in the entire northern third of the globe. A study published in 1994 by two Swedish scientists identified damming as "one of the most dramatic and widespread deliberate impacts of humans on the natural environment." According to their analysis, the number of large dams in the world increased sevenfold from 1950 to 1986, to about 39,000. By roughly the turn of the century, the World Commission on Dams reported that this number had grown to as many as 48,000. As a result, Stanford and Ward concluded that "[a]ltered ecosystems below dams and diversions are now the most prevalent lotic [riverine] environments on earth," and have a far greater impact on aquatic ecosystem health than chemical pollution.[64]

In the Colorado River, the decline of the pikeminnow is just one symptom of this disruption of the river's natural connections. This majestic fish once was abundant in the entire warm water portion of the Colorado River basin, from the upper Colorado and Green rivers through the lower basin and their respective tributaries, including the San Juan and the Gila sub-basins. Now, the species is confined to roughly a quarter of its historic range. Although accurate counts are difficult, according to the best available estimates, fewer than 9,000 of these fish remain, mainly in the Green River sub-basin. Scientists believe that fewer than 1,000 pikeminnows survive in the upper Colorado River sub-basin, mainly in the area of Grand Junction and the lower Gunnison River. The virtually isolated population in the San Juan River barely hangs on with an estimated nineteen to fifty individuals.[65] Although information on predam fish populations is scant, researchers believe that the pikeminnow probably was the "dominant fish" in the Grand Canyon before 1900. By 1977, when the first comprehensive research was conducted on fish populations in the canyon (a decade and a half after Glen Canyon Dam was completed), pikeminnow were gone, along with bonytail and roundtail chubs. The last wild pikeminnow was reported in the Gila River drainage in 1950. "The massive, salmon-like runs of the Colorado squawfish [pikeminnow] near Yuma, Arizona, where farmers would pitchfork thousands into horse-drawn wagons for use as fertilizer in the mid-1800s came to an end." By 1975, pikeminnow had disappeared from the entire lower basin.[66]

Like an artery, a river is usually considered a linear system, or part of a branched network of linear systems. In arteries, however, the flow of blood is constrained within largely impervious walls. Natural rivers are not confined within their channels. Rivers ebb and flow laterally, and form a network of connections to the adjacent riparian and terrestrial ecosystems and to the groundwater below. Artificial changes to the Colorado River and its watershed have disrupted those lateral and vertical relationships as well. Those changes are explored in the next chapter.

CHAPTER THREE

Only the Hills Will Know:
Changes in the Watershed

Driving downriver from Boulder City, Nevada, the small hometown of Hoover Dam, I headed south on U.S. 95, and then east on Nevada 163 toward the river. I was visiting places along the lower river into Mexico, trying to get a firsthand impression of the many ways in which we have modified the river. Suddenly, the dark, desert night gave way to the bright, flashing lights of Laughlin, Nevada. Laughlin is a miniature Las Vegas, complete with Harrah's, Golden Nugget, Flamingo, Riverside, Colorado Belle, and other huge casinos. Because I was trying to study the river and not the gaming industry, I crossed over the river to quieter Bullhead City, Arizona, which is much smaller and lacking in casinos. But this provided only a partial escape. Some casinos have motels in Arizona and free boat rides so you can sleep in one state and gamble in another. The river runs between straight, rock-armored banks, held in place beneath concrete walls that protect the casinos from the inevitable rise of the river.

The next day, I drove south along the Arizona side of the river. The Colorado River floodplain widens from what is typical up on the Colorado Plateau. There is considerable development, some old, some quite new. Strip malls, fast-food restaurants, subdivisions, motels, superstores, all in the dusty desert of the Mohave valley. A lot of the construction is new. South of Needles I passed expensive-looking new houses with docks and boat ramps leading right to the water from the back door, obviously in the path of the next floods.

At Lake Havasu, where the river is restrained behind Parker Dam, adobe homes are built right down to the reservoir. More big strip malls,

a Home Depot, a Wal-Mart, and the "Cowboy's Truck City RV Mart."
The RV Mart must do a very brisk business, based on the chain of riverside RV parks along much of the rest of the lower river. Water Wheel
Resort. Hidden Valley. Riverside Recreation Community. Lost Lake.
"Lost Lake" indeed; what about the lost river? In between the new
developments are fields of alfalfa, cotton, and other crops. A sign read:
"Do not enter when flooded." A Bureau of Reclamation official I inter-
viewed the previous day mentioned that we cannot just move all of these
people out of the floodplain. But we continue to move them *in* at a rapid
pace. In a period of less than three years in the mid-1990s, the U.S.
Army Corps of Engineers issued 151 permits for dredging, excavation,
grading, filling, and other activities that alter or destroy riparian habitat
along the lower Colorado to facilitate docks, boat ramps, subdivisions,
RV parks, utility corridors, marinas, and construction or maintenance of
flood control structures.[1]

All of this development severs or disrupts lateral and vertical habitat
connections that are important ecologically, as are the linear connections discussed in the previous chapter. Many of these changes are less
conspicuous to the casual observer than the obvious blockages caused by
huge concrete dams. These modifications extend laterally from the river
through its floodplain and riparian zones, into a series of connected
upland ecosystems, and downward into connected groundwater. First,
we will look at changes in important habitats at the water's edge and
their impacts on another species of endangered "big river fish." Then,
we will move upward in the watershed, and look at wider-scale changes
that affected fish and other resources in the headwaters and other tributaries of the Colorado River. Although most current restoration programs focus on the main stem of the river, significant habitat losses and
resulting impacts to fish have occurred in the upper reaches as well.

Habitats at the Water's Edge: Disruptions to River-Floodplain Connections

Along with the Colorado pikeminnow, another big river fish that
evolved to survive the fury of the Colorado was the razorback sucker
(*Xyrauchen texanus*). Like the pikeminnow, razorbacks are in serious
trouble throughout the length of their former range. Originally present
throughout the Colorado River basin from its headwaters in Wyoming,
Colorado, and New Mexico to the river's delta in Mexico, razorbacks are
now limited to twenty-five percent or less of their historic range.
Although the declining status of the species was known long before

(razorbacks were listed as endangered under Utah law in 1973 and Colorado law in 1979), the U.S. Fish and Wildlife Service did not list razorback suckers as an endangered species until 1991. Once numbering in the tens of thousands or more, razorback populations have plummeted dramatically. Where the largest populations survive, ironically in artificial impoundments like Lake Mead and Lake Mohave, only hundreds remain. Perhaps most troubling, what is left of the dwindling razorback population is downright geriatric. There is little evidence of "recruitment" of young fish—successful natural spawning, survival of larvae to the adult stage, and integration into the natural population. Razorback suckers can live for more than forty years. In areas in which populations remain, age estimates range from twenty to forty-four years old.[2]

In a small warehouse converted into a fish hatchery just outside of Grand Junction, Colorado, hundreds of thousands of tiny razorback sucker larvae swim in circles in tanks four feet in diameter. Biologists spawned these larvae artificially from native fish taken from the river, and ultimately will release the juvenile fish into the natural environment. In other parts of the basin, past efforts to stock young razorbacks failed abysmally. From 1981 to 1990, more than 13 million hatchery-spawned razorbacks were released into historic habitat in the Gila, Verde, and Salt rivers in Arizona. None are known to have survived in the long-term. More recent efforts to introduce razorback young into the upper Colorado River basin also largely failed. Although there are probably multiple reasons for these failed restocking efforts, biologists believe that the principal problem is predation by nonnative (introduced) fish species. So these painstaking efforts to spawn a new generation of razorbacks in artificial hatcheries succeeded mainly in producing high-cost fish food for exotic species of fish, and perhaps in giving the introduced species even a greater edge over their native competitors.[3]

Recently, biologists have been more successful in efforts to supplement adult in-stream populations by using natural and artificial off-stream ponds as rearing habitat.[4] Hatchery-reared fish are not introduced into the main stem of the river until they reach adult size, by which time they can successfully evade predation and compete more effectively for food, habitat, and other resources. Successful introduction and survival of some adult fish is important because it does help to prevent outright extinction of razorbacks in the river, and may constitute the last chance for this species. But this approach has problems as well. Even fish reared in artificial ponds show low survivorship, possibly due to their relative inability to adjust to the flow, turbidity, temperature, water quality, and food sources in the natural river.

Hatchery stocks also pose troublesome issues of population genetics (discussed in chapter 5).

Relying on artificially bred fish reared in artificial habitats also does not fully meet the goal of the Endangered Species Act, which is to restore natural, *self-sustaining* populations in relatively native habitats. This purpose flows logically from the statutory definition of "conservation," which is to restore species to the point where affirmative efforts provided for under the act no longer are necessary.[5] Meeting that more ambitious goal would require fish to spawn successfully under natural conditions, and a sufficient percentage of the offspring to survive to adulthood in the wild, without artificial off-stream rearing habitat, to replenish the population over time.

In the predam river, riparian wetlands, tributary mouths, backwaters, and inundated floodplain habitats were critical to survival of young fish. After spawning, larvae escaped predation and took advantage of preferred temperatures, food sources, and other conditions in those areas until they grew to adult size. Dams, water withdrawal structures, channelization, and levees eliminated many of these areas or cut them off from the main channel. As explained in chapter 2, disruptions in the natural flow and redistribution of sediment changed the process by which these backwater habitats form and reform. Apparently, reduction or elimination of backwater habitats impedes the ability of the native species to replenish populations naturally.

The upper Colorado River is not nearly as developed as the lower river. Grand Junction, Colorado, is the largest population center along the entire river, with a population in 2000 of about 42,000 people. But much of the upper Colorado flows through federal lands, and many parts of the river are locked in narrow, cliff-bound canyons. Human development occurs mainly within the broader, flatter zones of the river. To protect development in these areas, we build dikes, levees, and other control structures to prevent the river from intruding into former backwater habitats during spring floods. Those control structures compound the harm caused by water withdrawals. By taking so much water for irrigation and other uses, we reduce the natural flow of water needed to support backwater habitats adjacent to the main river. Thus, although scientists continue to blame the razorback's decline on a range of factors, including dams, water diversions, changes in temperature, and pollution, the most immediate barrier to species recovery appears to be the loss of floodplain and backwater habitats relied on by razorbacks and other native fish for spawning and rearing. Most of this lost habitat is not in the main channel of the river, but at the water's edge.

Several hundred miles downstream of Grand Junction, below the monster dams that now artificially define the Colorado River system, is a long stretch of river that runs relatively placidly through a broad expanse of semi-Sonoran desert. Before human development, dense riparian forests lined large portions of this reach with varying mixes of species, sometimes extending miles from the main river channel. In the Mohave valley, now the target of so much riverside development, Lieutenant Joseph Christmas Ives in 1857–58 portrayed a riparian zone varying from one to seven miles wide of "inviting meadows, with broad groves of willow and mezquite [sic]." In Cottonwood Valley, he found "graceful clusters of stately cottonwoods in full and brilliant leaf."[6]

In 1910, Dr. Joseph Grinnell from the University of California at Berkeley floated the lower river from Needles to Yuma, and later provided a more systematic description of the lush riparian vegetation that filled broad alluvial valleys. Grinnell described forests dominated by cottonwood and willow along with a diverse understory of shrub vegetation, and emergent marshes of cattail, bulrushes, tules, and cane to the south. These riparian zones provided some of the most important bird habitats in the Southwest. The higher terraces above the river hosted an expanse of deep-rooted honey mesquite and a shrub community of saltbush, inkweed, quail bush, and other drought-tolerant vegetation. Native cottonwoods and willows tolerated the frequent spring floods in which the river spread from its banks across the broad floodplain of the lower Colorado. They depended on the river's annual flood cycle to distribute seeds; spread soil, organic matter, and other nutrients; flush salts from alkaline soils; and otherwise nourish and renew the vitality of the riparian zone. Farther downstream in the delta, the spring floods spread water and nutrient-rich silt across broad swaths of flat terrain, and supported almost 2 million acres of riparian, freshwater, brackish, and tidal wetlands. This area once supported an estimated 200–400 species of vascular plants, and animals such as jaguars, beavers, deer, coyote, and large flocks of waterfowl.[7]

The native riparian forest was also the historic nesting habitat of a small songbird called the southwestern willow flycatcher (*Empidonax traillii extimus*), which the U.S. Fish and Wildlife Service listed as an endangered species in 1995.[8] Just as loss of riparian backwater habitat and the presence of exotic species are among the most significant problems for the razorback sucker, loss of this natural riparian forest habitat and replacement of much of what is left with invasive plants is a major underlying cause in the decline of the flycatchers.

The riparian forests of the lower Colorado suffered first at the hands of steamboat operators who plied the lower river in the middle part of the 19th century, before railroads overtook them as the dominant channels of commerce. The Yuma Crossing State Historical Site hosts a steamboat exhibit with huge pistons and a wheel seven or eight feet in diameter. Boats with names like *The Searchlight*, *The Mohave*, and *The Explorer* carried cargo from Port Isabel at the mouth of the Santa Clara Slough and Robinson's Landing in the main channel as far north as Callville (now submerged beneath Lake Mead), to avoid high Mexican import fees and overland freight costs from California. Steamboat companies cut riparian forests extensively as the only readily available source of fuel. By 1890, almost none of the mature native cottonwoods remained. Before the dams, however, those losses were temporary. Each spring flood brought seeds from upriver and conditions in which they could germinate and thrive, leading to natural replenishment over time.[9]

Later insults were increasingly difficult to reverse. Beginning in the late 19th century, settlers cleared large areas of riparian forest for agriculture, and more recently for residential and recreational development. By the mid-1980s, more than seventy-five percent of the Mohave, Parker, Palo Verde, and Yuma valleys had been converted to agriculture. Hundreds of river miles of formerly forested riparian habitat lie beneath Lakes Mead, Havasu, Mohave, and other reservoirs. And as discussed more extensively in the next chapter, in much of the remaining riparian zone the original plant community was replaced by invasive salt cedar (*Tamarisk* spp.). Out of an estimated 400,000–450,000 acres of riparian vegetation along the lower Colorado River before the dams, only 126,000 acres remain, and just 23,000 of those acres support native vegetation. The rest is dominated by salt cedar.[10]

Once the river's former fury was controlled by Hoover Dam and other structures, even more of the floodplain was put to use for agriculture and other development through channelization, dredging, riverbank straightening and armoring, construction of levees, and other engineered changes. By 1996, federal agencies had constructed 114 miles of levees and 168 miles of bank line stabilization along the lower Colorado River between Davis Dam and the southern international boundary with Mexico. In combination with regulated flows and diversions from dams, these engineered structures sever the riparian zone from the river by preventing seasonal inundation. Depending on location, water projects can either raise the riparian zone water table to the point where riparian vegetation is flooded or lower it below the root zones of native riparian species. Water structures thus have cut not only the linear but also the lateral and vertical flows of water, organic materials, and other nutrients on which the native riparian plant communities once relied. Groundwater pumping along parts of the lower river corridor exacerbates this problem. As a result, the vast majority of historic flycatcher nesting habitat is now gone. Many patches of remaining cottonwood-willow forests apparently are not suitable for flycatcher nesting because of an absence of adjacent areas of standing water or saturated soils.[11]

In part due to this massive disruption and loss of riparian forest habitat along the lower Colorado River and other regions in the Southwest, the southwestern willow flycatcher population has plummeted dramatically. In recent years, fewer than 500 flycatchers have been documented throughout the species' historic range, in 300–500 identified "territories." Many of the remaining territories contain small numbers of birds (five or fewer), often with no documented nesting success, and consist of small patches of highly fragmented habitats the distance between which impairs migration, success in locating potential mates, and recolonization. As a result, in 1997 the Fish and Wildlife Service indicated that "at such low population levels, random demographic, environmental, and genetic events could lead to extirpation of breeding groups and eventually render this species extinct, even if all extant sites were fully protected."[12]

Unfortunately, this small songbird is not alone in its plight. The Lower Colorado River Multispecies Conservation Program (discussed in chapter 7) is a collaborative, multiagency effort to mitigate impacts to various species that formerly thrived in this region. The program is designed to address the needs of twenty-seven species of birds, fish, mammals, reptiles, amphibians, and plants, culled from a candidate list of 149 "special status species" because they are imperiled in some way. So many species are at risk along the lower river because riparian habitats are particularly critical in a region that is otherwise so arid, with few places of refuge and nourishment for birds and other water-dependent species. Lieutenant Ives was forced by the slow progress of his boats across the mudflats of the lower river to ride overland to Yuma on a borrowed horse to arrive before the mail left. He commented on a country "quite destitute of vegetation," with occasional growths of mesquite, cottonwood, and willow along the river but otherwise bare hills and desert shrubs in the gravelly beds of the valleys.[13] Even today, the contrast is evident. Driving U.S. 95 from Parker to Yuma, Arizona (the straight route), you leave the green fringes of the Colorado and cross an open desert populated only by sparse shrubs separated by swaths of bare, hard soil.

Riparian areas are transition zones between terrestrial and aquatic ecosystems, with many relationships to the associated river. In the case of the Colorado, which in its natural state had little in-stream primary production, the main source of food energy fell or drifted into the river from the riparian zone. Riparian vegetation also shaded the river and moderated temperatures, stabilized the banks, bound the soil, and filtered and slowed the pace of runoff. The lower Colorado River between Grand Canyon and the delta is the largest riparian and wetland habitat in the Sonoran Desert, a slim ribbon of green within hundreds of miles of aridity. To put the importance of this riparian oasis in perspective, the Sonoran Desert it bisects covers about 120,000 square miles, is 870 miles long, and up to 400 miles wide.[14]

One key to recovery of the southwestern willow flycatcher and other species is restoration of sufficient portions of wetlands and riparian zones along the lower Colorado River corridor. Indeed, this is one of the main agenda items for the Lower Colorado River Multispecies Conservation Program. The severed connections between the river and its adjacent habitat do not stop, however, in the river's immediate riparian zone. As with all large rivers in major drainage systems, the health of the river is connected to the condition of the land in the entire watershed. Next, we look at changes in land use and habitats beyond the water's edge that also had profound impacts on the health of the aquatic ecosystems of the Colorado River basin, especially in headwaters and tributaries where native trout and other species of fish and wildlife have been affected.

Disruptions to Land-Water Connections in the Watershed

"America's Redrock Wilderness." This is one of the current slogans in the battle to preserve and protect America's wild and natural lands.[15] The redrock wilderness message is sold with images of magnificent but stark southwestern landscapes—sandstones in a rainbow of colors; deeply cut and steeply sloped arroyos and canyons; vast expanses of desert abounding in life, but only to those patient enough to look at the right times and in the right places. Without meaning to negate the importance or legitimacy of the issue, there is a serious irony here. Much of the American Southwest now known for these sparse but beautiful scenes was anything but bare redrock wilderness several hundred years ago, and in many places as recently as a century and a half. This does not mean that those wonderful places are less deserving of protection. But we should not assume that their current appearance is their only "natural" condition.

When European visitors—as Spanish conquistadors and missionary priests, early trappers and cattle ranchers—first crossed the Colorado Plateau and other parts of the Colorado River watershed, the landscape was far different from what we see now (and what tourists travel thousands of miles to visit). Rivers that now trickle or run dry for much of the year, now identified on maps as "intermittent," ran year-round (or nearly so) through healthy channels with vegetated riparian zones. From these riparian areas to the adjacent valleys and up to the mountain meadows, some parts of the Southwest were considerably more luxuriant and supported wildlife in larger numbers than exist today. Many of those changes have harmed fish and other wildlife and resources.

Early European explorers, whether in pursuit of beavers or gold, described lush valleys and meadows throughout much of the Colorado River basin. "Soils were undisturbed, most creeks were clear, and beaver were abundant." Early trapper James O. Pattie complained of being "fatigued by the difficulty of getting through the high grass, which covered the heavily timbered bottom" along upper portions of the Gila River in Arizona. The river at that point was a beautiful clear stream about thirty yards wide, filled with fish. Farther downstream he reported a river "skirted with very wide bottoms, thick-set with the musquito trees [mesquite], which bear a pod in the shape of a bean, which is exceedingly sweet."[16]

Although the recorded impressions are necessarily anecdotal rather than systematic, they are certainly contemporaneous and reflect a vastly different regional ecology than we see today. In a 1946 survey (published in 1950) of the recreational resources of the Colorado River basin, in language more eloquent and effusive than is typically allowed modern bureaucrats, the National Park Service reported:

> The stories that early explorers brought back from the western wilderness of the incredible lushness and wealth of vegetation and wildlife read like fairy tales to us now, who will never see birds in flocks of millions that darken the earth for hours as they pass, or grass as tall as a man's shoulder, thronged with endless herds of buffalo, extending to the horizon. Civilization has left for present generations so little trace of the overwhelming primeval abundance of the "wild west" that the imagination, even when aided by the early records, now finds it almost impossible to reconstruct the original picture.

Although the lowlands of the Colorado River basin were arid compared to other regions, the Park Service nevertheless wrote that "by contrast with conditions in much of the basin today, the forage and the wildlife habitat that it sustained were almost unbelievable." Some areas of the basin that are currently most arid, such as the upper Green River

Valley and Red Desert of Wyoming, had grass interspersed with the sage and supported bands of antelope and bison, sage hens congregating at water holes by the thousands, and elk in vast numbers. In the vicinity of what is now Fontanelle Reservoir on the Green, naturalist and author Ann Zwinger noted that the Hayden expedition described "broad river bottoms richly carpeted with good grass and big cottonwood groves." But, she lamented, "Now there is no shade whatsoever."[17]

Somewhat better watered areas like the valleys of the Duchesne, White, and Yampa rivers hosted elk, bison, sage hen, and mountain plovers in "uncounted thousands." Upland plover—by the mid-20th century threatened almost everywhere—was "rather common" in 1877. Grass in the Wasatch Range grew six to twelve *feet* high, but is now six to twelve *inches*, with desirable forage now replaced by weeds and shrubs that are far less palatable to resident wildlife, such as cactus and juniper and other shrubs. Elk were nearly exterminated from the Colorado River basin by 1910, and once abundant bighorn sheep had dwindled to a fraction of their former strength by 1900. Thousands of wild turkey once inhabited the forested mountains of Arizona, New Mexico, and southern Colorado, but now are seldom seen.[18]

Farther south in the basin, often in areas now set aside as national parks and monuments for their stark canyon beauty and with a primary focus on geology and scenery, the contrast between what was and what is sharpens in focus. Parts of Canyonlands National Park, now immense playgrounds of colorful sandstone, were "surprisingly fertile." The Henry Mountains of southern Utah, the last major mountain range in the continental United States to be charted, "still stood in a valley of grass." Antelope frequented the San Rafael Swell of south central Utah as late as 1895, but disappeared soon thereafter. The 1950 National Park Service report said: "Even the vast, unpopulated southern desert lands which are commonly imagined as eternally unchanging would in many places be unrecognizable to those who first discovered them."[19]

The same was true even in what are now known as some of the most barren deserts of North America, all the way to southern Arizona. Again, according to the National Park Service: "Those familiar with the bare, dry, sandy bed of the present intermittent Santa Cruz River at Tucson, in the heart of the Arizona desert, will find it difficult to believe that prior to 1776 water there was so abundant that travelers sometimes had to wait for weeks before it was possible to get horses across its swampy expanse." Tucson was established in what was then a swamp. The perennial river supplied water, the abundant vegetation provided wood, and the rich low-lying lands supplied meadowlands for stock. "Where now there is only powder-dry desert, the grass once reached as high as the head of a man on horseback." The river bottom was densely forested with a wide belt of giant mesquites extending miles beyond the riverbanks. The region teemed with quail, doves, peccaries, and beaver. After 1800, the valley was denuded of mesquites, the ironwood was stripped bare to feed lime kilns and build fence posts, and the wildlife that depended on this vegetation disappeared. The same story unfolded in "nearly every other mesquite valley of the Southwest."[20]

Scientists who have compared various early accounts with current conditions in southern Arizona concluded that the region's riparian zones were not just different but ecologically diverse: "These accounts suggest that there was considerable complexity in riparian zones in the 1800s. The valleys were wetter and more open than today, groundwater levels were higher, and channels were not as deep. But the precise conditions varied from place to place and from time to time."[21] Marshes, meadows, and riparian forests all occurred in different locales and at different times.

Dams, diversions and other engineered changes are directly responsible for some of the most severe changes to habitats closest to the river. When the Laguna Dam was built in 1910, for example, Dr. Grinnell observed that the water level was raised "conspicuously" for at least ten miles downstream. Cottonwoods were killed by the rising water levels around their trunks, and mesquites and other second-tier plants were drowned, replaced by mudflats growing invasive arrowweed. Along other stretches of the lower river, projects designed to divert water and control floods produced an opposite but equally damaging effect. Flood control projects such as riprapping to stabilize banks, channelization, and channel dredging lowered the adjacent water table dramatically. This desiccated areas that were periodically moist, draining backwaters and robbing riparian plants of their sustenance.[22]

Artificial water structures, however, do not explain all of the changes to the landscape of the Colorado River basin. Many of those vast changes preceded the major dams by decades or more. From the perspective of efforts to restore the Colorado River, both the reasons for and the impacts of this virtually basinwide change in land condition are significant. Here, the lateral connections flow in both directions, from river to uplands and back again. The effects of these environmental changes in headwaters and other tributaries, and in the lands they drain, can be attenuated far downstream in the main stem of the Green and Colorado rivers, which are the focus of the large-scale restoration programs discussed in most detail in later chapters. But scientists have long cited destruction of riparian and upland habitats as a major cause of impairments to fish populations in tributaries throughout the Colorado River drainage, including the lower Colorado River, the Gila River, the San Pedro River, the Salt River, and others. This is especially true where grazing and other land disturbance occurs in or near the riparian zone. In addition to the four big river fish that are the focus of so much attention now, affected native species in various rivers and streams include the woundfin, speckled dace, and spikedace, as well as various species of native trout. Efforts are under way in various Colorado River sub-basins to protect some such species, but those efforts suffer from continued changes in watershed health caused by 200 years of land use changes in the basin.[23]

Land Use, Climate, and Habitat Change in the Colorado River Basin

The first and probably the most significant artificial cause of terrestrial ecosystem change in the Colorado River region—indeed across the entire West—I will refer to by analogy as the "Second Mountain Meadows Massacre." For those not familiar with Utah history, the "first" Mountain Meadows Massacre occurred in 1857 in a high mountain meadow in southwestern Utah, in which a group of pioneers from Arkansas, Missouri, and Kansas were slaughtered en route to California. Accounts of the incident vary, but according to most current versions Mormon settlers tricked an encamped wagon train of emigrants to cede their weapons in return for protection from American Indians, who the emigrants mistakenly believed were responsible for an ongoing siege. The unarmed party then was summarily executed. The Mormon community may have been reacting to the imminent arrival in Utah of a large force of federal troops. Or, they may have considered this retribution for their own persecution at the hands of the citizens of Missouri and Kansas a decade before, or for more recent anti-Mormon incidents in Arkansas. They may have feared new harm at the hands of these groups, or simply overreacted to taunting by the emigrants. Whatever the explanation, the massacre stands as one of the ugliest chapters in the history of relations within the white population in the West.[24]

The Second Mountain Meadows Massacre is less shocking as an indicator of human nature, but equally telling as an example of the impacts we have had on the ecology and hydrology of the west. As described by the National Park Service: "In 1844, Mountain Meadows, on the summit of the Colorado River–Great Basin divide in southwestern Utah, was a beautiful valley of green grass that extended for miles, bounded by piñon- and juniper-covered hills. By 1877, the grass had been mostly replaced by desert shrubs as a result of sheep grazing, and the appearance of the landscape was so greatly altered that landmarks of 20 years before could no longer be identified. Today, the 'meadow' is dry, and gray with sage. Great gullies have drained away the water, so that even if ungrazed, the grass could not come back." The first Mountain Meadows Massacre was over in a matter of minutes. The second was almost as rapid from a relative perspective. Livestock were first moved into the region in 1863, and serious erosion began within just seven years.[25]

The onslaught of livestock grazing seriously affected the ecology of the western range, especially as grazing intensity increased astronomically during the late 19th and early 20th centuries. In just two years from 1907 to 1909, grazing of public lands increased from 600,000 cattle, 59,000 horses, and 1.7 million sheep to 1.5 million cattle, 90,000 horses, and 7.7 million sheep. In Utah, sheep herds nearly quadrupled from 1 million in 1885 to 3.8 million in 1900. In Arizona, cattle herds skyrocketed from 50,000 in the mid-1800s to 1.5 million in 1891.[26]

Congress recognized the rapid deterioration of western range conditions when it passed the Taylor Grazing Act of 1934, although those reforms ultimately did little to reduce that harm. The U.S. Forest Service issued a comprehensive report in 1936 documenting that grazing led to serious erosion, associated flooding and siltation of streams and reservoirs, and other adverse impacts to watersheds. A series of later government reports indicated that large percentages of public grazing lands were in fair to poor condition even after decades of attempted reforms. In fact, by the early 1990s the U.S. Bureau of Land Management and the Forest Service reported that riparian areas on public lands were in "their worst condition in history," and that grazing was the principal cause. Debra Donahue summed this up by arguing that "[g]razing's ecological impacts are more widespread than those of any other human activity in the West."[27]

In an analysis of proposed water projects in the Colorado River watershed, the National Park Service reported in 1950 that huge herds of livestock, stocked in concentrations far beyond the capacity of the land to feed, ate native grasses literally to the ground and trampled the remaining growth underneath their hooves. The soil below was left unprotected from sun, wind, violent summer thunderstorms, and then invasion by less desirable species of plants. Erosion accelerated with later construction of roads, ditches, canals, and diversion dams. Soil turned to dust, which blew hundreds of miles away. The resulting erosion was catastrophic in a region where topsoil may replenish at the rate of an inch in 1,000 years. With natural topsoil seldom more than seven inches deep, natural replenishment could thus take 7,000–10,000 years.[28] Many of the native plants cannot naturally reseed under the altered conditions.

In addition to ecological impacts to the western range, this increased erosion was important from the perspective of proposed dams and other water projects. One justification for many of the major dams, including Hoover and Glen Canyon, was sediment control. Although the river naturally carried massive loads of sediment, artificially increased erosion only exacerbated that situation in the main stem of the river. There is, of course, a serious irony here. From the perspective of current efforts to restore ecosystems in the main stem of the Colorado River in the Grand Canyon and elsewhere, a significant problem is the *absence* of enough sediment.[29] As discussed in chapter 2, most of the natural sediment flow downriver is trapped behind Glen Canyon Dam and other dams up and down the river. In chapter 5, we will look at efforts to modify dam operations to increase the degree to which the remaining sediment from tributaries (such as the Little Colorado River) can be mobilized to rebuild important beaches for river runners and critical backwater habitats for fish. As discussed above, however, changes in the riparian ecosystems of tributaries also had serious adverse effects on upstream fish and wildlife populations, many of which have not recovered to this day. Will efforts to restore watershed health in upstream tributaries reduce the flow of eroded sediment downstream, to the detriment of efforts to restore the Grand Canyon? This requires an historical look at the effects of grazing and other development on the streams and riparian zones in the Colorado River basin.

Travelers in the Southwest see countless dry riverbeds with steep, cut banks, some ten to 100 feet deep. It is tempting to view them as entirely natural and entirely in place, because they do resemble the much larger natural canyons and because they are so ubiquitous. Like the expanses of sage and sand amidst the red rock, many are of recent origin, but by now have reshaped our image of the region. We are drawn to them aesthetically because they *seem* like part of the arid Southwest, just the smallest attributes of canyon country. But whether they really "belong" requires some historical perspective. That outlook could affect restoration goals, as discussed in later chapters, if it modifies our concept of the state of the ecosystem to which restoration efforts might aspire.

The natural streams in much of the Southwest had relatively flat or gentle banks lined with marshes, fertile floodplains, and other riparian vegetation. Higher spring flows overflowed the banks and were absorbed gradually by streamside plants and the stable soils they bound. Riparian vegetation and soils absorbed runoff from even violent summer storms, and released it into the stream channel slowly. Those conditions provided favorable in-stream habitat for native trout and other species. According to some researchers, however, once cattle and sheep denuded the vegetation from the banks, there was nothing to stop the force of the silt-laden water. Erosion cut the banks until gentle streams turned into steep arroyos, often in relatively short periods of time. The original stream channels were not wide enough or deep enough to carry so much water, and quickly degraded in the path of the larger and faster floods. Compared to the gentle slopes of the original southwestern streambeds, arroyos are trenched with rectangular cross sections, steep sides, and water flowing only in the deep channel.[30] Perennial streams became intermittent not only due to human diversion, but because water no longer was retained in the stream bank soils and plants. The natural sponges were gone.

The Fremont River in south central Utah, which flows into the Dirty Devil and then the Colorado, is a good example of the rapid arroyo cutting that occurred during just a short period in the late 19th century. Originally the channel was 100 feet wide and coursed within a gentle floodplain. The fertile Fremont valley, like many parts of southern Utah, was settled during the latter half of the 19th century largely for agriculture. A large flood in 1896 eroded the channel twenty-two feet below its original level. A series of floods continued over the next several decades until the channel was converted to a trench approximately seventy feet deep and 1,320 feet wide. Irrigation works were destroyed, and fields and most of a town site eroded away.[31] Other southwestern waterways suffered similar fates. Oraibi Wash, once a small stream, was cut into a twisting gully eighty miles long and twenty to eighty feet deep. Chaco Canyon was a continuous, clear stream with no gully in 1849. By 1877, it was an arroyo sixteen feet deep and forty to sixty feet wide; and by 1930 it had grown to thirty feet deep and 200–300 feet wide. The stream now flows in the canyon only during flash floods. Kanab Creek did not begin to erode until 1883, but just three years later it was a gully sixty feet deep, seventy feet wide, and fifteen miles long. Kitchen valley, along a tributary to the Paria River, was among the latest areas in southern Utah to be settled with cattle ranches. Since 1915, a channel forty feet wide has been cut through the valley, destroying shallow lakes and meadows. The community of Paria, Utah, about fifty miles above the junction with the Colorado, was destroyed entirely just a decade after it was formed by flash floods that "washed away farmhouses and fields and converted the narrow stream channel into a wash that extended in places from rock wall to rock wall."[32]

Similar stories can be told of countless watersheds throughout the region. Franklin Flat in the San Simon Valley of Arizona was covered with "grassy, flower-spangled meadows" in 1885, with grass reportedly stirrup-high along a continuously flowing valley with practically no banks. By 1895, 50,000 cattle grazed in the valley. By 1934, the green meadow was replaced by drifting sand, and the now intermittent river "had cut a chasm through the heart of that once-fertile land." The Gila River in Arizona, the southernmost significant tributary to the Colorado, once was a large, permanent stream with clear water confined within a well-defined channel flanked by rich marshes and dense riparian growths of willow, cane, and other species. Now it is intermittent, with its lower reaches almost always dry, and bordered instead by "desolate wastes of sand and silt." The San Pedro River, once the Gila's only perennial tributary, flowed through a level valley with banks covered in willow, cottonwood, sycamore, and ash, interspersed with lush growths of native grasses. By 1900, the river bed had entrenched from ten to forty feet, the vegetation was largely gone, and cattle and horses had cut channels from the hills to the river.[33]

Harold S. Colton published an article in 1937 describing the recent and rapid history of erosion of the Little Colorado River.[34] The chronicler of the 1583 Espejo party, Luxan, wrote of a "fine, beautiful, and selected river almost as large as the Del Norte [Rio Grande], containing many groves of poplars and willows."[35] Based on personal interviews with pioneers who brought livestock to the lush valley in the 1870s and early 1880s, Colton reported that the area still hosted fine groves of cottonwoods and adjoining hillsides covered with grama grass. When the intense rains of the early 1890s arrived, however, the impacts of grazing pressure left the river and its channel and banks vulnerable to erosion. Many of the old cottonwoods were uprooted and washed downstream. The river overflowed its narrow banks, carrying massive amounts of sediment downstream. In just two years, the river had cut back fourteen feet. Ultimately, Colton lamented: "We can see the Little Colorado as it was fifty years ago; a narrow, perennial stream lined with cottonwoods and willows. We see it now in a wide, sandy bed, which is dry much of the year. The willows have departed and only a few gnarled cottonwoods remain of the once extensive groves. The surrounding hills that once bore a good stand of grama grass are now covered with a desert pavement of polished pebbles. Navajo sheep see to it that no young trees get a start."[36]

So although the human massacre at Mountain Meadows was a tragic anomaly caused by a peculiar conjunction of tensions and circumstances, the ecological massacre at Mountain Meadows was just one example of a persistent phenomenon that altered the face of the region. The Commission on Water Resources Policy's Basin Study of the Colorado published in 1950 identified the massive scope of the problem in every state in the basin. In Arizona, erosion plagued nearly two-thirds of the state, with gullying on 45 million acres, severe gullying on 32 million acres, loss of more than three-quarters of the topsoil on 3.5 million acres, and half on another 20 million. In the Colorado River watershed in Colorado more than three-fourths of public lands and half of all private lands were eroding. Gullying occurred on forty percent of public lands in New Mexico, with large arroyo cutting common. Small drainages that one could once step across were one-quarter mile wide and 100 feet deep. Sheet erosion had uncovered the roots of pinyons and junipers over wide areas of the state. About half of the land in the basin in Utah was erosive, with gullying on 11 million acres. The cause of all this erosion, according to the Commission, was overgrazing. A full eighty-five percent of the basin by then was used for grazing, still with too much stock for the prevailing conditions.[37]

The degree to which livestock grazing is responsible for ecological and hydrological changes in the western landscape, however, remains in dispute. For decades, some experts argued that increased erosion and sedimentation resulted from overgrazing, but others attributed it to natural geologic factors. The Spanish conquistadors first introduced cattle to what is now the American Southwest in the mid-16th century, and several other periods of heavy livestock use occurred under Spanish and later Mexican rule. Those uses did not lead to similarly intensive ecological changes. But that may simply reflect issues of scale. Earlier grazing densities and the total acreage affected did not remotely approach those that occurred during the great western cattle boom of the 1880s and beyond. Likewise, despite the immense amount of grazing that began in the late 19th century, increased erosion and other impacts occurred even in ungrazed areas. Clearly grazing contributed to ensuing problems at least to some degree: "Grazing unquestionably weakened the old plant communities, leaving them open to invasion; it unquestionably upset the balance between infiltration and runoff in favor of the latter." Just as riparian forests along the lower Colorado could recover

from extensive cutting by steamboat operators, though, western grasslands were resilient and capable of recovery even after extensive grazing.
Recovery from late 19th century grazing pressure was made more difficult, however, due to a change in climate. At the same time that dramatically increased grazing pressure rendered many areas of the western
range vulnerable, the weather changed in ways that generated more
intensive floods.[38]

Much of the documented arroyo development occurred shortly after
cattle and sheep were introduced to the western range in such large
numbers, with visible accompanying impacts on vegetative cover, accelerated water runoff, and soil erosion. But for at least two reasons, grazing cannot be the only reason for the dramatic increase in soil erosion
and arroyo cutting beginning in the late 1800s. Arroyo development is
not just a modern phenomenon. Evidence of prehistoric arroyo cutting
and refilling is preserved in sedimentary materials, suggesting patterns
of erosion and fill that followed large-scale climatic changes. Erosion
occurs when increased precipitation causes more runoff, and the climate
in recent centuries has been relatively more arid.[39] Moreover, as discussed above, large herds of cattle were introduced to areas of Arizona
and New Mexico by Mexican ranchers in the late 1700s without the rash
of erosion caused a century later. Likewise, not all areas were affected
equally by the introduction of cattle by 19th century ranchers.

More likely, the significant increases in erosion and arroyo cutting in
parts of the Southwest during the late 1800s and early 1900s resulted
from the concurrence of intensified grazing and an increase in the inci-
dence of intense rainfalls in parts of the basin after periods of soil-
parching drought. Specific storms were identified as the source of widespread erosion from the late 1860s through the early 1900s. Periodic
changes in climate are part of natural climatic variability. The period
from 1906 to 1930 was one of the highest runoff periods in half a millennium, while the period from 1870 to 1894 was unusually dry. Increased grazing and the resulting loss of protective vegetative cover may
have made many of the watersheds far more susceptible to the effects of
the intensified rainfall and runoff that followed.[40]

Thus, it was probably the combination of severe overgrazing, other
intensifications in land use, and the onset of a cycle of intense rainfalls
that caused the epidemic of soil erosion, stream sedimentation, and
channel cutting that beset much of the Colorado River basin around the
turn of the 20th century, and that contributed to the impairment of
western stream ecosystems. However, at least one other relatively less
studied but equally disputed factor may have contributed to the prob-

lem. As with livestock grazing, the underlying culprit was the assumption of western settlers that the natural resources of the region knew no limits, and their failure to recognize the importance of one particular "cog" in the western ecosystem. Overtrapping of beavers in the early 19th century may also have contributed to the deterioration of western headwater stream channels.

Beavers and Watershed Health

Beavers dammed the rivers of western North America long before there was a Bureau of Reclamation. But they did so in a very different way. Rather than designing huge structures placed at a limited number of strategic locations, beavers built tens of thousands of much smaller dams in the headwaters of the many small drainage systems that fed the region's larger rivers and streams. The very smallest headwater streams generally are too small to dam, and spring floods in larger rivers wash beaver dams downstream. Most beaver dams are on intermediate headwater streams, but have significant downstream influence as well.[41]

Like the marshes and floodplains that lined the river valleys, and in conjunction with the high marshes and meadows that bordered the mountain streams, the beaver dams buffered flows so that large volumes of water from both winter snowmelt and intense summer thunderstorms would release slowly, stored in the multichambered sponge of soil, plants, and beaver dams. As noted by the National Water Commission, "The importance of beaver in the basin has never adequately been determined, yet beaver may be very helpful from the standpoint of sediment and water control in many small tributary streams." Moderated flow conditions and the beaver dams themselves also caused more sediment to be retained in headwater streams and riparian zones, rather than being flushed violently downstream. In effect, complexes of beaver dams slow the rate of water flow and sediment transport. One study measured fifty to seventy-five percent lower concentrations of suspended solids in stream water below beaver complexes. Even recently abandoned beaver dams have beneficial impacts on the riparian environment. Sediment collects where the pond used to be, leaving fertile soils along the banks where plants quickly take root and thrive, further stabilizing the soil and preventing erosion. Under these conditions, intense, erosive floods typical in the Colorado River basin likely were far fewer and far less severe.[42]

As with the impacts of grazing on vegetation and streams, human depletion of the West's beaver population was remarkably swift. French and American fur trappers began to work the Colorado River watershed in 1824, after most of the Rio Grande watershed had been trapped out. Trapping parties were remarkably efficient, even accounting for some degree of exaggeration. James O. Pattie boasted that his party of eight men would trap as many as sixty beavers in a single morning. Encouraged by high prices to feed the hat markets of a style-conscious Europe, trappers had every incentive to seek out beavers throughout the West. But as with so many other wildlife populations, the apparently limitless resource quickly became scarce under intense trapping pressure. By 1860, just decades after the trapping invasion began, a precolonial North American beaver population estimated at between 60 and 400 million was nearly extinct. Beaver recovery efforts begun in the early 1900s have restored populations to between 6 and 12 million, and Ann Zwinger commented that they are so frequent along the upper Green River that it is difficult to believe they were once trapped out of the whole drainage. But beaver populations remain a fraction of their original abundance.[43]

The overall impact of beaver eradication on western streams and rivers remains in dispute. Some scientists argue that beaver populations had largely recovered by the 1850s, decades before the increase in arroyo cutting occurred.[44] Other scientists believe that beaver eradication was potentially significant to the decline in western watershed health. Suzanne Fouty, a hydrologist at the U.S. Forest Service, studied beaver complexes from Arizona to Oregon. Fouty rejects the very premise of the debate over whether increased erosion and arroyo cutting in the Colorado River basin and elsewhere in the West were caused by intensive livestock grazing, a change in climate during the late 19th and early 20th centuries, or the loss of beavers. She believes that all three acted in tandem to help bring about the transformation of the western range—but the point of her research is that of the three, the role of beaver eradication has been least understood and most underestimated.

According to Fouty's thesis, obliteration of beaver populations by trappers left the region vulnerable to increased erosion and flooding. Had this loss not been followed almost immediately by the intensification of livestock grazing and then close on its heels by a shift in precipitation patterns, perhaps beaver populations would have recovered with no major impact, like the cottonwood forests of the lower Colorado after they were cut by steamboat companies. But overgrazing made the region more vulnerable to the ecological calamity to follow by denuding protective plant cover and by creating pathways for runoff and erosion. The decade from 1835 to 1845, immediately after the most intensive trapping period, turned out to be among the wettest in the last millennium. When the rains came, the existing beaver dams breached, and those dams were not replaced because the workers were gone. These natural buffers no longer stemmed the flow of storms, leading to massive erosion, flooding of stream and river channels, and significant transformation of western riparian ecosystems. Likewise, had beaver populations remained healthy when the impacts of overgrazing occurred and the storms arrived, Fouty surmised, perhaps the impacts would have been significantly smaller.[45]

The silver lining to this theory that beaver loss contributed, perhaps significantly, to the impacts of overgrazing and an increase in intense storms is that reintroduction of beavers to western stream systems may help riparian restoration efforts as well. In one study in southwestern Wyoming during the 1980s, scientists reintroduced beaver to restoration study areas and evaluated the impacts on ecosystem restoration success. Where beavers were reintroduced or allowed to recolonize an area naturally, water levels remained relatively constant throughout the year rather than following the now familiar pattern of torrential spring floods followed by trickle flows during the summer months. These conditions promoted rapid recovery of willows and other riparian vegetation, such that "full riparian recovery was underway." Suzanne Fouty's research provides similar evidence. Fouty created exclosures in which she either removed grazers or added beavers. Removing the grazers showed little impact on already affected areas; adding beavers, and necessarily the dams they built, served to trap and retain large amounts of sediment in the system. Likewise, Fouty underscores that even pristine-looking western rivers and streams are not "natural" or "healthy" from an ecological-historical perspective. "We are all so conditioned by what streams are supposed to look like," says Fouty. "You have seen post cards; you've seen movies. Everything you've ever seen in a river is a river that has been so highly damaged and that's what we are using for our model."[46]

The beavers may provide several lessons relevant to our exploration of restoration goals and strategies in later chapters. When the causes of aquatic ecosystem impairment are multiple and interrelated, solutions directed at single causes will likely be incomplete, and might fail altogether. Comprehensive restoration strategies may involve ecosystem components, such as beavers, that bring back some of the conditions and processes that contributed to natural ecosystem health. Finally, just as we might falsely view expanses of redrock wilderness in the Southwest as "natural," modern conceptions of southwestern stream channels as steep-banked arroyos and cut channels probably miss the mark relative to prevailing ecological conditions before the arrival of western settlers.

Identifying appropriate restoration goals will require some inquiry into predisturbance hydrological and ecological conditions, choices about what past conditions can be and should be achieved, and possibly trade-offs between resources in various parts of the ecosystem. To get a preliminary sense of some of these complexities, it is useful to review some of the possible implications of the changes in watershed health discussed above for Colorado River restoration efforts.

Implications of Watershed Health for Colorado River Restoration

Although the linear connections described in chapter 2 are relatively intuitive, the lateral and vertical connections described in this chapter are also important to the health of various components of the Colorado River ecosystem, along with its contiguous and associated riparian and upland habitats. Water, sediment, nutrients, and other resources flowed from the predam river to the adjacent marshes and riparian zones. Those riparian areas, in turn, shared energy and resources with adjacent upland habitats and beyond, and the wildlife that depended on them. But the connections worked in both directions. Vegetation adjacent to the waterways and upland bound the soil and buffered storm flows, protecting the river from the effects of the region's highly variable climate and hydrological conditions.

Like the linear connections portrayed in the last chapter, dams and diversions severed some of these lateral connections between land and water. Reduced flows and "regulated" flows designed to move consistently throughout the year rather than according to the natural annual hydrograph curtailed or eliminated the lateral movement of water into the adjacent terrestrial habitats. Human activities throughout the region over the past one and a half centuries, however, also seriously disrupted the connections flowing from land to water, resulting in significant changes to the hydrology and ecology of the Colorado River and its tributaries. Those activities include logging in the headwater forests, grazing in mountain meadows and tributary riparian zones, and more recently, building homes, shopping centers, RV parks, and resort casinos virtually to the water's edge. All of these activities contributed to an increase in the speed and volume of runoff to the region's rivers and streams, and impaired habitat for native fish. At the same time, these changes significantly altered the region's riparian forests and marshes, its grasslands, meadows, and other habitats that once hosted far greater numbers and diversity of wildlife.

The overriding message is that watersheds are not just bodies of water, but are connected intimately with the entire associated land mass. The hydrologist's definition of "watershed" is "[t]he entire surface drainage area that contributes water to a lake or river." Ecologists now conceptualize watersheds as not one but a collection of "ecosystems composed of a mosaic of terrestrial 'patches' that are connected (drained) by a network of streams." Under this view, river systems are not just two-dimensional (linear and lateral) but four-dimensional in nature. They are longitudinal or linear (upstream-downstream); lateral (river-floodplain to riparian zone and beyond); vertical (groundwater-surface water); and temporal (all three spatial dimensions change over time). Thus, for the Colorado as with all major rivers, the health of the river depends in part on what occurs on associated watershed lands. River restoration efforts cannot end at the water's edge, or even in the relatively limited riparian zone. We cannot view rivers as lines of water flowing downstream, disconnected from their sources of food and energy. In restoring the Colorado River and other water bodies around the country, we must consider connections in all dimensions.[47]

But we also must take steps to explore more rigorously the predisturbance conditions that should inform decisions about restoration goals. Of the now-extinct passenger pigeons that reportedly blackened the midwestern skies with their numbers, more than a half century ago Aldo Leopold wrote: "Men still live who, in their youth, remember pigeons. Trees still live who, in their youth, were shaken by a living wind. But a decade hence only the oldest oaks will remember, and at long last *only the hills will know*." A few pages later, in an even more melancholic passage, he warned: "Perhaps our grandsons, having never seen a wild river, will never miss the chance to set a canoe in singing waters."[48]

The generation that knew the passenger pigeon is now gone, as are the generations that knew the wild rivers of the American Southwest, at least rivers that were truly wild in an ecological sense. Of the native Colorado River basin, perhaps only the oldest cottonwoods now remember. Even they will soon be gone, as the average cottonwood lives slightly more than a century.[49] Then, only the red rock hills and cliffs will know the splendor and the abundance of the grasslands that once lay at their feet and the herds of antelope and bison that fed there, of the forests and meadows in the mountains and hills above them, the thickets of willows and groves of cottonwoods that lined the rivers and streams that carved their paths through them, and the massive flocks of birds that flew above them.

It may be that some of those changes are, by now, irreversible, a topic to which we return in later chapters. The degree to which we can

"restore" portions of the Colorado and other watersheds to conditions
that predated western settlement, and whether those changes are desirable, is scientifically complex and infused with value judgments we have
not yet made. Without understanding what the region was before we
built the dams, introduced livestock, and made so many other changes,
however, it is difficult even to make those choices.

Time will tell whether our grandchildren will see the rivers of the
Southwest and their adjacent habitats restored to something closer to
their wild condition, and have a chance to set canoes in waters that sing
once again. That will happen only if we shed our perception of the
Southwest as a land of barren red rock and sparse vegetation, of dry
streambeds meandering through rocky and sandy channels beneath
steep arroyos, and of flashy casinos built to the edge of an armored,
channelized riverbank to protect us from the river's natural flux. To
restore the Colorado and the other great rivers of the Southwest, we
first need to restore our vision and collective memories of what they
once were. Then maybe we will know as well as the hills.

So far, we have looked at the ways in which human development has
changed various physical aspects of aquatic ecosystems in the Colorado
River basin. We also caused fundamental changes in the ecology of the
Colorado River system, however, by artificially introducing significant
biological changes, especially through the intentional or accidental
introduction of nonnative species. Those changes are discussed in the
next chapter.

CHAPTER FOUR

Tree of the People: Tree of Life

The Grand Canyon Lodge, at Bright Angel Point on the canyon's north rim, sits atop one of the world's most magnificent cliffs. Not far from here, Clarence Dutton wrote: "In all the vast space beneath and around us there is very little upon which the mind can linger restfully. It is completely filled with objects of gigantic size and amazing form, and as the mind wanders over them it is hopelessly bewildered and lost. . . . Everything is superlative, transcending the power of the intelligence to comprehend it."[1]

The lodge also sits at the edge of the lush Kaibab Plateau, interspersed with dense forests and open meadows, and described by Dutton in similar superlatives as a "paradise of the explorer who, weary of the desert, wanders with delight among its giant pines and spruces, and through its verdant but streamless valleys." Before the livestock grazing era, the Kaibab Plateau was home to an immense quantity and diversity of indigenous wildlife, and even now supports large populations of deer and other animals. Down below, the Colorado River and the canyons through which it flows host some of the most unique ecosystems in the Southwest, in part because the region lies at the junction of three major habitat zones—the Colorado Plateau to the north, the Mojave Desert to the south and west, and the Sonoran Desert to the south. There is a wealth of natural history to be described in the vicinity of the Grand Canyon. At the park's North Rim visitor center a few yards away, however, the focus is almost entirely on scenery and geology.[2]

The most notable exception in the Grand Canyon Lodge is the large statue of "Brighty" the burro, in the observation room overlooking the canyon. Brighty was the original plumbing system for the lodge, hauling enough water to supply hundreds of guests and employees. (Water is now piped from Roaring Springs, partway down the canyon, to both the north and south rim visitor complexes.) Although he was made famous by the children's book *Brighty of the Grand Canyon*, it is ironic that a burro is the most prominent animal highlighted on the North Rim. Grand Canyon historian Stephen J. Pyne wrote: "Brighty did for the Grand Canyon what Smokey Bear did for forest fire. The baby boomers who grew up with it learned that, like the burro Brighty, the Grand Canyon should remain 'wild and free.'"[3] But neither Brighty nor any of his burro cousins are indigenous to the region, and no longer do they run wild or free. "Wild" burros in fact were escaped burros that became naturalized to the canyon and other parts of the west. "Naturalization" refers to the establishment and successful reproduction in the wild (without further human intervention) of a nonindigenous species, as when daffodils reproduce and spread across a field after a few bulbs are planted nearby.

Although introduced burros are not key to the restoration efforts that are the principal focus of this book, they illustrate the kinds of impacts that exotic species can have in this region. They also highlight the difficult problem of deciding what restoration goals are appropriate for large ecosystems that change significantly over time. Both of those interrelated issues are addressed in this chapter.

Wild Burros and the Goals of Environmental Restoration

Burros are not only out of place in the Grand Canyon, they did not evolve in the wild anywhere in the Western Hemisphere. Native to northwestern Africa, burros were brought to North America as domesticated beasts of burden. Miners introduced them to the Grand Canyon beginning in the 1860s. When some escaped into the wild, burros proved quite fertile, and reproduced and spread rapidly under conditions of ample forage and few successful predators amid the steep cliffs. New herbivores were introduced, but no new predators.[4]

Beginning in the 1920s, after concluding that burros were damaging vegetation, polluting water, and degrading wildlife habitat, the National Park Service began a burro control program. Rangers shot thousands of burros, but populations recovered after each effort. Finally, to the dismay of animal rights activists as well as some ecologists and advocates for park history, the Park Service issued a policy in the 1970s to eliminate burros entirely from Grand Canyon National Park. By the early 1980s, most of the animals were removed and relocated, the remaining burros were shot, and reportedly a single animal remained in just one of the three former herds (thus making further reproduction unlikely).[5]

The nature and extent of harm to the Grand Canyon ecosystem actually caused by introduced burros is the subject of dispute. An article published in the *Journal of Wildlife Management* in 1958 argued that native vegetation suffered "[w]herever burros are concentrated."[6] According to some ecologists, burros cropped plants to the ground, dug out roots with their hooves, extirpated grasses and some shrubs from heavily grazed areas, cut trails and thereby increased erosion along steep slopes, and contaminated springs and other water sources with their feces. Robert H. Webb and colleagues documented some of these impacts by a process of repeat photography. These scientists photographed a series of Grand Canyon sites a century after photos were taken at identical locations by Robert Brewster Stanton's railroad party, and compared the views to make observations about changes in ecology, topography, and other physical landscape features. Webb's researchers documented plant species that appear to have been locally extirpated by burro grazing, although other species such as Mormon tea were relatively less affected.[7]

On the other hand, some paleoecologists, scientists who study the ecology of foregone epochs by evaluating fossils and their locations relative to physical surroundings at the time, note that other herbivores such as the Shasta ground sloth, Harrington's mountain goat, and ancestral horses occupied essentially the same ecological niche during the Pleistocene epoch as wild burros now do. Those earlier species either died out due to a change in climate and forage or were hunted to extinction by early human populations from Asia. The naturalized burros, then, arguably fill the same ecological role as their Pleistocene antecedents. There are two competing theories about why the Pleistocene herbivores vanished, and the validity of the "same niche" perspective may depend on which of these explanations is more correct. If herbivores disappeared from the region due to a shift in climate and redistribution of vegetation, perhaps the remaining flora is inadequate or too sensitive to overexploitation to support similar populations of modern herbivores. Based on an analysis of fossilized dung, all components of the diet of Pleistocene herbivores remain along the modern river corridor, although that analysis does not demonstrate whether these plants occurred in the same density or condition. If the first human migrants to the region hunted prehistoric grazing populations to extinction, the ecological niche filled by those animals remains vacant. If so, allowing even an introduced population to thrive might not be all bad, and might in fact constitute a sort of restoration.[8]

The differing perspectives are highlighted by the long-standing debate between ecologists Paul S. Martin and Steven W. Carothers, both respected researchers of the Grand Canyon and its ecosystems. Martin notes that changes in the region over the past 40,000 years, as evidenced in the fossil record, dwarf the small changes depicted in Webb's repeat photography over the past century. Then, as now, the same or a very similar assemblage of plants existed, and was sufficient to support a diversity of herbivores. Why, he asks, should we object to the presence in the canyon of wild burros—whose genealogy can be traced to early North American horses—if they graze the same forage as analogous herbivores did millennia ago? Among the possible ecological benefits he identifies are that burros eat alien as well as native plants, burro carcasses could help support efforts to reintroduce native scavengers such as California condors, and burros could provide food for wolves should they be reintroduced to the region.[9]

Carothers counters that burros in the Grand Canyon were "overgrazing habitats and inflicting long-lasting damage to the ecosystem." Both plant and small mammal populations were suppressed in areas grazed by burros compared to those that were not. Carothers believes that our mandate is to restore natural ecosystems as closely as possible to conditions unaltered by humans. Burros are introduced rather than native, notwithstanding their distant evolutionary connections with species that existed in a far different ecosystem and climate ten millennia or more in the past. He is willing to compromise, however, by supporting a proposal to maintain burro populations on the contiguous Hualapai Indian Reservation, in part to benefit the tribe and in part as food for reintroduced California condors, which once inhabited the region as well. If this distinction is accepted, of course, wildlife and other ecological management decisions will be made according to artificial (national park/Indian reservation) rather than ecosystem boundaries.[10]

The debate over the fate of wild burros in the Grand Canyon ecosystem is a microcosm of a wider controversy throughout the West, but also illustrates that law and politics sometimes play major roles in establishing restoration goals, an issue to which we will return in later chapters. Burros and wild horses receive special legal protection under the Federal Wild Free-roaming Horses and Burros Act. Congress passed the act in 1971 to protect "all unbranded and unclaimed horses and burros on public lands of the United States." The act directs the U.S. Bureau of Land Management "to protect and manage wild free-roaming horses and burros as components of the public lands," and provides additional protection to wild horses and burros that stray onto private lands. As explained by the U.S. Supreme Court in upholding the statute against its most significant challenge to date:

> In passing the Wild Free-roaming Horses and Burros Act, Congress deemed the regulated animals "an integral part of the natural system of the public lands" of the United States . . . and found that their management was necessary "for achievement of an ecological balance on the public lands." According to Congress, these animals, if preserved in their native habitats, "contribute to the diversity of life forms within the nation and enrich the lives of the American people."

The court also cited Congress's directive that the responsible federal officials "protect and manage (the animals) as components of the public lands . . . in a manner that is designed to achieve and maintain a thriving ecological balance on the public lands."[11]

In addition to its ecological rationale, Congress sought to protect wild horses as "living symbols of the historic and pioneer spirit of the West." Although this perspective on the historical place of wild horses and burros in the western range is certainly valid, Congress's ecological judgment about the role of these animals is questionable. Spanish conquistadors brought both wild horses and burros to North America, which suggests that they should play no role in the restoration of a natural ecosystem inhabited by native species. Although wild burros had never existed in the western hemisphere, the same species of horse did roam North America up until 12,000 years ago. They either were among those large mammals hunted to extinction by early Native Americans or they died out due to climate change. Nevertheless, they were not part of the ecosystem when Europeans arrived.[12]

The debate over the place of naturalized grazing species in the West suggests broader questions about ecological "restoration" in what is now a significantly modified landscape, sometimes with far different species of plants and animals. The National Park Service based its decision to eliminate burros on the philosophy that the park ecosystem should be restored as closely as possible to "the condition that prevailed when the area was first visited by the white man." But why should the sudden appearance of Spanish conquistadors serve as the baseline for ecological restoration, as opposed to the condition of the canyon when settlers from Asia arrived several thousand years earlier? As Webb noted, in evaluating what this managed ecosystem should look like, "one must decide what perspective to take: a limited view based on a few generations of experience, or a longer view based on millennia. Whichever perspective one chooses, humans cannot easily be dismissed as an unnatural force in the environment." If the goal is to restore ecological conditions as closely as possible to what they were before the arrival of Europeans, wild horses and burros have no place on those portions of the range that are set aside for ecological restoration. But if we adopt a more functional strategy, in which stable ecological *conditions* are sought, wild horses and burros might have a legitimate role as surrogate species that fulfill the same or similar ecological roles as the extinct ungulates that roamed portions of the region, especially if native predators and scavengers such as wolves and condors are reintroduced as well. The U.S. Fish and Wildlife Service and a cooperating coalition of public and private organizations began to reintroduce condors to the wild in 1992, and by April 2000 more than 60 flew the skies above Arizona and California.[13]

These scientific and philosophical questions become even more complex in the case of other exotic species that have multiplied throughout the Colorado River basin. The Pleistocene species that Brighty and his progeny may have "replaced" are now gone, permanently. Other exotic species in the basin, both terrestrial and aquatic, displace native species that still exist. The proliferation of this wide variety of exotic species constitutes the third major ecological transformation of the region, and the one that may be the most difficult to redress. We will look first to introduced species of plants, then to a seemingly innocuous species of bird, and then to the wide range of exotic fish that now pervade the Colorado River basin.

Salt Cedar and Southwestern Riparian Ecosystems

In *Death Comes to the Archbishop*, her novel of Catholic missionary priests in the Southwest after the Mexican-American War, Willa Cather used tamarisk (Latin genus name *Tamarix*, commonly known as salt cedar) as a metaphor for the harsh southwestern deserts and the people and all other life that survived a sparse existence there:

> On the south, against the earth wall, was the one row of trees they had found growing there when they first came, —old, old tamarisks, with twisted trunks. They had been so neglected, left to fight for life in such hard, sun-baked, burro-trodden ground, that their trunks had the hardness of cypress. They looked, indeed, like very old posts, well seasoned and polished by time, miraculously endowed with the power to burst into

delicate foliage and flowers, to cover themselves with long brooms of lavender-pink blossom.

At the same time, however, she invoked tamarisk as a symbol of human settlement, even some comfort, in this harsh desert:

> Father Joseph had come to love the tamarisk above all trees. It had been the companion of his wanderings. All along his way through the deserts of New Mexico and Arizona, wherever he had come upon a Mexican homestead, out of the sun-baked earth, against the sun-baked adobe walls, the tamarisk waved its feathery plumes of bluish green. The family burro was tied to its trunk, the chickens scratched under it, the dogs slept in its shade, the washing was hung on its branches. Father Latour had often remarked that this tree seemed especially designed in shape and colour for the adobe village. The sprays of bloom which adorn it are merely another shade of the red earth walls, and its fibrous trunk is full of gold and lavender tints. Father Joseph respected the Bishop's eye for such things, but himself he loved it merely because it was the *tree of the people*, and was like one of the family in every Mexican's household.[14]

There is irony in Cather's depiction of ancient tamarisks as the "tree of the people" in the small villages of what is now New Mexico and Arizona. Like Brighty and kin, tamarisk is not native to the region. In fact, it most likely did not even exist there during most of the period covered in *Death Comes to the Archbishop*, the main story line of which begins in 1851 and ends early in 1889. According to most sources, various species of tamarisk were introduced to the New World through nurseries as an ornamental plant, first in New York City and Philadelphia in the 1820s, and by the middle of the 19th century in California. The "earliest definitive record of tamarisk in the Southwest" was in Galveston, Texas, in 1877, where it had already naturalized, and in an 1892 photograph in Arizona's Salt River drainage.[15]

Willa Cather was not ignorant of natural history. In acclaimed works such as *O Pioneer!* and *My Antonia*, she portrayed with significant realism the character of both the native midwestern prairies and the region as it was altered by pioneer agriculture. How, then, could she have made such a significant error about the natural history of the American Southwest? There are at least two possible explanations. First, Cather researched *Death Comes to the Archbishop* from Santa Fe during the 1920s. By then, tamarisk was a well established invader, and she

probably observed thickets of the encroaching shrubs during her journeys through the region. In other respects, she similarly described the
region as it was likely to have been *after* the arroyo-cutting floods of the
late 19th and early 20th centuries: "The very floor of the world is
cracked open into countless canyons and arroyos, fissures in the earth
which are sometimes ten feet deep, sometimes a thousand."[16]

A second possible explanation is that Cather accepted at face value
references to tamarisk in the journals of early Spanish missionaries. The
Dominguez-Escalante party, whose story is told in more detail in chapter 6, camped at a place near the current Utah-Arizona border they
named "Arroyo del Taray" (now known as Fort Pierce Wash). "Taray" is
defined as "tamarisk" in most modern Spanish-English dictionaries, but
it is not identified or defined as such in earlier dictionaries and botanical works published in Mexico. Some cite this as evidence that the
Spaniards brought the species to the Southwest much earlier. One botanist
wrote off this reference as a rare anomaly in the region. But that would
not explain how this Old World genus arrived in one or more isolated
locations in the Southwest. Tamarisk was notably absent from the journals and other papers of any of the other early explorers across the Colorado River and other parts of the region. Other scholars noted more
plausibly that the Spanish word "taray" may not refer to tamarisk. Early
Spanish explorers might have used the word to describe a riparian willow (*Salix taxifolia*) that is native to riparian zones in Mexico and the
southwestern United States, and is known in parts of Mexico as "taray,"
or "taray de rio."[17]

Whatever the reason, Cather was not alone in describing tamarisk as
an aesthetically pleasing part of the natural order of the Southwest. In a
wonderfully poetic essay in which she described the Colorado River
basin as a "sumptuous feast," Ellen Meloy included among the "thick-
headed pinyon-juniper broccoli" and the "tangy salad of hackberry, coyote willow, and other riparian greens" a healthy serving of "cottonwood
celery and tamarisk slaw." (Among other "exotic" foods in the banquet
she facetiously described "catfish bouillabaisse," "carp carpaccio," and a
"Russian olive.") Ecowarrior icon Edward Abbey described tamarisk
along the banks of the Colorado as if it were part of the natural landscape, delighting in his view of a great blue heron flapping its wings
"among the lavender plumes of a tamarisk tree," and joyously describing sandbars where "lacy blossoms of young tamarisk wave in the breeze."
Renowned nature photographer Elliot Porter, whose images eulogized
Glen Canyon before the dam in one of the Sierra Club's most famous
coffee table bibles, extolled the "velvety lawns of young tamarisks

sprouting on the wet sandbars." Historian Donald Worster reflected, perhaps accurately as a matter of human but not natural history: "Once, men and women recollected, the West had been a land of canyons leading on to canyons where tamarisk and cottonwoods rustled in a slight breeze blowing up at twilight."[18]

This is another example of the phenomenon discussed in chapter 3, in which our concept of what is "natural" can be shaped by current conditions. The ecological reality, however, is far more complex and controversial. Tamarisk now covers large swaths of the riparian zones in the Southwest, but the ecological and hydrological impacts of that invasion and the desirability of corrective measures remain in dispute.

By as early as 1920, some species of this prolific plant had spread to most of the major drainage basins of the Southwest. From there, it proliferated northward into the Great Basin, Rocky Mountains, and parts of the Great Plains, in many areas significantly displacing willows, cottonwoods, and other native vegetation. By 1929, tamarisk thickets had become a nuisance along irrigation canals and a "jungle" along the Gila River in Arizona. Recent estimates suggest that tamarisk covers up to 1.6 million acres from northern Mexico to central Montana, and from central Kansas to central California.[19]

In the Colorado–Green River system, the invasion began around 1925 and spread throughout the basin within two decades. The first known naturalized specimen in the region was collected by a botanist at Kanab, Utah, in 1909; by 1925 books reported tamarisk along the Virgin River, and it expanded rapidly thereafter. Along the main stem of the Colorado, apparently tamarisk was limited initially by the river's raging annual floods, which scoured new plants from unprotected banks. Survivors remained only behind boulders or other obstructions, or in backwaters with low flow velocities. No tamarisk is depicted in the photographs taken by the Stanton expedition in the 1890s. In their pioneer botanical survey of 1937 (discussed more in chapter 6), Elzada Clover and her assistant Lois Jotter found tamarisk at only a few sites in the Grand Canyon. In other reaches of the river, however, Clover and Jotter reported tamarisk as among the most prominent species. Similar patches appeared at the junction of the river with tributaries such as the San Juan and Green rivers beginning in the 1930s, and the plants began to spread more widely by the 1950s.[20]

After Glen Canyon Dam evened out flows in the river below, the tamarisk invasion intensified dramatically. In the late 1960s, Dr. Paul Martin found tamarisk "abundantly distributed" due to an "explosive spread" through the region; the intruder filled the banks along much of the available habitat in the canyon by the late 1970s. Tamarisk thickets thrived around Lake Mead and downstream riverbanks now protected from a once volatile river.[21]

A number of attributes can allow tamarisk to outcompete native plants, although that does not occur in all places and under all conditions. Like native cottonwoods and willows, tamarisk is a "phreatophyte." Phreatophytes thrive primarily in floodplains and the shorelines of lakes and reservoirs, where roots can easily reach riparian groundwater which is connected closely with surface water. Tamarisk does best where the groundwater table is no more than fifteen feet below the surface, and usually does not survive where the water table exceeds twenty-five feet. In some cases, however, where groundwater levels decline gradually, tamarisk roots can descend to nearly 100 feet.[22]

Although all phreatophytes produce large volumes of seeds that spread primarily by wind and water and that germinate readily in most riverside soils, tamarisk is particularly aggressive, producing a huge number of seeds over a long period of time. The flowering season for tamarisk extends from April through September or October, much longer than for competing species, although individual tamarisk seeds remain viable for only a couple of weeks. A mature plant can produce a half million seeds in one year. Thus, tamarisk seeds will germinate across a wide range of time after floodwaters clear the banks of existing trees and then recede, especially when other plant seeds are not present. Tamarisk seeds can germinate not only on moist soil but also in flowing water, and plants become established when the germinated seeds are stranded on a shore or sandbar. As noted by one analyst of the tamarisk invasion, "[t]he natural flood regime of rivers in the Southwest, which is characterized by receding snowmelt flows in late spring and early summer, is therefore ideally suited to colonization by tamarisk."[23]

Tamarisk also tolerates adverse conditions such as saline soils better than do native phreatophytes. Tamarisk has been found growing in Death Valley, where the groundwater has dissolved solids (salt) levels approaching five percent, although the plants do not thrive at those levels. In fact, tamarisk actually limits the ability of its competitors to take root by exuding salts from its leaves, thus increasing salinity in surrounding soils. Tamarisk is long-lived, and survives both drought and flood better than native species. Mature plants can survive submergence for up to seventy days. It even recovers more quickly in the wake of fires (via surviving shoots) than do native plants, which are not well adapted to fire. In addition, tamarisk was transplanted from its native Old World environments without the natural controls that limit its spread in its native habitat, such as insects, diseases, and other pests. This allows tamarisk to expand with no effective checks other than habitat availability.[24]

Nevertheless, some experts believe that tamarisk would not have spread so successfully absent dramatic artificial changes in stream regimens and streamside habitats. Where native phreatophytes such as cottonwood, willow, rabbitbrush, and greasewood remain, smaller numbers of tamarisk may invade but become just part of the otherwise native riparian plant community. Where the native plants are artificially cleared from drainage channels, or where reservoir shores and stream banks are altered significantly due to dam construction, tamarisk rapidly fills the newly created void with dense thickets that later-arriving natives cannot penetrate. Studies attribute the spread of tamarisk to reservoir construction and altered stream flows, which artificially generate extensive seedbeds that are perfect for tamarisk.[25]

Other researchers, however, argue that the tendency of tamarisk to displace native plants is exaggerated. Rather, they suggest that tamarisk only takes root on bare, disturbed, or highly saline soils that are otherwise unsuitable for native plants, or along eroded streambeds where the water table is lowered and higher terraces formed. Tamarisk can tolerate drier soils and lower groundwater than native species such as cottonwoods and willows. In addition, tamarisk can curb erosion, allow soil to accumulate, and otherwise "prepare" sites for native species that establish later but eventually will replace the tamarisk, in part by growing taller and shading out the tamarisk.[26]

The ecological impacts of tamarisk also remain in dispute, suggesting other difficult choices about restoration goals that we will revisit in later chapters. Large monocultures of tamarisk reduce the natural biodiversity of the region, supplanting not only native plants but also the insects, birds, and other species adapted to particular plant hosts. For example, many frugivores (animals that primarily eat fruit) and insectivores that inhabit native riparian communities avoid tamarisk thickets almost entirely because key food sources are absent. Some researchers found higher bird density and diversity in native plant groups compared to those dominated by tamarisk. Native herbivores such as bighorn sheep and deer disfavor tamarisk as forage, and tamarisk seeds are too small to provide useful food for birds or rodents.[27]

Again, however, other scientists disagree either in principle or in degree. Although tamarisk may not equal native plants in habitat value, it does provide nesting habitat and some sources of food, depending on the species of bird, amphibian, or small mammal. Where tamarisk provides *some* habitat value on a site where native plants cannot establish or survive anyway, arguably more total habitat is created. And in some places, especially where tamarisk occurs in mixed stands of native and introduced species, studies show improved bird diversities and densities. As a result, the U.S. Fish and Wildlife Service is reluctant to sanction complete tamarisk removal where it provides some habitat for endangered southwestern willow flycatchers and other species. In the Grand Canyon, black-chinned hummingbirds nest almost exclusively in tamarisk. Nonetheless, two researchers who evaluated tamarisk value for wildlife concluded that native riparian communities are generally superior for wildlife, with the possible exception of habitat for doves and bees.[28]

Tamarisk thickets can also alter significantly the physical characteristics of rivers and streams, although a range of factors can affect river morphology, making it difficult to reach conclusions about causes and effects from particular sources of change. But once again, views vary about the relationships among tamarisk (and other riparian vegetation) and river channel size and shape. Some scientists argue that tamarisk increases channel erosion by blocking natural overflow routes, which routes the erosive force of floods onto stream banks. This, they argue, can pose serious flood control problems by reducing channel capacity. William Graf cited two floods in Arizona in 1978 that caused agricultural losses of more than $9 million and urban losses of $75 million, which he attributed to the effect of tamarisk on channel capacity. In the upper Green River, others argue that tamarisk has reduced average channel width by seventeen percent, and that tamarisk root systems prevent redistribution of sediment by stabilizing riverbanks. But some researchers argue that changes in sediment delivery and river flows have much larger impacts on river morphology.[29]

The degree of environmental harm or benefit attributed to tamarisk affects riparian restoration efforts, to which we will return in chapter 7. As with wild burros in the Grand Canyon, choices must be made about whether tamarisk removal is desirable or appropriate as part of river restoration, and according to what criteria. Even if those decisions are resolved, questions remain about the extent to which tamarisk removal is possible given its massive current proliferation. Early programs of cutting, burning, and poisoning failed because the species regenerates so rapidly. More success has been possible with a combination of mechanical removal below the root crown, herbicide application, and long-term inundation, although biological control methods are being developed as well.[30] As we will see in more detail in later chapters, establishing restoration goals requires some sense of what is possible and what is not, given the magnitude of existing environmental change.

Tamarisk is not the only invasive species with detrimental impacts to the southwestern willow flycatcher and other birds along the lower Colorado River. Another culprit is a small species of bird that engages in an unusual form of parasitic competition.

Brown-Headed Cowbirds, Habitat Change, and Interspecies Competition

The Bill Williams River, the only significant tributary to the Colorado between the Gila and the Virgin (the confluence with which now lies beneath Lake Mead), enters from the east through a beautiful cattail marsh in the Bill Williams National Wildlife Refuge. Lieutenant Joseph Christmas Ives visited the confluence twice, once with Lieutenant Whipple's expedition in February 1853, and again on his own in the same month in 1857. The first time the river was thirty feet wide and several feet deep. Four years later, due to natural variation in climate, it was a "very narrow gully, through which a feeble stream was trickling," its former mouth overgrown with willow thickets. Just downstream, Ives reported extensive shoals, bars, braided channels, and islands, which suggests the capacity of this seemingly small side stream to carry large volumes of water and sediment to the river below.[31]

During my exploration of the lower river, I longed to explore this lush-looking river upstream. I regretted not having done so after I learned that the refuge still hosts, in healthy, reproducing populations, virtually every indigenous species that has not been extirpated from the lower Colorado River region. But because I had an appointment at a restoration site downstream, I perused the visitor center instead. There were tanks with small razorback suckers and bonytail chub, but what most caught my eye was an exhibit on a small variety of blackbird called the brown-headed cowbird (*Molothrus ater*). While I studied this information a woman emerged from her office to ask if I had any questions. To my good fortune, she turned out to be Dr. Kathleen B. Blair, the refuge ecologist and a font of knowledge with an enthusiastic willingness to share her insights with an interested guest on an otherwise slow December day at the refuge.

Cowbirds hardly seem like a formidable foe. They are not much bigger than southwestern willow flycatchers, and they sport neither hawkish claws nor powerful long beaks. But they can eradicate other bird species in a more subtle and equally effective way. Rather than building their own nests, females lay their eggs in the nests of other birds, such as warblers, vireos, finches, and flycatchers, in a process known as "brood parasitism."[32] I know I am inappropriately anthropomorphizing by comparing this reproductive method to our own tendency to foul the nests of others, rather than recognizing it as a successful strategy of reproduction and survival. Still, it is difficult to *like* a bird that would do such a thing.

Female cowbirds lay thirty to forty eggs per season, but place only one or two in each host nest. This means that they "borrow" between fifteen and forty nests a year, and may push one host egg out of the nest or crack a hole in it for every egg they place in the nest. The unsuspecting host mother incubates the eggs and feeds and raises the cowbird chicks, which often consume the lion's share of the food because they tend to be larger and grow faster than the host chicks, take up space, and sometimes push the host chicks out of the nest.[33] Imagine the surprise of the host mom when she sees the "ugly duckling" she has raised. Although it is tempting to blame the consequences of this parasitism on the cowbirds, they do only what their evolutionary history teaches them. We are really to blame for changing environmental conditions in ways that altered species distribution and abundance and exaggerated the impacts of this reproductive strategy.

Field guides show brown-headed cowbirds as ubiquitous throughout Mexico, the United States, and southern Canada.[34] Technically, then, they are not an exotic species. But while cowbirds are native to North America, they were rare along the lower Colorado River until the turn of the 20th century. As suggested by their unusual name—cowbirds look nothing like cows—these were originally "bison birds" from the Great Plains. Cowbirds followed bison herds, which roamed large areas in search of fresh forage. Just as the impacts of bison grazing were spread over large areas, allowing time for recovery before the herd returned, cowbirds affected local bird populations temporarily. Host species also had time to develop, through a process of coevolution, defenses to the cowbirds' brood parasitism.[35] And because cowbirds are relatively shy birds (admittedly an odd notion for a bird that invades the nests of other species), in their original habitats they did not penetrate large forests, thus leaving many nests alone.

Human settlers changed ecological circumstances in two ways that dramatically affected the cowbird-host dynamic. First, flocks of cowbirds moved to modified habitats with cattle as hosts after western settlers hunted the bison to near extinction. As a result, cowbirds are now a resident species whose impacts are concentrated on individual host populations for longer periods. Second, habitats that once consisted of large forest patches are now divided into much smaller patches with a lot of "edge" between forests and open fields, allowing cowbirds to invade a much higher percentage of nests. With these changes, cowbirds affect host nest bird populations much more acutely.

The cowbird example raises somewhat different restoration issues than suggested by our discussion of burros and tamarisk. No one brought brown-headed cowbirds to the Colorado River basin for anthropocentric reasons, only to have them escape beyond our effective control. Instead, by altering habitats and other conditions, we created circumstances in which the mix and interaction of species changed. The common backyard American robin was native to North America. It was not the ubiquitous, thriving presence it is now, however, until we converted millions of acres of forests to suburban lawns. In the case of brown-headed cowbirds, by extirpating the bison and fencing their once sweeping range, we eliminated the natural conditions in which the effects of the cowbirds' reproductive strategy of brood parasitism were held in check. At the same time, we created new conditions in which this species could, and did, thrive, to the detriment of other resident species.

There are other examples of this phenomenon in the Southwest. Mesquite (*Prosopis julifora*) and one-seed juniper (*Juniperus monosperma*), for example, are both native plants to the Southwest. But over the past century, they invaded areas formerly covered by other species, due to grazing, fire suppression, and other major changes in land use brought by European settlers. Mesquite invasion from Texas to Arizona affects more than 70 million acres formerly covered in grasslands. Expansion of juniper range is somewhat less extensive, but has also reduced the acreage of grasslands in the Southwest.[36] As was true for cowbirds, the significant human disruption was not the introduction of new species. It was changes in land use and habitat conditions that altered the relative success of different existing species. That suggests different challenges for restoration efforts. If we decide that it is desirable to restore the western range and other ecosystems to something closer to the mix of species that occurred under some set of previous conditions, we cannot simply remove the newly successful invasive species. More likely, we will need to restore the conditions in which the earlier mix of native species evolved.

In other cases, however, the introduction of exotic species in the Colorado River basin has been far more purposeful. Such is the case with fish and other aquatic life, communities of which have experienced an even more epidemic transformation throughout the system. Those changes may be even more difficult to reverse, and are even more fundamental to restoration of the Colorado River.

Effects of Introduced Species on Native Fish of the Colorado

The humpback chub (*Gila cypha*), in name and in morphology, is the paradigm of the big river fish adapted to the volatile conditions of the predam Colorado River. It was the last to be identified as a distinct variety, in fact the last major fish species identified in North America. In 1942, ichthyologist Robert Rush Miller found an unfamiliar specimen in the archives of Grand Canyon National Park's natural history collection. The sample had been captured and preserved by a park naturalist a decade earlier, and might have languished in the archives had Miller not stumbled on it while doing other work. Miller published a paper four years later identifying the humpback chub as a unique species endemic to the Colorado River. Steven Carothers and Bryan Brown later wrote, "How much longer the humpback would have remained unknown to science if Dodge had not made the effort to pack the fish out of the canyon and properly curate it will never be known." Early specimens of humpback chub apparently were confused with bonytail chub and other similar species. River runners Elsworth and Emory Kolb, for example, reported large congregations of "bony tail" near the confluence of the Little Colorado River—a place now known for its humpback population—and photos confirm them to be humpbacks.[37]

The relatively late identification is also understandable because humpbacks often inhabit deep, inaccessible, swift-flowing canyon reaches. Fish are not easy to see or catch in those remote, turbulent conditions. Inundation of some of the deep canyons by Lake Mead and other reservoirs also eliminated some likely habitat before humpbacks were identified as a distinct species. The Fish and Wildlife Service listed humpback chub as endangered just twenty-five years after it was identified as a species, and remaining populations face serious threats. According to the Service, humpbacks now occupy slightly more than two-thirds of their historic range. Because of the late identification as a species, however, after extensive human changes had already occurred, the original habitat range is poorly understood and likely underestimated.[38]

The name "humpback" derives from a dorsal "hump," a large mass of flesh that develops as the fish matures, and which scientists believe provides stability during heavy flow conditions. Other adaptations to aid swimming in the swift river include a narrow, flattened head; a tapered, hydrodynamic body; a deeply forked tail fin; and large, fanlike fins. These fish also evolved features to withstand the abrasive effects of the sediment-laden water, such as thick skin, relatively few, deeply embedded scales, and a sensitive sensory system to navigate and locate food in the turbid river.[39] In theory, those morphological advantages should equip the humpback to outcompete introduced fish that evolved under different, less severe conditions. That theory, however, withered in the face of the Normandy-like invasion of nonnative fish in the Colorado River basin, a phenomenon that was part of a much broader pattern of artificial, often intentional environmental change beginning in the mid- to late 1800s.

Smithsonian scientist Spencer Fullerton Baird became the first Commissioner of the U.S. Fish Commission, predecessor to the U.S. Fish and Wildlife Service, in 1871. Western settlers wanted more diverse (and presumably familiar) fish species for food and recreation, and Baird and colleagues responded with extensive programs to plant species such as common carp, channel catfish, smallmouth bass, and largemouth bass in western rivers.[40] As a result, it is now difficult to distinguish eastern, midwestern, and western fisheries that once boasted far more natural diversity. We homogenized America's rivers much in the way that the real estate, fast-food, and retail industries converted once varied American communities into cookie-cutter patterns of suburbs and strip malls. In the process, we decimated native fish populations just as we destroyed local shops and cafes.

Commissioner Baird did not intend to wreak havoc on the ecosystems of western rivers; he certainly does not hold singular blame; and ecological understanding at the time was limited. As a scientist, Baird identified a large number of North American fish species, and some of his scientific successors generously forgive his errors: "If he could only know of the adverse effects of such actions on indigenous species, some of which he described, we are convinced he would feel remorse."[41] Baird's programs and their progeny, however, resulted in some of the most dramatic ecological transformations in history.

Fish transplantation and stocking were key components of a national fishery program that relied heavily on hatcheries. Managers shipped eggs collected from native stocks to distant watersheds around the country. Beginning in the late 19th century, for example, federal and state agencies transported tens of millions of chinook salmon eggs from the first Pacific salmon hatchery on the McCloud River in California to the eastern United States and to foreign countries. Because Californians were disturbed at the loss of this natural resource, fishery managers replaced them with nonnative shad and other East Coast species. As noted by fish biologist Jim Lichatowich, "Californians readily traded chinook salmon for shad as part of a wholesale restocking of freshwater ecosystems." To maintain salmon hatcheries at optimal capacity, managers traded eggs without recognizing that salmon stocks evolved within particular watersheds, particularly for purposes of return spawning migrations.[42] As we will see in later chapters, genetics can be very important in ecosystem restoration.

Salmon transplantation was part of a larger Progressive Era program of environmental manipulation grounded in the belief that humans could, and should, maximize the efficient use of natural resources, including entire species of plants and animals, for human benefit. Ann Zwinger referred to this as "a human triumph over *the things nature forgot*, a kind of Manifest Destiny for the natural world."[43] By 1895, thirty species of fish plus the American lobster, eastern oyster, and soft-shelled clam had been transplanted from the East to the West Coast. Carp were brought to the United States from Germany, and introduced to U.S. rivers on a virtually nationwide scale, to the dismay of many current biologists. As I stood on a bridge overlooking the Gunnison River with Chuck McAda of the Fish and Wildlife Service, several large dark shadows swam below. When I asked what they were, he replied with great disgust: "Oh, probably just demon carp."

The Colorado River system was no exception in this wholesale transformation of the fish assemblages in America's rivers, lakes, and reservoirs. In its 1945 Colorado River planning report, the Bureau of Reclamation's description of ecological values, two sentences long, focused entirely on introduced recreational fish: "The cold, clear mountain streams abound in trout, the most common varieties being rainbow, eastern brook, native rainbow, and Loch Leven. Bass, crappie, and bluegill prefer the lakes and reservoirs to the moving waters of the streams." Likewise, the 1950 basin study by the President's Water Resources Policy Commission discussed trout, bass, and even sunfish, but made no mention of native species.[44]

At least sixty-seven and as many as seventy-five nonnative fish species have been introduced into the Colorado River basin over the last century, with forty-two now established in the upper basin and thirty-seven in the lower basin. There is anecdotal evidence that carp and catfish were already widespread in the basin by the late 1880s. Although some species were released accidentally, fishery officials intentionally introduced many as game, or as forage to feed the game species. Threadfin shad, for example, were planted in Lake Powell to fuel the popular recreational bass fishery. Exotic fish from the Mississippi River basin, the Midwest, and the Old World now inhabit the lower basin.[45]

The U.S. Forest Service stocked exotic fish in Grand Canyon beginning in 1919, followed by the National Park Service in 1920, and the Arizona Department of Game and Fish in 1964. Species included trout, coho salmon, carp, catfish, and bass. Other species, mainly forage fish such as golden shiners, red shiners, woundfin dace, Rio Grande killifish, and fathead minnow, were introduced either intentionally as food for the sport fish or accidentally via bait buckets. But fishery managers did not stop with introduced fish. They also added other food sources to stimulate the new sport fishery. In 1932, 50,000 freshwater amphipods were planted in the moss at Bright Angel Creek. In 1965, managers planted shrimp, snails, leeches, caddis flies, damselflies, and mayflies at Lees Ferry, essentially as trout food.[46]

Introduced species quickly had serious effects on native fish throughout the Colorado River basin. In the headwaters of the Green River, the original Green River cutthroat trout (*Salmo clarki pluriticus*) nearly disappeared due to competition and interbreeding with introduced rainbow trout, among other causes. Named after William Clark of the Lewis and Clark expedition, *Salmo clarki* once was the widest ranging trout species in the West, but now are severely diminished and, in some cases, endangered. The Arizona native trout (*Salmo gilae*) began to decline before 1900. By 1950, it was largely gone due to competition and interbreeding with introduced rainbow trout. Similar fates befell other native trout, which were already stressed due to watershed changes discussed in chapter 3.[47]

In the lower basin, reservoirs now are dominated by threadfin shad, largemouth bass, black and white crappies, sunfish, and striped bass—all exotics. The remaining stretches of river are filled with common carp; red shiner; channel, bullhead, and flathead catfish; mosquitofish; mollies; and to emphasize the exotic, African cichlids. Two biologists noted recently that "[t]oday, the Lower Colorado River has the dubious distinction of being among the few major rivers of the world with an entirely introduced fish fauna." Seven of nine species of freshwater fish native to the lower river are listed as endangered, and most are gone.[48]

In the modified Grand Canyon ecosystem, only four species of native fish are found regularly, compared to at least twenty-five reported exotics. Two species of trout (rainbow and brown), carp, channel catfish, and some species of forage fish are now ubiquitous throughout the system, sometimes in very large schools. By contrast, only three out of the eight native fish that inhabited the Colorado River in Grand Canyon (bluehead and flannelmouth suckers and speckled dace) maintain healthy, reproducing populations. Bonytail chub, roundtail chub, and Colorado pikeminnow have been extirpated from the canyon. Razorbacks are likely to share the same fate, and experts believe that extirpation is virtually inevitable. Humpback chub sit somewhere between these two extremes. Although humpback populations appear to have declined steadily until about 2002, recent data suggest that the population may be stabilizing if not recovering somewhat.[49] We will revisit the possible reasons for this recent trend and the prospects for recovery in chapter 6.

Was this decline in the native fish fauna of the Grand Canyon and elsewhere in the river caused by the drastically changed river conditions brought about by dams, the presence of exotic species, or both? The answer will have important implications for which restoration strategy or strategies are most likely be most effective, if one goal is to recover native fish populations.

Some biologists have placed most of the blame on the dams. In what W. L. Minckley labeled an "ecological catastrophe," the demise began shortly after 1900, as the river was progressively harnessed and regulated for water, power, and flood control. Bonytail began to decline noticeably in the lower river throughout the first half of the 20th century. They were gone from the lower river by 1950, and experienced major declines elsewhere along the river beginning in the 1950s, after the construction of main stem dams. Minckley noted that the native fauna appears to have collapsed downstream to upstream, in the same direction in which we progressively built the dams.[50] If that is true, the native species are not likely to recover so long as the dams remain, or at least if they are operated as they are now.

At the same time, other experts remind us that native fish populations in parts of the Colorado River began to decline after exotic fish species were introduced, but *before* those reaches of the river were significantly modified by dams, diversions, or other structural changes. Channel catfish and carp were introduced to the Colorado River in the late 1800s, and by 1963, even before Glen Canyon Dam was finished, were already the most common fish in many reaches of the river. Edward Abbey wrote proudly of how easily he caught catfish in Glen Canyon, using rotten salami for bait, when he floated the river just before the dam was completed. Likewise, during his famous 1963 trek, Colin Fletcher caught catfish and carp in the Grand Canyon using his hiking staff and some tackle. Several largely marine species in the lower river, including machete, striped mullet, spotted sleeper, and woundfin, were extirpated before 1900; roundtail chub and pikeminnow followed shortly thereafter. A 1926 collection from the Gila River yielded seven species, four of which were not native.[51] If exotics are the main culprit, strategies to remove or reduce populations of those species (discussed in later chapters) might do more good than dam modification or removal in restoring native fish.

More likely, however, the situation is more complex, and impairments to native fish in various parts of the basin resulted from some combination of the structural and watershed changes discussed in chapters 2 and 3, along with the introductions of nonnative species. Under this theory, introduced species already began to affect native fish before the major dams were built, through a combination of predation and competition. Dam construction then exacerbated this situation by creating more favorable conditions for the introduced species relative to the natives.

Humpback chub, for example, are highly vulnerable to predation before they reach adult size. Although survival rates under wild conditions apparently vary and are not well understood or documented, studies indicate that only about one out of every 1,000 larval humpback survives to adulthood. One main reason is predation, and although larval fish obviously were eaten by native fish during the evolutionary history of the species, the feeding habits of the voracious new predators are not similarly ingrained in the survival strategies of the young natives. In the Grand Canyon, for example, introduced brown trout, rainbow trout, channel catfish, and black bullheads are major predators of juvenile humpbacks, in amounts that can jeopardize entire annual spawning populations. The Fish and Wildlife Service believes that catastrophic events are the biggest single threat to the ultimate viability of the species, and that predators could remove entire age classes of native fish.[52]

The same is true for other species. Experts believe that "predation by non-native fishes is the single most likely factor precluding recruitment of razorback suckers in nature," and that efforts to restock razorbacks from hatchery populations failed largely due to predation by catfish and other introduced species. Introduced mosquitofish decimated Gila topminnow populations within two years by eating their eggs and young. As explained by some experts, native fish of the Colorado evolved to deal with a single main predator (Colorado pikeminnow), but within a period of fifty years were overwhelmed by several dozen new predators they did not recognize, and against which they had not evolved effective defenses.[53]

The more subtle effects of competition, however, also can take their toll on populations of native fish. Introduced fish may have superior survival skills, having evolved in conditions of higher species diversity and more variable conditions. To compete successfully under those conditions, those species had to develop more aggressive behavior, or strategies to protect their eggs and young. As such, they can outcompete natives for food, space, and other resources. Highly specialized natives that evolved in relative isolation, by contrast, had no need to develop the ability to compete with these hardy generalists.[54]

Although predation and competition already affected native fish, dams and other modifications in the main stem induced changes in flow, water chemistry, temperature, and other conditions that exacerbated the effects of interspecies interactions, and minimized or reversed the evolutionary advantage enjoyed by indigenous species. Increased erosion harmed trout and other fish in headwater tributaries, as discussed in chapter 3, but main stem species evolved to thrive in warm, sediment-laden waters. Studies conducted after Flaming Gorge Dam began operation showed that cold, clear water released from the dam caused the disappearance of pikeminnow, humpbacks, and razorbacks from a 65-mile reach below the dam. The same is true downstream of Glen Canyon and other major dams. Water released from the dam is generally too cold for spawning by native fish, and the only viable spawning habitats are in tributaries or near warm springs. If larval native fish descend into the main river while they are too young, they often die of thermal shock from the cold water. Others succumb to predation. Because of erratic swimming behavior in the colder water, and because clear water is better for sight predators such as trout, native young lost the natural advantage they enjoyed by evading predators in the murky waters of the predam river.[55]

If the native fish of the Colorado do suffer from these combined sources of harm in a complex set of interactions, choices among restoration approaches will be even more difficult. It is clear that the presence of exotic species is a significant part of the problem. But efforts to remove or reduce those invasive species may turn out to be difficult or impossible, and whether those efforts will benefit native species significantly remains uncertain. In addition, not everyone agrees that we *should* eradicate all of those species, such as the trophy-sized trout sought by anglers in the tailwaters of Glen Canyon, Flaming Gorge, and other major dams, or the giant striped bass that are fodder for fishing tournaments in Lake Mead. In fact, based on some of the place names encountered up and down the river, the new species are viewed with some degree of pride. "Bullhead City." "Catfish Paradise." Never "Razorback City" or "Humpback Paradise." Those competing sets of personal values are the subject of even more perplexing debates. They require us to choose which species should remain, and which should go.

Are Exotic Species All Bad? Choices Among Restoration Goals

Shortly after describing tamarisk in *Death Comes to the Archbishop*, Willa Cather exalted a different image of the Southwest's native riparian splendor, for its own virtues and as another metaphor for life in the desert: "Beside the river was a grove of tall, naked cottonwoods—trees of great antiquity and enormous size—so large that they seemed to belong to a bygone age. They grew far apart, and their strange twisted shapes must have come about from the ceaseless winds that bent them to the east and scoured them with sand, and from the fact that they lived with very little water,—the river was nearly dry here for most of the year." Naturalist Ann Zwinger knows the virtues of cottonwoods as well as anyone: "An old cottonwood is protector of anthills, shelter for wren, shade for small plants, platform for osprey, perch for heron, shovel for Fremont Indian, all things to all creatures." Edward Abbey too sang their praises:

> Long enough in the desert a man like other animals can learn to smell water. Can learn, at least, the smell of things associated with water—the unique and heartening odor of the cottonwood tree, for example, which in the canyonlands is the *tree of life*. In this wilderness of naked rock burnt to auburn or buff or red by ancient fires there is no vision more pleasing to the eyes and more gratifying to the heart than the translucent acid green (bright gold in autumn) of this venerable tree. It signifies water, and not only water but shade, in a country where shelter from the sun is sometimes almost as precious as water.[56]

Tamarisk may well be a "tree of the people," introduced to the region by people for decoration and promised agricultural and other benefits. Just as clearly, though, in the desert Southwest cottonwoods represent the "tree of life." According to Hopi legend, a man named Tiyo made the first successful navigation of the Grand Canyon by floating through in a hollowed-out cottonwood trunk. In a similar but more recent and hotly disputed legend, prospector and Civil War veteran James White claimed to have floated the canyon in a small raft built of cottonwoods before Powell's famous journey two years later. In both cases, the cottonwoods literally sustained life. Even as metaphors the illusion rings true. But now, thicket upon thicket of the tree of the people displace hundreds of miles of the tree of life. Describing the trees inundated by Lake Powell, Russell Martin lamented "the branches of the old cottonwoods now looking like the desperate hands of the drowning."[57]

In his excellent analysis of the Pacific Northwest salmon crisis, *Salmon Without Rivers*, biologist Jim Lichatowich wrote: "When humans try to control an ecosystem, they must decide what will live and what will die." Later, more facetiously, he commented: "Rivers needed to be tilled by the hand of man to suit his needs. Only those fish favored by humans, those deemed capable of efficiently producing commodities, should be allowed to inhabit rivers."[58] In the Colorado River, for many years managers have chosen trophy trout for the tailwaters below the dams and giant bass in the reservoirs. These fisheries boast tremendous economic and recreational value, attracting millions of anglers every year, and hundreds of millions of dollars spent on food, lodging, equipment, and the like.

The values of the imperiled indigenous biota, by contrast, are far more difficult to translate into tangible dollars. Harm to endangered species can foretell deeper symptoms of ecosystem harm that could ultimately threaten much more than individual species. Former Secretary of Interior Stewart Udall observed: "Endangered species are . . . the messengers of change, and we must heed their messages." They are parts of integrated systems, and each part serves an integral if not fully known purpose. But each species, many biologists note, also harbors within its gene pool precious gifts that cannot be ignored in the quest for pure maximization of financial returns, gifts that may return more of real value in the longer term: "Native species constitute a dictionary from which words may be chosen to compose management prescriptions for the future." Such gifts may include genetic raw materials for breeding stock or medicines. Individual species may hold the solutions to scientific puzzles, which may never be solved once the keys are discarded. Noted environmental philosopher Holmes Rolston III: "Destroying species is like tearing pages out of an unread book, written in a language humans hardly know how to read, about the place where they live." Others still argue the value of diversity for its own sake. "What we must avoid is a domesticated, homogenous earth; for many, it would be a far less fascinating place to live."[59]

In deciding whether, how, and how much to restore the ecosystems of the Colorado River, choices loom large. Shall we retain the naturalized ecosystems brought about by two centuries of major human change, and simply try to make them as functional as possible? Or should we try to reverse that change to the fullest extent possible, eradicating the new to restore the old? And if so, what point in ecological time defines our restoration goals? But those choices also affect our use of the river for even more significant economic needs, such as water and electric power. Those choices, and the processes by which they are made, are the focus of the rest of this book. We begin in the next chapter by exploring ongoing efforts to resolve the difficult conflicts between water development and endangered species recovery in the upper Colorado River basin.

CHAPTER FIVE

Down the Great Unknown: Environmental Restoration in the Face of Scientific Uncertainty

By August 13, 1869, John Wesley Powell and his party had spent three months exploring some of the least-known remaining territory in North America, the canyons of the Green and upper Colorado rivers. Although trappers and others had floated parts of the Colorado, this was the first attempt to traverse the river through all of its deepest and most dangerous canyons. Virtually nothing was known about this stretch of the river and the country through which it flowed. Cartographers could not even place the junction of the Green and the Grand rivers to within 100 miles.[1]

By the time Powell reached the head of the Grand Canyon, he and his men had discerned at least some of the river's mysteries, such as which kinds of canyon rock presaged dangerous rapids and which foretold smooth water. Yet, despite three months of intensive experience in which life or death could turn on the accuracy of their predictions, Powell and his companions knew little of what lay ahead. In perhaps the most famous passage from his re-created journals, compiled from notes taken during two separate Colorado River expeditions, Powell wrote: "We are now ready to start on our way *down the Great Unknown*. . . . We have an unknown distance yet to run, an unknown river to explore. What falls there are, we know not; what rocks beset the channel, we know not; what walls rise over the river, we know not. Ah, well! we may conjecture many things."[2]

We might make a similar statement about efforts to restore the ecosystems of the Green and Colorado rivers. Talented and dedicated scientists have spent years studying the river's ecosystem, hydrology, and geology, and much has been learned. Yet what remains unknown is formidable, and much is left to future study and analysis. How much water do the fish need to survive and reproduce naturally, and when? Will different flow regimes or changes in water temperature help exotic species more than native fish? Is more water alone sufficient without restoring other essential habitat components, for example, by reconnecting the river with its natural floodplain and renewing the natural flow of sediment downriver? Scientists continue to learn more about those issues, but clear and complete answers remain elusive because of the complexity and variability of the system and the difficulty of conducting experiments in remote, rugged areas. At the same time, populations of endangered species continue to decline.

As geophysicist Henry Pollack explained brilliantly in the context of global warming and other current issues, uncertainty is fundamental to the scientific process.[3] Scientists iteratively form hypotheses based on existing data, test those ideas by conducting experiments and seeking additional information, and revise their hypotheses accordingly. Often, different scientists posit competing theories or explanations about phenomena that are understood incompletely, or for which adequate data are not available. Sometimes, different studies suggest multiple possible conclusions. This ongoing scientific process is healthy, because it is how we learn. But at the same time, we often must make important public and private choices in the face of that scientific uncertainty. Likewise, sometimes we must determine when and how past decisions should change as scientific knowledge and other factors evolve, especially when new or modified decisions and policies alter longstanding legal and economic expectations.

This chapter examines the manner in which we make restoration decisions in the face of scientific uncertainty, using endangered species recovery efforts in the upper Colorado River basin (figure 5.1) as our main example. Restoration decisions obviously affect the fate of endangered native fish and other components of the Colorado River ecosystem. Yet those choices also affect legal rights and obligations of people who use and rely on the river, including rights established in the Colorado River Compact and other components of the Law of the River described in chapter 1. A key legal and policy question, then, is who should bear the burden of scientific uncertainty? When we do not understand fully the environmental impacts of certain actions (building dams, diverting water, stocking or removing alien species, changing water flow or temperature), we must decide whether those activities can proceed until environmental harm is proven, or whether they must be prohibited until proven safe. Once those decisions are made and legal rights and obligations are fixed, we continuously face choices about when shifts in scientific theory or information justify changes in that legal and economic stability. Of course, those decisions also turn on evolving public opinions, politics, and other factors, which we address beginning in chapter 6.

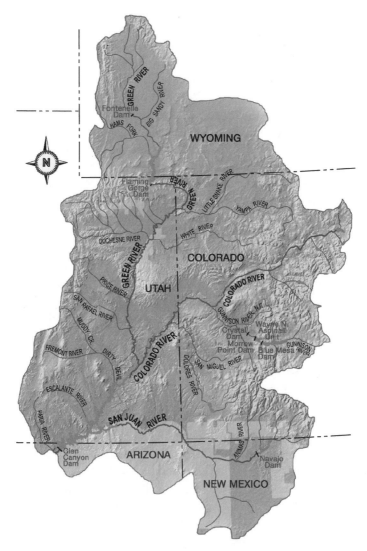

Figure 5.1. Upper Colorado River basin. *Courtesy of the U.S. Bureau of Reclamation*

We will begin by looking at the legal and political context in which decisions were made to build the major dams and to take other actions that now impair the Colorado River ecosystem, when few environmental laws or regulations were in place. Then, we will explore how decisions affecting river development and restoration changed over time, as Congress adopted laws that require federal agencies and others to evaluate the environmental impacts of their actions more carefully before decisions are made. Ironically, in some cases environmental restoration efforts may be impeded as much as they are aided by that new legal regime, at least as those laws have been interpreted and implemented in the upper Colorado River.

Ghosts of the Green River—Native Fish Eradication and Early Environmental Law

Powell's journals barely mention fish or other ecological attributes of the river, except in connection with the expedition's dwindling food supplies. In 1895, just three decades later, the United States Fish Commission issued a report identifying the unique fish fauna of the Colorado River.[4] By then, however, the commission had already begun to stock much of the basin with alien fish. Despite their new knowledge about what species inhabited the ecosystem, scientists knew little about how the system *worked*. They had no way to know how the native fish would respond to the introduced species. Stocking began four decades after Charles Darwin published *On the Origin of Species* and Gregor Mendel conducted his famous experiments with peas that gave birth to the new science of genetics. But understanding often lags behind knowledge, and no one yet realized the implications of Darwin's and Mendel's ideas. Modern ideas of population biology and ecology would not be developed more fully for decades.

Perhaps more important, new scientific information does not necessarily translate to changes in society's values, attitudes, or priorities. The Fish Commission responded to public pressure to expand the kinds of fish available for food and sport, and to include varieties that were familiar to new settlers from Europe or back east. The public valued catfish and brown trout and carp, and believed they should be introduced into western rivers to *improve* the environment. Under the prevailing Progressive Era philosophy, we used scientific knowledge and technical prowess to make nature more "efficient" and more "productive" for human use.

It was more than half a century before scientists began to realize that alien species and other human threats jeopardized the continued existence of the unique fish of the Colorado River. In 1961, renowned fishery biologist Robert Rush Miller warned of the serious plight of the basin's endangered fish: "Perhaps nowhere else in North America has the upset of natural conditions been more strikingly reflected by biotic change than in the arid Southwest." But Miller's warning did not translate into immediate action. At about the same time, a consortium of government agencies was preparing an intensive and deliberate strategy to transform the natural species mix in the Colorado River. Construction of Flaming Gorge Dam was nearing completion. The Bureau of Reclamation and others promoted fishing and other recreational uses as a major public benefit of the dam, along with water storage, flood control, and hydroelectric power. The last thing they wanted was a reservoir filled with "trash fish," or even introduced carp and perch. For the reservoir they intended gigantic lake trout and kokanee salmon, specimens to mount on cabin walls. In the cold, clear tailwaters downstream they preferred enormous rainbows that would attract fly-fishers from around the country. Catch and release, maybe, but better a record rainbow than an ugly sucker.[5]

Fearful that the disheveled local riffraff would inhibit or displace the new beauty queens, the Wyoming and Utah fish and game departments, joined by the U.S. Bureau of Sport Fisheries and Wildlife (heir to the U.S. Fish Commission and predecessor to the U.S. Fish and Wildlife Service [FWS]), hatched a plan to apply the fish poison rotenone throughout 444 miles of the Green River to the end of Flaming Gorge. Robert Rush Miller and Carl Hubbs, who had studied the fish of the Southwest since at least 1915,[6] led the resistance to the proposal by the American Society of Ichthyologists and Herpetologists. These scientists worried about the impacts of rotenone on native fish populations, which had already declined precipitously. The group feared an even more rapid expansion of carp, catfish, and other exotics.

But other groups advocated different preferences for the newly modified river. The Sport Fishing Institute supported the eradication program to improve sport fishing. The Bureau of Sport Fisheries and Wildlife believed that the fisheries would change with or without the poisoning, due to dams, diversions, and pollution. Why not help the process along and replace a fishery of questionable value with one that would be used by thousands of people a year? Local populations and state and local officials overwhelmingly approved. A headline in Rock Springs, Wyoming, touted: "Successful Stream Rehabilitation Rids Green River of Coarse Fish."[7]

In a quasi-military assault, the agencies set up fifty-five rotenone "drip" stations along the river, dispensing more than 21,100 gallons of rotenone over three days with the help of more than 100 people supported by vehicles, airboats, and one helicopter. Judged by military criteria, the operation was a success. The agencies generally achieved their goal of eliminating fish in the river, paving the way to restocking the ecosystem with preferred species.

In one major respect, though, the eradication program did not go as planned, and the "collateral casualties" included native fish downstream of the target zone. Along with the drip stations, the agencies established one "detoxification" station just north of the boundary of Dinosaur National Monument. Rotenone application initially was scheduled to occur after the gates of the new Flaming Gorge Dam closed, so that most of the poison would remain in the new reservoir until it was detoxified. Because final construction was delayed, however, the agencies decided to continue the program but to detoxify rotenone in the river upstream of the national monument. Unfortunately, the poisoning went much more smoothly than the detoxification. Rotenone concentrations in the river were higher than expected, in part due to declining river levels. So more detoxification agent (potassium manganate) was needed than was available. Bad weather made it difficult to apply the crystalline substance, resulting in incomplete treatment. National Park Service officials reported dead and dying fish in the national monument, and fish kills as far south as the Yampa River, what biologist Paul Holden (from whose account of the incident I borrow here) called the "ghosts of the Green River."

Scientists had only a limited basis to predict the consequences of the largest fish eradication effort attempted to date. No fish were found in the lower Green River in Wyoming in 1963. The agencies stocked Flaming Gorge with rainbow trout and kokanee salmon, although lake trout ultimately became the trophy species of choice in the reservoir. But by the following summer, native flannelmouth sucker and mountain whitefish recolonized much of the area, followed later by introduced carp and Utah chub. Although the eradication program may have paved the way for new game fish, it also opened up habitats formerly used by native fish for species that were neither native *nor* desired, such as white suckers. Paul Holden wrote: "The words of Miller and Hubbs thus had a ring of truth; poisoning without consideration for ecological consequences caused concerns for both native fishes and non-native salmonids." Within a year, however, scientists found humpback chub and other native species in Dinosaur National Monument. The dam itself, with its accompanying changes in flow, temperature, and habitat, posed far more dramatic threats to native species than the one-time application of rotenone. Whatever the cause, though, native species never returned to the river above Flaming Gorge.[8]

The Flaming Gorge fish eradication saga is a good example of how, until the 1970s, federal agency actions could be taken with such little evaluation of potential environmental impacts. Compared to today, the speed with which a small coalition of federal and state agencies planned and executed such a dramatic program to poison fish through hundreds of miles of one of the West's great rivers is nothing short of amazing. No permits were needed. No detailed scientific analysis was conducted. No alternatives were considered, at least in any formal way. And although the agencies granted opposing scientists one meeting to express objections, there was no broader mechanism for the public to understand the basis for the decision or to express their views, pro or con. Perhaps most significantly, no one filed a lawsuit to stop the project. "Father Knows Best" was still popular on television, and we clung to the Progressive Era belief that at least some scientists and engineers knew best.

Ironically, though, this incident helped lead to a major shift in our views of how much scientific and public input was needed before taking actions with significant effects on the environment. Secretary of Interior Stewart Udall issued a directive demanding greater caution when taking actions that might endanger unique or threatened species:

> Whenever there is a question of danger to a unique species, the potential loss to the pool of genes of living material is of such significance that this must be a dominant consideration in evaluating the advisability of the total project.

> I am taking measures to assure that future projects are reviewed to assure that experimental work is taken into consideration, and that possible deleterious effects are evaluated by competent and disinterested parties.[9]

Several years later, in 1966, Congress passed the predecessor to the modern Endangered Species Act (ESA), and a year later the FWS listed humpback chub and Colorado pikeminnow (then still known as squawfish) as two of the first endangered species under the new law. In just four years, the native fish of the Colorado morphed from public enemies suitable for poisoning to the beneficiaries of a new federal protection program. Holden quipped: "[M]any biologists who worked on the Green River project eventually held positions in which they were responsible for protecting the very species they had once worked to destroy."[10]

The legal landscape within which native Colorado River fish were managed changed rapidly. In 1969, Congress passed the National Environmental Policy Act (NEPA). NEPA prohibits any federal agency from conducting any "major federal action significantly affecting the quality of the human environment" without first preparing an environmental impact statement (EIS). Agencies must study the possible environmental effects of programs in advance, and evaluate all feasible alternatives to see if project purposes can be met in less damaging ways. NEPA is also an environmental sunshine law. Federal officials must announce in advance programs that might have adverse environmental impacts, and open those proposals to public scrutiny. Outside scientists, environmental groups, and other citizens can comment on proposed actions, and agencies must consider those comments and respond to differing views before making a final decision. If an agency fails to prepare an EIS, or if an opponent believes an EIS is deficient, the project can be challenged in court.[11]

Had NEPA been in place at the time, the Green River poisoning program could not have occurred with such speed and lack of deliberation. An EIS would have been prepared. The agencies would have been obligated to identify potential impacts on native fish, to consider possible alternative courses of action, and to open their analysis to public scrutiny. The scientists who urged caution and other members of the public would have had a formal opportunity to comment, and the agencies would have had to explain which advice it accepted (or not), and why. The decision would not have been left to one small set of agency officials with a single goal of promoting sport fish.

As interpreted by the U.S. Supreme Court, however, NEPA would not have required the agencies to adopt a different course of action in the face of the competing sets of opinions that might have been expressed in this new, open process. In NEPA, Congress established a national policy to "create and maintain conditions under which man and nature can exist in productive harmony, and fulfill the social, economic, and other requirements of present and future generations of Americans." But the Supreme Court later decided that although NEPA demands serious scrutiny of environmental impacts and alternatives, it does not dictate particular results.[12] If the agencies believed it was more important to develop new sport fisheries than to protect native fish, they would have been free to do so.

Three years after passing NEPA, however, Congress adopted a more potent tool for protecting Colorado River fish and other threatened and endangered species, the ESA of 1972. Like NEPA, the ESA included important "look before you leap" requirements. Federal agencies that plan any action that might jeopardize the continued existence of a listed threatened or endangered species must first "consult" with the FWS (or, for marine species, the National Marine Fisheries Service). As used in the ESA, "consultation" requires a formal back-and-forth process in which an agency proposing an action first must prepare a Biological Assessment identifying threatened or endangered species that might be affected. The FWS then prepares a Biological Opinion evaluating the impacts of the proposed action on the species. Where the action would jeopardize the continued existence of the species or population, or destroy or adversely modify its critical habitat, the FWS must either prohibit the action or identify "reasonable and prudent alternatives" to avoid those impacts.[13]

Unlike NEPA, the ESA is more than just a "study first" and public disclosure law. It prohibits actions that harm listed species or their habitats, although with a number of exceptions. Agencies must "insure that any action . . . is not likely to jeopardize the continued existence" of endangered species or to destroy critical habitat. The agency may undertake the identified reasonable and prudent alternatives, or forgo the action altogether. The ESA also makes it unlawful for any person (not just agencies) to "take" any endangered species. "Take" is defined broadly to include a wide range of potentially damaging activities: "harass, harm, pursue, hunt, shoot, wound, kill, trap, capture, or collect," although various exceptions again apply. For example, the FWS may issue an "incidental take" permit if the agency finds that the take is incidental to an otherwise lawful activity and meets other requirements to protect the species.[14]

The Supreme Court confirmed the potential power of the ESA in the famous snail darter case, the high-stakes collision between the relatively new statute and the virtually completed Tellico Dam on the Little Tennessee River. When the lawsuit began, the dam was nearly eighty percent completed and the government had already spent $53 million on the project. Congress approved appropriations for the project after the ESA was in effect, which sponsors argued effectively repealed or modified the ESA regarding that project. Nevertheless, the court upheld an injunction against further work on the dam pending ESA compliance. Chief Justice Burger (typically one of the justices least likely to vote in favor of environmental protection) wrote that "[o]ne would be hard pressed to find a statutory provision whose terms were any plainer than those in § 7 [the consultation provision]" of the ESA.[15]

Thus, although the ESA does allow for some exceptions, it probably would have prohibited the Green River eradication program. In the ESA, Congress articulated a clear choice for the nation about the value of endangered species and biodiversity relative to economic development and other choices, although we will see that fulfillment of those species protection goals depends on the manner in which the law is implemented.

NEPA and the ESA were prominent in implementing the first and second environmental strategies discussed in chapter 1, mitigation and prevention. Some criticize these laws as serving no purpose other than delay, but in some sense that was the point. In our fervor to move our economy forward, to clear, grade, build, develop, log, mine, drill, dam, divert, farm, manufacture, dig, dredge, fill, cut, whatever, we acted with precise attention to engineering, but little or no consideration of environmental impacts or long-range consequences. The new environmental laws constrained those actions. Despite its lack of a clear ban on damaging actions, NEPA probably accomplished more than is commonly recognized by slowing the pace of decisions and allowing other ideas to develop in the period of respite. New harm can be prevented if agencies accomplish the same good in less damaging ways. Even when a proposed action is taken, agencies must adopt measures to ensure that impacts are minimized and some form of compensation is provided.[16]

NEPA thus shifted the burden of proof from opponents of hasty actions to the agencies to show that proposals were properly evaluated and aired fully. The ESA imposed on agencies and proponents of development projects the duty to show that proposed actions, or "reasonable and prudent alternatives" to those actions, would not harm threatened and endangered species. Good trial lawyers know that shifting the burden of proof can dictate the outcome of lawsuits. Why, then, more than three decades after the new laws were passed, do four species of native Colorado River fish remain endangered (Colorado pikeminnow, humpback chub, razorback sucker, and bonytail chub), with one on the brink of extinction and efforts to restore the others proceeding at a painfully slow pace?

One answer lies at the intersection of environmental law and science, and in the manner in which the ESA has been applied to Colorado River restoration programs. In some ways, the burdens of proof imposed by the new laws have been reversed, perhaps illegally, and the risk of uncertainty remains on the fish rather than the proponents of development.[17] As a result, the new environmental laws have served just as much to impede species recovery and broader restoration efforts as to restrict harmful new development. The third strategy of the modern environmental era, restoration, has become an unintended victim of the reforms designed to facilitate mitigation and prevention. To understand how this happened, we return to the Law of the River, which, as explained in chapter 1, created legal and economic expectations against which species recovery and environmental restoration efforts in the basin must be balanced.

The Law of the River and Changing Environmental Awareness

By late 2004, the Colorado River basin was in the midst of a record drought five years long and counting. Runoff during water years 2000–2004 was the lowest of any five-year period in the observed record, averaging just less than 10 million acre-feet (maf) a year.[18] I arrived at the December 2004 annual meeting of the Colorado River Water Users Association in Las Vegas a day early to witness the impacts of the dry spell on Lake Mead. The Las Vegas Bay boat ramp was closed due to low water. The Lake Mead National Recreation Area map identified several access points that were also closed, and warned of "shifting sand bars and low water." The bay looked like a wet mudflat, and inflows from the Las Vegas River smelled of sewage. A sign read: "Notice: The water in this stream contains contaminants found in runoff from urban areas and treated sewage. Human contact not recommended." The lake still seemed huge, and to put matters in perspective, still held around 14 maf of water. But the bathtub ring, the bleached white stripe showing the reservoir's high water mark, was high up in the cliffs, in sharp contrast to the dark canyon rock above. According to the gentleman at the gift shop, the reservoir was down ninety-six feet.

At the Water Users Association meeting in Caesar's Palace, veterans of the river's periodic cycles made their best efforts to appear calm. Colorado Supreme Court Justice Gregory Hobbs, a highly respected expert in Colorado River law and policy, assured the audience that the compact was written and the dams were built to accommodate this kind of shortage. The negotiators discussed the risk of drought and struck a balance. Indeed, he argued, in some ways shortages are *good* for the basin, like a dose of foul-tasting medicine. "In scarcity is the opportunity for community," he advised, but the long-term lessons were clear: "Store, store, store in the good times for the times of scarcity."

A panel of speakers from the upper basin, however, bemoaned the already serious impacts of the drought on water users. Poor range conditions in Utah caused $200 million per year in losses to livestock. Some reservoirs in Colorado were down to "dead storage" (below levels at which water can be used given the location of the withdrawal structures), and would take several years to refill even with higher precipitation levels. Elk died in Wyoming because inadequate forage forced them to graze on lichens with high concentrations of toxins. Colorado estimated economic losses of $2 billion, with impacts to businesses from farming and ranching to commercial rafting.

A real sense of the stakes could be felt when the moderator asked the upper basin panelists if they feared a "call on the river" from the lower basin states. In western states that use the prior appropriation doctrine, the "first in time, first in right" system explained in chapter 1, a "call on the river" occurs when less water is available than needed to meet the paper water rights of all users along a particular river. Senior users—those who have put water to beneficial use the longest and hence hold the earliest priority dates—place a "call" on those with junior rights to close their gates until enough water is provided to the seniors. Within the upper basin states, for many streams only those with the most senior priority dates were actually receiving water. A lower basin "call" on the Colorado River, the first in the history of the compact, would portend even more serious consequences.

Recall that under the basic deal struck in 1922, most of the risk of drought falls on the upper basin. The lower basin (especially California) agreed to limit its Colorado River water use to specified amounts to allow the upper basin time to develop its water share, a goal that remains only partially met three-quarters of a century later. In return, the upper basin guaranteed to deliver a rolling ten-year average of 75 maf of water measured at Lees Ferry. The most basic rationale for Glen Canyon Dam was to allow the upper basin to use its water, with a hedge against drought. Water stored in Lake Powell is used to meet the delivery obligation, allowing the upper basin states to continue to use water upstream even in dry years. If the reservoir is too low, however, the lower basin theoretically can require the upper basin to meet its delivery obligations from upstream flow, even if that means upper basin uses are not met.

Lake Powell still held a two-year supply of water as of December 2004, but was at less than half of its capacity (forty-six percent). The combined storage in all of the major upper basin reservoirs (Powell, Navajo, Flaming Gorge, and Blue Mesa) was at a worrisome fifty-four percent.[19] Upper basin representatives reacted with varying degrees of concern. Larry Anderson, director of Utah's Division of Water Resources, expressed little fear "so long as there is water in Lake Powell," which of course begs the question. If the drought persists, he acknowledged the possibility of a call on the river, but expressed faith that the hydro-logical pendulum would swing back to wetter conditions. Colorado's Hal Simpson indicated that his state already was devising a system to identify all Colorado River water users in his state, and to plan ways to curtail use if a legitimate river call is made. Patrick Tyrell of Wyoming was more dramatic, referring to even the prospect of a river call as the "dark side of the compact."

What does this have to do with protecting native fish hundreds of miles upstream from Lees Ferry? The drought underscored the signifi-cance of the 1922 compact for virtually every water management deci-sion in the basin ever since, including choices about how to implement the ESA. Under the Law of the River, in return for accepting the risk of drought, the upper basin states bought the right to develop water over time, without fear of losing their allocation to the rapidly growing lower basin, *or anyone else*. The compact provided the certainty by which states could forge long-term water policy with stable expectations.

But, as explained above, the ESA established new requirements that could prevent the upper basin from using all of its allocations under the compact. Water withdrawals in the upper basin rely on decisions by the Bureau of Reclamation, which are federal agency "actions" for purposes of the ESA. Water contracts, reservoir operating plans, and other deci-sions require ESA consultation, even for projects built before the ESA was passed.[20] When the Bureau issues a contract that allows a farmer to take water from the river, ESA consultation may be required to deter-mine the resulting impacts on endangered fish. If the FWS determines that the proposal would jeopardize the continued existence of a threat-ened or endangered species, the withdrawal could be prohibited or con-ditioned on implementation of reasonable and prudent alternatives, that is, mitigation measures.

The battle line was drawn. When the FWS initiated ESA consulta-tion regarding new upper basin water uses in 1977, the history of which is explained in more detail below, the upper basin states cried foul. The federal government could not, they argued, change the rules so dramat-ically in the middle of the game. The Law of the River also rested, how-ever, on other key factors that would change significantly over time, including our scientific understanding of how much water is likely to be available, and what additional demand for that water might arise. All of this suggests that perhaps the legal foundation on which management of the Colorado River rests needs to be revisited to account for new scien-tific understanding and current economic, political, and environmental realities.

The Law of the River, Evolving Scientific Understanding, and Changing Demand

Schoolchildren learn that you can determine the age of a tree by counting growth rings in the trunk. Each ring reflects one year of growth. But each ring holds even more information. The *width* of each annual ring indicates how much growth occurred during that year, and the width therefore varies from year to year. Although many factors can affect tree growth, the biggest single factor for many species, not surprisingly, is water. In wetter years, trees grow more and produce wider rings. Scientists can calibrate the relationships among water, other environmental variables (such as temperature), and tree growth with measured data for the same species from recent years. By measuring growth rings from trees at properly selected locations around the basin, they can then estimate the amount of water available during that year, and thereby the total runoff in the Colorado River basin.[21] This new information about how much water was available in the basin hundreds of years before we began to measure the river directly calls into question one of the major factual assumptions relied on during compact negotiations.

The basin states negotiated the Colorado River Compact at the end of the wettest ten-year period on record (1914–1923), during which average annual flows reached nearly 19 maf per year. The negotiators relied on the period from 1896 to 1921, when flows averaged almost 17 maf, and believed that they could safely dole out at least 16.5 maf of water per year. The rest of the 20th century, however, turned out to be much drier. The average river flow from 1896 to 2004 was less than 15 maf. The average annual river flow from 1922 to 1982 was just over 14 maf. During the "wet" years of the early 1980s, the ten-year rolling average flows approximated the 16 maf assumption for the first time since the compact was written.[22] The 1999–2004 drought highlighted the risks of shortage, especially as demand for water increases in the region.

For the relatively young field of hydrology, two decades of data on a single river may have seemed adequate at the time the compact was negotiated. Now we know that a couple of decades is a mere blip in hydrological time, and that such a small record of data can mask much larger, longer term fluctuations in climate and river flows. In 1976, Charles Stockton and Gordon Jacoby studied 450 years of tree ring records from thirty locations in the Colorado River basin and calculated a longer term average of 13.5 maf per year. This analysis also identified periods of longer and more severe drought than anything we saw in the 20th century. A more recent study based on expanded data and newer analytical methods produced a somewhat higher long-term average, but still significantly less than compact negotiators assumed. This analysis also suggested that more severe droughts have occurred over the past several hundred years than the 1999–2004 drought that caused so much concern among basin water users. The authors of this recent analysis warned that the "long-term perspective provided by tree ring reconstructions points to looming conflict between water demand and supply in the Upper Colorado River basin."[23]

Some analysts argue that the 16 maf assumption was the product of a scientific "error," indeed, that "[i]t was a very costly error for the West." Others claim that the commissioners were "well aware" of the risk of drought when they negotiated the compact, and took that into account as they allocated the associated risks. Still others assert that negotiators knew that the information they used omitted data from the drought that had occurred around the turn of the century, but ignored this information in negotiating the agreement. There is even some indication that the upper basin states were intentionally given false, or at best misleading, scientific information to induce them to agree to the deal. At best, scientists at the time were well aware that the available hydrological record was not entirely reliable, and did not represent a complete cycle of climate in the Southwest.[24]

The stakes were monumentally high. The delegates sought a flexible and permanent solution to allocation of the Colorado River, one that would stand the test of time even as circumstances and conditions changed. When taking a calculated risk, sometimes what we do not know is more important than what we do, especially when adopting a water "constitution" with an indefinite life span. The hydrological information used to support the compact underscores the perils of making far-reaching decisions that will affect the fates of millions of people across a huge region based on incomplete and uncertain science, or at least without understanding and taking that uncertainty into account.

Global warming now adds to this uncertainty about the reliability of basin water supplies. Some scientists already suggest a correlation between global warming and declines in annual snowpacks in the West. Predictions based on scientific models vary on whether climate change will cause more or less precipitation in the basin, but some studies suggest more rain and less snow, which would reduce runoff into reservoirs. Assuming a slightly different set of changes, some parts of the basin may be wetter, and other parts may be drier. Other models predict less runoff in all scenarios, with potentially significant reductions in reservoir storage and resulting inability to meet the allocations and deliveries required by the compact.[25]

Other uncertainties about water uses in the basin, and politics, also led the compact commissioners to ignore some water needs altogether. Despite Mexico's claims to water for irrigation and other uses, the commissioners denied Mexico a seat at the table, arguing that the negotiations involved a purely domestic matter. Some parties, including Herbert Hoover, believed that Mexico was legally entitled to nothing. The compact ultimately addressed this issue by providing that if water was guaranteed to Mexico by later treaty, it would be supplied equally out of the allocations to each basin. In the 1944 treaty the United States guaranteed Mexico 1.5 maf per year, thus adding 750,000 acre-feet to the upper basin's annual delivery obligations.[26]

Compact negotiators gave American Indian tribes even less consideration. Fifteen years earlier, the U.S. Supreme Court had ruled that in setting aside reservations for the support of various tribes, the federal government impliedly "reserved" sufficient water to meet the needs of those communities. This created a major exception to the prior appropriation doctrine in western water law. Appropriative water rights are measured by the amount necessary to meet the beneficial use to which the water is put. An appropriative right does not exist until water is diverted and put to beneficial use, and seniority under the "first in time" doctrine (a "priority date") is usually determined by when the water was first used. Federal reserved rights, by contrast, need not be quantified until some later date. But the priority date remains the time that the federal government reserved the lands for the specified use. Because Indian reservations predate so many western water rights, the later quantification of a reserved right with an early priority date can seriously affect existing water users. Yet despite the presence of the large number of American Indian tribes along the Colorado River, the commissioners made no effort to evaluate tribal needs. Instead, they left the matter largely open in the compact by providing: "Nothing in this compact shall be construed as affecting the rights of Indian tribes."[27]

This casual treatment of tribal water rights came back to haunt the states in 1962, when the Supreme Court ruled that some American Indian tribes along the river corridor were entitled to nearly 1 maf of water a year. The court also ruled that federal reserved rights apply to other federal land reservations such as national parks, forests, and recreation areas. Since then, various settlements of tribal water rights in the basin have doled out nearly another million acre-feet of water per year. But the biggest wild card remains in the deck, not yet fully played. Recently, the Navajo Nation filed a lawsuit designed to force the federal government to recognize and quantify their Colorado River water rights. Outstanding claims for the Navajo and other tribes could amount to nearly another 3 maf.[28] Water allocations quantified for those needs also must be taken from the apportionments given to the individual basins and states.

Perhaps the biggest source of legal uncertainty about future water needs and allocations, however, brings us back to the fish. In negotiating the 1922 compact, the commissioners gave no thought to ecological needs, and societal values had not yet shifted to the point where those needs would have been given much credence. Now, through the ESA and other environmental laws, society has expressed a new set of preferences that were not evident when the compact was negotiated. How much water is needed to satisfy those needs is the subject of considerable scientific uncertainty. When the compact was negotiated, the basin states were in a position equivalent to that faced by Powell and his party when they entered the "Great Unknown." They faced a long, uncertain, and very bumpy metaphorical ride down the rapids of the Colorado. To date, as explained in the next section, upper basin recovery programs have imposed almost all of the risk of this uncertainty on the fish, and virtually none on those who wish to continue to develop upper basin water rights.

Who Bears the Risk of Uncertainty?

The federal government first listed humpback chub and Colorado River squawfish (now known as pikeminnow) as endangered species on March 11, 1967. Another full decade passed before the FWS began to take any serious steps to protect those species in the upper basin. From 1977 to 1981, the FWS wrote "jeopardy opinions" for all major water projects proposed by the Bureau of Reclamation in the upper basin, reflecting the judgment that further water depletions would jeopardize the continued existence of those species. All of those depletion projects proceeded, however, because the FWS identified "reasonable and prudent alternatives" that would protect the endangered fish and their habitats. The FWS ordered the Bureau of Reclamation to release more water from Flaming Gorge and Blue Mesa reservoirs to offset the effects of the additional withdrawals from the river.[29]

But not all of the proposed new depletion projects were tied to Bureau dams. Moreover, analyzing the effects of each individual withdrawal did not foster a comprehensive approach to identify and address the cumulative effects of all of the existing and proposed water projects.

In 1979, the Bureau of Reclamation agreed to fund a comprehensive investigation known as the Colorado River Fish Project. The FWS agreed to provide the information needed to protect the endangered fish and to allow river development to continue as planned. But it was difficult to predict how long it would take to develop a scientifically based recovery effort.

Under the ESA, the FWS must reject a project if no reasonable and prudent alternatives are identified that will avoid jeopardy to a listed species. Stopping projects altogether, however, would propel the ESA headlong into the well-entrenched Law of the River, under which upper basin water users are allowed to continue to develop their water rights. This tension came to a head in 1981, when the FWS agreed to allow project sponsors to avoid a jeopardy finding, and to proceed with new or expanded water projects, by making a one-time contribution of funds into the recovery program. The established price was ten dollars per acre-foot of water diverted, with periodic adjustments for inflation, based on estimated recovery program costs of $25 million. In hindsight, this figure was grossly understated. Indeed, it is difficult to envision how the agencies predicted the cost of the program when they had so little idea what the recovery effort would entail.

By 1984, the FWS had issued biological opinions concluding that nearly 100 specific water projects were likely to jeopardize the continued existence of endangered Colorado River fish. The upper basin states complained that the deal reflected in the Law of the River had been broken. Congress declined to weaken the ESA in response to these claims, and development interests sought a more cooperative approach that would reconcile the ESA, state water rights, and the Law of the River. This led in 1987 to the Upper Colorado River Endangered Fishes Recovery Program (known by its two component parts as the Recovery Implementation Program–Recovery Action Program (RIP/RAP)), which essentially adopted the earlier "pay-as-you-go" approach on a much larger scale. Water users avoid jeopardy opinions for water withdrawals by paying a one-time fee. The price rose to $14.36 per acre-foot, again adjusted annually for inflation. Once more, the basis for this calculation was not clear, when the FWS still had little idea what a recovery effort would entail.[30]

The underlying conflict avoidance focus of the upper basin recovery program is no secret. The stated program purpose is "to recover the endangered fishes while providing for existing and new water development to proceed." These goals are promoted by a management committee consisting of various affected interest groups, a process that has been criticized as well as praised, and to which we return in chapter 6. Cooperation and dispute avoidance are certainly laudable goals. The much harder question, though, is who bears the risk of program failure, the water users or the fish? ESA expert Federico Cheever, although strongly supporting broadly based recovery plans under the ESA, warned that this program "subordinates the section 7 consultation process" and "creates a significant danger of regulating species out of existence." Other analysts note that the fate of the fish has become secondary to the political process.[31]

In some ways, the philosophy of the recovery program makes a lot of sense. An individual water withdrawal might not cause significant harm to the ecosystem. But the cumulative impacts of multiple depletions, along with other ecosystem impairments, threaten the Colorado River fish with extinction. Similarly, a series of uncoordinated mitigation efforts tied to individual water depletions in the upper basin probably will not correct a problem with so many interdependent causes. It makes more sense to develop a comprehensive, ecosystem-based recovery and restoration program.[32]

This kind of coordination is used in other environmental programs, such as wetland mitigation "banks." The Clean Water Act and related federal regulations prohibit wetlands filling if a "less damaging practicable alternative" will accomplish project purposes. Developers must mitigate unavoidable harm by restoring similar wetland values and functions, preferably in the affected ecosystem. Wetland filling often occurs in the same fragmented way as water depletions along the Colorado River: an acre here for a new condominium, another acre there for a strip mall. It may make more ecological sense to restore an entire wetland system than a disconnected series of smaller patches adjacent to each development. Mitigation banks allow developers to contribute money toward broader restoration efforts, and then to draw "credits" from the bank instead of mitigating losses from each development separately. A fundamental premise of mitigation banks, however, is that the currency (restored wetlands) must have real value before it can be used. Some wetland restoration projects succeed, others fail, but the difference remains poorly understood. If a developer can fill a healthy wetland in return for uncertain future restoration, she will enjoy the development benefits now with no similar assurance that wetland losses will be offset by effective restoration.[33]

The Upper Colorado Recovery Program takes the opposite approach. Project proponents pay a one-time fee now, and are allowed to proceed with development projects, long before valuable restoration currency is in the bank. This compromise rests on the agency's *prediction* that the planned recovery efforts will work: "[P]articipants accept that certain positive population responses to initiatives are not likely to be measurable for many years due to the time required for endangered fishes to reach reproductive maturity, limited knowledge about their life history and habitat requirements, sampling difficulties and limitations, and other factors."[34] The per-acre-foot mitigation fees estimated in the mid-1980s did not reflect the risk that program costs might increase above the rate of inflation, that the program might have to change dramatically, or that recovery efforts might fail.

As of now, population trends for the four listed species are not encouraging in the upper basin.[35] If the program does not work, the water will already be diverted and allocated, although the FWS did retain some authority to account for this uncertainty. The Service has "ultimate authority and responsibility for determining whether progress is sufficient to enable it to rely on the RIP as a reasonable and prudent alternative and identifying actions necessary to avoid jeopardy." The FWS can define additional actions as reasonable and prudent alternatives, and under more serious conditions it can apply those requirements retroactively and apply depletion charges to projects that predated the agreement. It can also abandon the process and apply the ESA to individual *future* projects according to new requirements. The program agreement never says, however, that existing water depletions can be withdrawn. And because water is perhaps the most critical limiting resource needed to recover fish populations, each new water withdrawal actually increases the risk of program failure.

Just how much of this risk is shifted from water developers to the fish depends in part on the definition of "success." The FWS "will use accomplishments under the RIP as its measure of sufficient progress," but the agreement does not define "sufficient progress." The program includes a list of specific activities, such as installing fish passages, purchasing riparian properties or easements, restoring floodplains, conducting scientific studies, educating the public, and acquiring water rights to dedicate to in-stream flows. Each of these efforts is desirable to improve fish habitat. They do not, however, define the recovery goal itself, which is essential to provide program accountability.[36] The program could perform magnificently in accomplishing individual tasks, but fail miserably in recovering the species. To understand the additional uncertainties in judging program success or failure, we need to understand more about the law, science, and policy of establishing species recovery goals.

The Related Roles of Science and Policy in Species Recovery Goals

Some things we will never know for sure, and the lore of the Colorado is full of such mysteries. As Powell and his men neared the point at which they would emerge from the "Great Unknown" and leave the Grand Canyon, three in the party chose not to continue and hiked out at a place now called Separation Rapids. Two mysteries surround this departure. Powell wrote that the three left because they lost their nerve, and were not willing to risk another set of dangerous rapids. Others in the party claimed that the three left in disgust due to Powell's imperious leadership style and other grievances. The three men were later murdered when they reached the top of the canyon, but the precise perpetrators and circumstances remain a mystery. Early accounts blame local American Indians, but documents written at the time and since uncovered by historians suggest that suspicious local residents were to blame.[37]

We may never solve all of the canyon's mysteries definitively, but no key decisions turn on the answers. And in the case of river lore, a little bit of mystique is intriguing. Some important decisions about the river's future, however, require us to grapple with mysteries of another kind, and for which the stakes are higher. One is the establishment of "recovery goals" for the four species of endangered big river fish, in which the FWS defines the number and size of populations and other conditions necessary to downlist a species from endangered to threatened, or to declare success by delisting a species altogether.

In establishing recovery plans, the FWS must articulate "objective, measurable criteria which, when met, would result in a determination . . . that the species be removed from the list."[38] Objective recovery goals serve a valid purpose by providing criteria against which progress can be measured and allowing reconsideration of restoration strategies where appropriate. It is misleading, however, to suggest that recovery goals are purely "objective" in the sense that any particular population target or other defined goal will ensure species viability. As with almost any predictive scientific process, defining recovery goals can depend on competing scientific assumptions and policy choices, many of which are not presumptively "right" or "wrong." Those choices can be more or less conservative in terms of the degree of certainty that a species is likely to remain viable if the stated recovery goals are met. Therefore, the manner in which the recovery goals are derived also affects the degree of risk imposed on project proponents or the fish. Once species delisting occurs, most of the constraints imposed by the ESA are lifted. There is a danger, then, that recovery goals can give a false sense of victory.

The FWS issued recovery goals for the four species of endangered big river fish on August 1, 2002, about three and a half decades after it first listed humpback chub and Colorado pikeminnow as endangered. For each species, the FWS established criteria for both downlisting and delisting. It based those criteria on a minimum number of distinct fish populations in particular parts of the river system, and minimum population sizes and biological characteristics for each of those populations. For example, downlisting of pikeminnow can occur when three separate populations of particular sizes and characteristics occur over a five-year period in the Green River, the Colorado River, and the San Juan River. The Green River population must be "genetically and demographically viable" and "self-sustaining," and must exceed 2,600 adults. This is the number that the FWS estimated as the "minimum viable population size needed to ensure long-term genetic and demographic viability." The FWS also identified a list of management tasks that must be completed prior to downlisting or delisting, regardless of population size and number. Those include, for example, changes in flow regimes downstream from major dams, restoration and protection of habitat, and additional studies. Finally, the FWS defined minimum monitoring periods within which the numeric goals must be met to account for fluctuations in population size and uncertainties in measuring numbers of fish in the river. This use of multiple methods and different kinds of criteria (population size versus management task) can help to offset some of the uncertainty inherent in the use of numeric population goals alone.[39]

In the first policy choice that could increase the risk of species extinction, the FWS narrowed the goals for species recovery relative to the broader ecosystem protection mandate in the statute: "Th[e] definition of recovery falls far short of requiring that a species must be restored to its historic range and abundance before it can be considered recovered or delisted. It also falls far short of requiring the restoration of a species to all suitable remaining habitat, unless this is necessary to sufficiently reduce the species' susceptibility to threats to a level at which the species is no longer threatened or endangered." The upper basin program clearly focuses on individual species recovery rather than broader ecosystem restoration. Although some have criticized the ESA for its focus on single species recovery, this may reflect how the law is implemented rather than how it is written. The overall purpose of the ESA is "to provide a means whereby the ecosystems upon which endangered species and threatened species depend may be conserved."[40] The FWS has ample basis, therefore, to interpret the law as providing a more comprehensive restoration mandate.

This decision to focus only on assurance that individual species will not go extinct, rather than restoration of the ecosystems on which those species rely, has practical as well as general policy implications. Most notably, the FWS concluded that pikeminnow recovery is "considered necessary" only in the upper basin, an implicit decision that it is acceptable to forgo the return of pikeminnow to the entire lower Colorado River basin. Similar limiting choices were made for each of the four species. Congress obviously did not intend restoration of the United States in the 21st century to anything close to a predevelopment condition. Equally clearly, choices must be made about how much habitat to restore for particular species, and where. However, few members of the general public likely knew that the FWS, under the veil of supposedly objective biological calculations, was making basic value choices that would eliminate large portions of the river from species recovery efforts. The recovery goals were open to public comment, but the notice published in the Federal Register hardly provided clear warning that major public judgments of this kind were at stake: "The draft recovery goals for each species provide objective, measurable recovery criteria for downlisting and delisting that identify levels of demographic and genetic viability needed for self-sustaining populations and site-specific management actions/tasks needed to minimize or remove threats."[41] More transparency on the part of the agency is appropriate when such fundamental policy choices are being made.

As explained further below, by limiting species recovery to portions of the basin, the FWS also increased the risk that the species might go extinct at some time in the future, if changed conditions cause serious damage to any of the upper basin populations. This requires an understanding of the rapidly evolving science of conservation biology, and how the concept of *chance* affects our ability to establish "objective and measurable" recovery goals.

The Role of Chance in Species Viability

Near its southwestern corner, the Colorado Plateau drops to a series of basins and ranges in the Mojave Desert extending westward past Las Vegas and into California. This is the home of the desert tortoise (*Gopherus agassizii*), another rare species that pits human development against biodiversity protection. The FWS listed the desert tortoise as threatened within much of its natural range in 1990. Individuals are killed by poaching, collection as pets, collisions with vehicles, trampling by livestock, disease, and predation. The species also suffers from habitat loss and fragmentation.[42] But, as with many threatened and endangered species, the desert tortoise faces another dangerous enemy: chance.

In spring and summer 2005, wildfires swept through parts of the region. Fires damaged thousands of acres of dwindling desert tortoise habitat, including more than ten percent of the Red Cliffs Desert Reserve in southwestern Utah.[43] With native plant species, desert habitats in this area experienced fires every thirty-five to fifty years. Invasion by exotic cheatgrass accelerated the fire cycle to an average of three to five years. Because fires generally prevent native species from going to seed, more frequent burns favor more cheatgrass, which in turn causes tortoise habitat to burn more frequently.

It is appropriate that our example involves habitat near Las Vegas. To population biologists, floods and fires are known as "stochastic events," which means that they occur periodically, but at uncertain intervals. Although we may know something about the odds, we cannot predict precisely when, where, or how frequently these events will occur any more than we can predict whether the roulette wheel will land on red or black. But in planning for ecosystem protection, we should anticipate that they *will* occur, at some times and some places. How frequently the events occur, how significant they are, and whether they help or hurt make a big difference in whether a population or a species survives.[44]

Desert tortoises live up to fifty or sixty years in the wild, but have a low reproductive rate. Only about a third of all tortoise eggs hatch, and only two percent of those live to reproductive age. Because of their longevity, however, even this low birth and survival rate might be enough to sustain a tortoise population. Disrupt that level of success with, say, a sudden increase in fires compared to conditions in which tortoises evolved, and the same population might not produce enough offspring to sustain itself. But if there are multiple but connected populations of tortoises in a patchwork of related habitats, one population wiped out or depleted by fires or some other chance event might be repopulated by tortoises in an adjacent habitat. Scientists refer to these related populations as "metapopulations."[45]

As noted by one population biologist, the life of any individual is "a continual roll of the dice," and the same is true for populations or species.[46] The more favorable the conditions, the more likely an individual will survive and reproduce. The more individuals in a population, the more likely the population will sustain itself under a wider range of conditions. The more separate but connected populations, the less likely that some chance event will extirpate a regional population or drive an entire species to extinction. Several related questions are important to determine when efforts to protect or recover desert tortoises will succeed. How much habitat is needed, and with what attributes, to support self-sustaining populations? How many tortoises make it likely that a given population will sustain itself over a particular period of time? How many populations, and of what sizes, make it likely that a metapopulation, or the whole species, will survive over a defined period of time?

Unless communicated clearly, single numeric population goals can be misinterpreted by nonscientists as definitive measures of success or failure, rather than as probabilities that a population or species will persist for a certain period of time. No matter how supportive the habitat, and no matter how large the population, there is always some chance that a population might be wiped out, perhaps by the sudden infestation of a parasite or a fire or drought that decimates the food supply. Certainty is not just elusive, but impossible. Scientists can only use the best available information to estimate the likelihood that, under particular conditions, a given population, metapopulation, or species will survive over a certain period of time.[47]

Decisions about chance (the acceptable probability of local extirpation or species extinction) and other key interrelated judgments of science and policy were made in developing recovery goals for the Colorado River fish. The pivotal component of the Colorado pikeminnow recovery goals, for example, is that individual breeding populations in the upper Colorado River basin be maintained at 2,600 adults, which the FWS calculated as the "minimum viable population" (MVP) needed to ensure long-term genetic and demographic viability. Why not 2,500, or 2,700, or 15,000? Although the details are not important per se, and although the FWS employed an accepted scientific method to calculate these population goals, it is important to realize that the numbers can vary quite significantly depending on a range of uncertain assumptions used in the methodology, and that those choices reflect implicit policy decisions about what level of risk of extinction or extirpation is "acceptable." Biologist Mark Shaffer asked, "Does a 95% probability of persistence for 100 years make extinction sufficiently remote, or all too imminent?"[48]

A "demographically viable, self-sustaining population" means enough individuals, taking into account all environmental variables and uncertainties in a particular habitat, to make it likely that the population will survive. "Demographic uncertainty" describes randomness in the survival and reproduction of individuals. Not all individuals reproduce successfully. Not all offspring survive to adulthood. "Environmental uncertainty" means that variations in weather, food, predators, competitors, and parasites all can affect the chance of offspring survival. A "catastrophic event" such as flood or drought could wipe out all young in a particular location. Because none of those factors can be controlled in the wild, the more individuals of both sexes, and the more matings, the greater the probability that enough young will survive to perpetuate the population.[49]

All of those factors, though, ensure only that enough individuals survive in the short term, under relatively stable conditions. The necessary adaptation to survive in the Colorado River, which has undergone such dramatic changes over time, requires a diverse gene pool so that individuals with certain characteristics survive as conditions change. Very small populations, in the wild or in hatcheries, reduce the genetic variation necessary to increase the chance of long-term survival of a population or species. Reduced genetic diversity in small captive stocks, maintained either to breed domestic animals or to replenish wild species, often results in reduced growth rate, adult size, survival, and reproductive success.[50]

In calculating the MVP of 2,600 adults, the FWS relied on several assumptions about the minimum numbers of individuals needed to establish viable, self-sustaining populations. To determine the "genetic effective population size," the number of individuals that contribute genes to the next generation, the FWS assumed that fifty individuals are needed to avoid inbreeding depression (which concentrates similar genes in a small population), and that 500 are needed to reduce long-term genetic drift (which concentrates random genetic mutations in small populations). The FWS acknowledged that we know too little about pikeminnow to calculate a species-specific genetic effective population size, so it used a default assumption of 500 "commonly used for fishes." This assumption is not necessarily "wrong," but some researchers suggest an effective population size of 5,000 (ten times higher), and warn that the lower figure can lead to significant loss of genetic variation after only a few generations.[51]

But 500 fish does not mean 500 successfully reproducing (what scientists call "genetically effective") fish. To calculate this higher figure, the FWS assumes a ratio of three male pikeminnows to every female. Based on this imbalance in the pikeminnow dance card, the FWS calculates that 666 total fish are needed to maintain 500 reproducing fish. It is very difficult to measure what percentage of fish breed every year, a figure that varies greatly among species. Because we have no information specific to pikeminnows, the FWS used an "overall average ratio for fish." But other researchers suggest that breeding success in pikeminnows is probably much lower, based on the closest real information we have for humpback chub in the Grand Canyon. Using a different assumption would nearly *triple* the number of total fish needed to ensure 500 successful breeders, from 2,600 to 6,600.[52]

One group of scientists who have been working with Colorado River endangered fish for many years believes that the assumptions used for these recovery goals could be seriously misplaced, and that using more conservative but plausible assumptions could dramatically change the MVP size adopted in the pikeminnow recovery goals. If we use the minimum effective population size of 5,000 proposed by some researchers, and the different ratio of total to genetically effective fish proposed by other scientists, we would require an MVP of more than 75,000 rather than 2,600 fish![53]

Given the uncertainty inherent in estimates of the minimum numbers needed to ensure viable populations, another way to avoid extinction is to protect metapopulations. As shown for desert tortoises, metapopulations are multiple, connected populations of individuals, assuming that each population is an independent breeding unit, and that migration between populations is possible, so that individuals from one population may replenish another in the event of a catastrophic decline. For pikeminnow, the FWS relies on three remaining populations in the Green River, upper Colorado River, and San Juan sub-basins. But this decision also depends on assumptions. For example, the FWS does not believe that sufficient habitat remains in the upper Colorado River to support more than 700 fish, rather than the 2,600 individuals it identified as the MVP size. Although that might be true, it also does not support the idea that the upper Colorado population will provide the requisite safeguard against catastrophic loss in the larger Green River population. Assuming replenishment from the San Juan River, where fewer than fifty pikeminnows were found as of 2002, is even more questionable. That small population is susceptible to genetic drift, inbreeding depression, and other problems.[54] Replenishment from the San Juan River population also assumes, with no current evidence, that pikeminnows can cross almost 200 miles of Lake Powell.

Whatever recovery goals are chosen, determining success or failure also assumes we can measure accurately how many fish remain in parts of the river at any time. Humpback chub inhabit deep, fast-flowing canyons, where it is difficult even to keep a boat steady much less to count fish. With capture and tagging methods scientists sometimes catch the same fish repeatedly, and access to many areas of fish concentration is difficult to impossible depending on flow conditions. One agency biologist reported at a program meeting that "[a]fter 20 plus years of study, we did not have a clear understanding about the status and trends of the [humpback] population." Other biologists noted that differences in methods can yield very different results: "It's hard to tell what's due to changes in the fish and what's due to changes in the biologists."[55]

So our ability to determine when the recovery goals are met may be just as uncertain as the goals themselves, leaving agency officials and the public frustrated about what decisions can be made, when, and with what confidence. Given the unresolved tension between the ESA and the Law of the River, water users want to know when the listed species will be sufficiently recovered so that the FWS can lift the jeopardy opinions that cast a cloud over the ability of the upper basin states to develop their full allocation of water. Uncertainty in the scientific analysis is understandable, as is the quest for stability in the legal process. The difficulty is in how we try to reconcile those apparently competing objectives.

Law, Science, and Uncertainty

Pioneering conservation biologist Michael Soulé noted that politicians and engineers often are frustrated with the inability of ecologists to provide simple, clear answers, but quipped that "the quest for a simple bottom line . . . is a quest for a phantom by an untrained mind." This conflict in expectations reflects what scholars who have studied this problem call a "culture clash" between law and science, which frustrates both scientists who present information and those who need to interpret and apply that information to make decisions. Sheila Jasanoff wrote that "science seeks truth, while the law does justice; science is descriptive, but the law is prescriptive; science emphasizes progress, while the law emphasizes process." David Faigman explained: "Science explores what is; the law dictates what ought to be. Science builds on experience; the law rests on it. Science welcomes innovation, creativity, and challenges to the status quo; the law cherishes the status quo." Although law seeks a reasonable degree of certainty, science is characterized by uncertainty, contradictory studies, and competing hypotheses and explanations. Law and science also operate on very different timetables. Although legal disputes must be resolved fairly based on the best evidence available at the time, science is characterized by evolutionary revision of existing hypotheses punctuated by occasional major paradigm shifts, but with the understanding that absolute truth is elusive and nothing is ever "proven."[56]

In a typical courtroom battle involving science, we use cross-examination, rebuttal, and similar procedures to help judges or juries decide who is *more likely than not* to be correct. In civil cases between private parties, it is logical for the law to choose certainty and finality over an impossible quest for absolute truth, so long as the process is fair to both sides. If two private parties need to resolve a dispute about whether a product was designed safely, we might need to choose between the opinions of different experts in product design and production. We cannot choose one hypothesis tentatively, with a procedure to revisit the verdict periodically as engineering paradigms, or even society's values, change. The parties need to get on with their lives, knowing whether they won or lost, even if they don't like the answer. But juries also implicitly make policy judgments in reaching verdicts about scientific information. Should we err on the side of product safety, or do consumers accept some risk when they purchase a potentially dangerous product?

In civil litigation, the jury system provides an accepted mechanism by which difficult, sometimes subjective judgments are made. If the public disapproves, elected officials can adopt new rules. The questions are even harder when we deal with public issues, such as how much water states can take from the Colorado River, and how much should be left to protect fish or to restore ecosystems. The stakes are higher, and those decisions affect more people (and fish) for longer periods of time. An open, democratic process is even more necessary so that the public will accept the resulting choices as valid. When such decisions are made by agencies, they should be completely transparent, with policy choices distinguished clearly from issues of "objective" science. Where the stakes are highest, some have suggested that those policy decisions should be made by elected officials rather than unelected bureaucrats, a theme to which we will return in the next chapter.[57]

In part, public acceptance of changing decisions based on evolving science also requires better government communication and public understanding of the nature of the scientific process and its inherent uncertainty. The recovery goals are an excellent example. The significance of the uncertainty reflected in the goals depends in part on how the minimum population size adopted in the recovery goals is interpreted by interested and affected members of the public, and by the agency itself. One of the founders of the theories used by the FWS, Michael Soulé, wrote that anyone who accepts any single figure as absolute "deserves all of the contempt that will be heaped upon him or her." But members of the lay public often misinterpret single point numbers such as the minimum population of 2,600 as having specific, almost definitive significance (the "right" answer), rather than reflecting inherent probabilities and uncertainties.[58]

In the text explaining the recovery goals, the FWS acknowledged the uncertainties inherent in the calculated minimum population size, and to some extent the recovery goals account for that uncertainty by adding other criteria (such as minimum management measures and long-term monitoring requirements) to the numeric goals. But this does not compensate for the fact that the deterministic numbers bear legal significance on which water users in the basin will rely. If the numbers are met for the prescribed period of time, and if the other criteria in the recovery goals are met, water users will expect downlisting or delisting to occur.

None of this, however, gets us to the even more difficult questions and uncertainties involved in determining *how* recovery goals or other restoration objectives for the Colorado River will be met. Actual restoration efforts are complicated by even more uncertainty about the relationship between changes in river management and the health of the aquatic ecosystem. Here, arguably the very legal regime designed to slow down new agency actions that might cause more harm to the river, and to force more consideration of environmental impacts and alternatives, impedes restoration efforts just as much, or more. To understand this irony, and to get a preview of the difficult choices addressed in the next chapter, we return briefly to the recovery goals, and then to Flaming Gorge.

Uncertainty in Restoration Strategies: A Preview

In a postscript to the recovery goal saga, a federal court later invalidated the recovery goals for the humpback chub in response to a lawsuit brought by several environmental groups.[59] The court rejected the goals not based on the uncertain scientific assumptions discussed above, but because the FWS did not provide estimates of how much time and money it will take to implement the various measures included in the goals, as required by the ESA. Although the groups asked the court to order the FWS to issue new, improved goals, the court held that the agency issued the goals *voluntarily*, and therefore was under no obligation to do so again. But the FWS announced that it will begin work on new recovery goals in 2007. This is not the first example of controversy leading to significant delays in the upper river recovery process.

Flaming Gorge Dam was completed in 1962, and full operation began in 1967. Biologists knew early on that construction and operation of the dam harmed the native fish of the Colorado. The FWS requested consultation under section 7 of the ESA in 1980, along with similar consultations on water projects throughout the upper basin. More than a quarter century later, the Bureau of Reclamation and the FWS continue to study the effects of these projects under both the ESA and NEPA, but the water projects proceeded. Dam operations were modified in some ways early on to mitigate impacts on fish and other resources. The Bureau established minimum release requirements in 1974, and in 1979 began to release water through a multilevel outlet structure so that warmer water would reach fish downstream. Both of those changes, however, were designed to help introduce trout rather than endangered natives.[60]

The Flaming Gorge ESA consultation led to a series of tests and studies about the effects of different flow releases on endangered fish downstream from the dam. Around the same time, the FWS issued a draft conservation plan proposing modified flows for all streams in the upper basin. The upper basin states argued that the FWS based its proposed flows on historic conditions rather than on documented science about how much water the fish really needed. In effect, the states argued that water projects could proceed without proof that the species would be protected (as required by the ESA), but that the FWS could not propose restoration flows to protect the fish absent adequate scientific proof. Again, choices were necessary about where to assign the risk of uncertainty, this time about the efficacy and impacts of proposed restoration strategies.

Jack Stanford, a highly respected independent scientist, was asked to review the proposed FWS flow recommendations. Stanford supported the general approach of the recommendations with some proposed modifications. Perhaps more significant, however, he urged an iterative, experimental approach to restoration efforts as a way to deal with the problem of uncertainty, as an alternative to the philosophy of doing nothing until definitive scientific answers are available:

> [A]ctions are often needed before scientific consensus can be achieved. . . . Uncertainties must be confronted by obtaining additional and more comprehensive information about how the endangered fishes function in the Upper Colorado River Basin ecosystem. Regardless of our ability to firmly demonstrate population dynamics of the endangered fishes, they are

rare, and further scientific study predicated on forecasting the future will not make them more abundant. . . . I make recommendations that couple action (implementation of flow regimes) with additional study to resolve the uncertainties discussed previously.[61]

We will discuss this concept of learning by doing, or "adaptive management," in more detail in the next chapter.

Because of its perceptions about inadequate information and the related political controversy, however, the FWS adopted Stanford's suggested strategy only in small part. Although providing for temporary shifts in dam releases to more closely mimic natural conditions, the FWS called for yet another study. Research continued for seven more years, leading to a phenomenally well documented report with more recommendations to change the timing, magnitude, and temperature of water released from the dam. The new report proposed increases in peak flows in spring and early summer to mimic the natural hydrological cycle, with a somewhat complex set of flow targets for different reaches of the river under different conditions, and warmer water temperatures during summer and fall.[62]

Even after all this study, the recommendations were not ready to be implemented, and the Bureau continued to operate the dam under the 1992 criteria. The Bureau evaluated the 2000 flow recommendations further in an EIS conducted pursuant to NEPA, to compare the pros and cons of two options: existing operations under the "temporary" criteria established in 1992 (dubbed the "no action" alternative because it would reflect no change in dam operations), and revised operation of the dam to achieve the 2000 Flow and Temperature Recommendations while maintaining and continuing all authorized purposes of the dam. The Bureau released a final EIS in September 2005, a quarter century after ESA consultation began. Finally, in February 2006, the Bureau issued a decision announcing that it would implement the new flow recommendations.[63] How much this will help the endangered fish remains unknown.

The irony here is the sharp contrast between decisions that preceded construction of Flaming Gorge and those that followed. When the dam was under construction, use of a massive dose of rotenone to rid the river of pesky native fish proceeded with little study of impacts or alternatives. Objections were brushed aside, no real public participation was provided, one set of preferences was paramount (recreational fishing), and a decision was made and implemented in a matter of months. In response to this and similar cases of haste, and the resulting environmental harm, Congress passed laws designed to slow the process down, to ensure that all issues and values are debated fully, and that decisions are made only with the best possible information. Those procedures apply just as surely to efforts to restore threatened and endangered species and their habitats. And those who fear that changes in dam operation to promote restoration efforts might impair their use of water and power use those same statutes to maintain the status quo as long as possible, using the inevitable scientific uncertainty as a tool for delay. As geophysicist Henry Pollack warned about this tactic, "Waiting until uncertainty is eliminated before making decisions is an implicit endorsement of the *status quo,* and often an excuse for maintaining it."[64]

On the other hand, proposed environmental restoration strategies clearly should not be exempt from requirements for full study and public review. Careful deliberation is equally appropriate before we undertake expensive restoration steps that might harm other important interests and values, because the relevant science is uncertain and evolving. We do not understand fully which impairments are most responsible for harming the fish, so we might implement expensive and damaging programs to no avail. We do not understand what conditions are most likely to help the endangered species, so efforts proposed with the best intentions might be misguided. Worst of all, some scientists question whether actions proposed to aid endangered fish will do more to harm rather than help those species, because they might provide more help to exotic competitors or predators than to the natives. But inaction until all of those issues are resolved definitively risks extinction for several species of big river fish. Efforts to evaluate all options carefully, and to navigate through the wide range of competing interests at stake, can help us to strike the right balance between the risk of inaction and the risk of ill-conceived action. We explore those trade-offs in the next chapter, in the context of an equally challenging effort to restore the stretch of the Colorado River between Glen Canyon and Hoover dams.

CHAPTER SIX

Casting of the Lots: Conflicting Methods and Goals in Environmental Restoration

On October 15, 1956, President Dwight D. Eisenhower pressed a ceremonial button in Washington, D.C. Secretary of Interior Fred Seaton telephoned onlookers in Kanab, Utah, who radioed the command to the nascent construction site for the Glen Canyon Dam. Workers depressed a plunger and an explosion rocked the canyon walls, beginning one of the largest construction projects in world history. There was little uncertainty about this effort. Although the dam would serve multiple purposes, it promoted a singular set of values. The reservoir would store as much water as possible. Turbines buried deep within the dam would generate as much electricity as feasible. The dam would promote development and allow the upper basin states to deliver the water required by the compact while still developing their own water supplies.[1]

More than four decades later, on March 26, 1996, Secretary of Interior Bruce Babbitt turned a valve and sent 45,000 cubic feet per second of water through the jet tubes of Glen Canyon Dam. This was not a purely ceremonial act (Babbitt turned the real valve), although it was certainly symbolic, shown live on the *Today Show* and *Good Morning America*. It was a grand experiment, perhaps the biggest ever conducted on how to restore an aquatic ecosystem whose biological resources were impaired by artificial changes in a river's flow. Three years later, Secretary Babbitt proclaimed the experiment a major success.[2] But that affirmation was premature, and more political than analytical.

The bigger difference between building the dam in 1956 and trying to repair the ecosystem it impaired forty years later is that the flood experiment was not driven by a monolithic set of values. The scientists and officials who conducted and monitored the experimental flood understood the Abe Lincoln principle: "You can't please all the people all of the time." Even absent environmental considerations, operating a dam involves trade-offs among different objectives. If you store more water behind the dam, perhaps to help the upper basin meet its water delivery obligation during a later drought, less storage space remains if the ensuing winter brings a bounty of snow. Torrential spring runoff will threaten dam safety if you cannot release water quickly enough to keep up with the flow. If you release as much water as possible to protect the dam, more than you can use to spin the turbines to generate electricity, you will not maximize power generation with the available water. If you dump all of that water all at once, you might cause flooding downstream, threatening new riverfront condos.

Environmental constraints make your job even more difficult. One group of scientists who were involved in or studied the experimental flood commented that no single way of operating the dam would improve conditions for all downstream resources, and that difficult choices among competing values are needed.[3] As shown in chapters 2 and 3, dams cause significant changes in riverine and riparian environments. Whether those changes are "good" or "bad" depends on whom you ask. The Colorado River once flowed through the pristine meanders of Glen Canyon, enjoyed by a relatively small group of self-reliant boaters and hikers who sought solitude and splendor in a booming modern world. Lake Powell now provides easy access to millions of people who prefer hotels, houseboats, and campgrounds, all of which boosts the region's economy. Downstream, the dam stemmed the spring floods that once replenished beaches and aquatic habitat, but scoured vegetation from the river's edge. We improved conditions for introduced trout at the expense of native fish, and swapped camping beaches for marshes and riparian forests. One main purpose of the experimental flood was to learn the effect that different flow patterns would have on different resources and uses of the river and its surrounding ecosystems.

In the last chapter, we explored several of the difficult choices decision makers face in planning and implementing large-scale environmental restoration programs in the face of scientific uncertainty. In this chapter, we look at two related sets of issues, largely in the context of ongoing efforts to restore the aquatic ecosystems of the Colorado River in the Grand Canyon.

First, it is often not clear which restoration strategies will help or hurt various target resources. That is primarily a scientific inquiry, in which the adaptive management process described preliminarily at the end of chapter 5 can be useful. Adaptive management suggests that if we ask the right questions using carefully designed and controlled experiments, perhaps we can determine which changes in dam operations and other management actions will help or hurt species or other resources, and how. An early proponent of the concept, Kai Lee, defined adaptive management as "treating economic uses of nature as experiments, so that we may learn efficiently from experience." Adaptive management, however, is not simply trial and error in the hope that we will ultimately get it right and miraculously save species and restore ecosystem health. The approach relies on testable hypotheses in which individual variables are controlled to the extent possible in a complex environment, so later efforts can draw more reliably on the lessons learned. Ecologists and adaptive management pioneers Carl Walters and C. S. Holling distinguished between pure trial and error, in which early choices are haphazard but improve from increased information, and adaptive management, in which each experiment is explicitly designed to generate the information necessary to make later choices.[4]

Second, when not all of the resources that are valued by different groups of people benefit equally from various restoration efforts, other difficult choices must be made about which resources should be promoted, and to what degree. Sometimes we must choose between native resources and new but still valued resources supported by modified environments, or the resources we extract or generate from those environments. This second set of choices cannot be made by scientists alone, or with purely scientific techniques.[5] As with the decisions discussed in chapter 5, both of those sets of related choices must be made openly and with adequate public notice, education, and input.

Let's look first at the maze of laws and regulations that agency officials must navigate to guide those decisions with respect to the Colorado River from Lake Powell to Lake Mead and beyond. One barrier to making clear decisions in the face of tough choices among competing values has been the set of conflicting, often contradictory statutory mandates that apply to Grand Canyon and other Colorado River restoration efforts. Then we'll explore some of the many scientific uncertainties that complicate decisions about the efficacy of different restoration strategies. Finally, we will look at the collaborative process designed to address tradeoffs between competing restoration goals, and ask how well it has worked.

Restoration Decisions in the Face of Conflicting Laws

My map of southwestern Utah identifies a point of interest called "Casting of the Lots." Strange name for a tourist attraction, but it has no connection to the gambling industry in nearby Nevada. The reference dates to October 1776, when the party of fathers Dominguez and Escalante stopped at this remote place to make a pivotal decision about their direction and their goals. The expedition left Santa Fe on July 29, 1776, with a primary charge of identifying a new overland trail to Monterey, California, and a secondary goal of exploring locations for new missions. Rather than taking a direct route, the expedition roamed more than 1,700 miles across significant portions of the Colorado Plateau and the Great Basin.

The fathers eventually aborted their efforts to reach Monterey. Concerned about the onset of winter, the party turned south to find a shorter route from Santa Fe to the tribes to whom they promised to return: "This delay . . . could be very prejudicial to the souls which . . . yearn for their eternal salvation through holy baptism." Apparently, the two leaders considered their secondary task as missionaries more important than the primary charge of route finding to California. This was the service to which they devoted their lives, in vows to a higher authority. But not all in the group agreed, perhaps fearing the wrath of their superiors if they returned without having reached Monterey. To prove that they had not chosen their path out of fear or "despotic will," Dominguez and Escalante agreed to "cast lots" to decide which way to go. To them, this was not a game of chance but an act of faith. God would make their choice by guiding the manner in which the lots fell.[6]

This was also a choice between two sets of core values. The Spanish conquistadors were bent on establishing an empire in the New World, and sought gold and other riches as well as glory.[7] The missionary priests with whom they were entwined, however, suffered years of deprivation in the southwestern deserts for a very different purpose—to save souls.[8] Sometimes, missionaries and conquistadors could reconcile the two sets of values. Spaniards who thirsted for empire drew moral justification from the concurrent mission of conversion. Where those values conflicted, they turned to God to resolve the impasse. The casting of the lots ratified the decision to continue south and back to Santa Fe. After additional trials and detours, the weary travelers located one of the few places at which men could ford the Colorado River in relative safety, a place later known as "Crossing of the Fathers" until it was submerged under Lake Powell.

We face a similar conflict among core values in our decisions about the river. To some, including early reclamation advocate William E. Smythe, the task of taming the Colorado also served a higher purpose, the "conquest of arid America." Smythe and others believed our destiny was to forge an irrigation empire in the Southwest. And like the Spaniards, they believed that their mission reflected "man's partnership with God." With an equal religious fervor, others now strive to save the wild river, the life it sustains (to the romantically inclined, the river's "soul"), and the pilgrims who journey there for spiritual sustenance. Although Smythe wrote that "the glories of the Garden of Eden itself . . . were products of irrigation," singer Katie Lee lamented that in damming Glen Canyon politicians and bureaucrats "orchestrated the drowning of Eden." Environmental historian Roderick Nash described this tension as "the legitimate ambivalence American culture feels between wilderness and civilization."[9]

Unlike the Spaniards of the late 18th century, however, we cannot seek divine guidance to plot our route or to choose among competing values. We live in a republican democracy in which we elect representatives to make choices by passing laws and making other decisions that, we hope, reflect the will of society. Of course, one advantage of casting lots is that it produces a definitive answer. Congress often seeks compromise among multiple interest groups, and therefore does not always speak with a clear and singular voice. As shown below, with respect to our management of the Colorado River, over time Congress legislated a set of conflicting, inconsistent directions in a maze of separate statutes. Those inconsistencies continue to complicate restoration efforts.

Environmentalists hail the debate over the Colorado River Storage Project Act of 1956 as a key event in the birth of the modern environmental movement. David Brower of the Sierra Club and others launched one of the first national environmental campaigns to prevent construction of two dams within Dinosaur National Monument. In a compromise the wisdom of which environmentalists still debate, Congress elected not to build dams in Dinosaur in return for an agreement to allow other projects, including Glen Canyon Dam (figure 6.1). The opponents did not quibble over the operating rules for the dams that would be built, however, and how they would affect the river's environment. The battle was over wilderness, not fish.[10]

In the 1956 law, Congress adopted a priority of purposes for which the dams were to be managed: water first, power second, environmental values a distant third. The stated project purposes were to regulate water flows, to store water for "beneficial use," to "reclaim" arid lands, to control floods, and "for the generation of hydroelectric power, as an incident of" other listed purposes. Generation of hydroelectric power was secondary. Congress underscored this point by directing that hydroelectric facilities be operated "so as to produce the greatest practicable

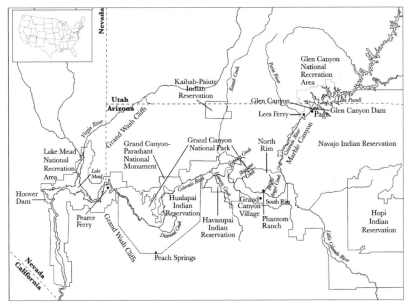

Figure 6.1. Grand Canyon reach of the Colorado River basin. *Courtesy of the U.S. Geological Survey.*

amount of power and energy," but only if power generation would not interfere with the compact and other aspects of the Law of the River. In a third section, Congress also "authorized and directed" the secretary of interior to develop public recreation facilities on lands adjacent to the new dams, and to allow public use "by such means as are consistent with the primary purposes of said projects." Finally, Congress called for "facilities to mitigate losses of, and improve conditions for, the propagation of fish and wildlife." Whatever environmental harm the dams caused was a necessary evil to be mitigated.[11]

A decade later, in the Colorado River Basin Project Act of 1968, Congress added additional guidance on priorities for water releases as it approved additional upper basin water projects. Releases from Glen Canyon Dam must be used first to meet the upper basin's share of the U.S. treaty obligation to Mexico, and second to meet the upper basin's deliver duty to the lower basin. Then, "to the extent surplus water exists," it may be used for additional uses in the lower basin, to maintain Lake Powell and Lake Mead at approximately equal levels, and to "avoid anticipated spills from Lake Powell."[12] The requirement to roughly equalize storage at Lake Powell and Lake Mead reinforces the lower basin's rights to water. The upper basin may not hoard water in Lake

Powell at the expense of water use and power production from Lake Mead. The ban on "anticipated spills" prevents releases at levels that exceed the power-generating capacity of the hydroelectric plant. Spills might be necessary to protect the dam during floods, as occurred during the wet years of the mid-1980s. Otherwise, spills "waste" water that could be used to generate electricity.

Again, no one was thinking about the environmental effects of the dams. Environmental lobbyists were busy with an even higher profile fight over two proposed dams in Grand Canyon National Park, a battle they ultimately won.[13] But the resulting rules for reservoir operation would affect environmental restoration efforts decades later. Do higher flows to restore the river constitute a prohibited "spill" of water destined to spin turbines? In 1970, the Department of Interior issued "long range operating criteria" to implement Congress's priority scheme. Every year the department issues an annual operating plan, based on water levels in the dams and weather forecasts, identifying specific rules for how much water will be released and when during the next "water year," which runs from October 1 to September 30.[14]

Glen Canyon Dam was not long in place before scientists began to realize its effects on the Colorado River ecosystem. Just a decade after the dam began to fill, scientists had already identified most of the major impacts the dam would have on the downstream river and riparian zone, although they would learn many more details in the years to come.[15] This was just about the same time that Congress reacted to broader public concerns about the environment by passing laws such as the National Environmental Policy Act (NEPA) and the Endangered Species Act (ESA), described in the last chapter. These laws ran headlong against the Law of the River and affected restoration efforts in many ways, just as they did with respect to Flaming Gorge Dam.

In 1977, the U.S. Fish and Wildlife Service (FWS) requested consultation under section 7 of the ESA because of the dam's impacts on endangered populations of humpback chub and Colorado pikeminnow. The next year, the FWS determined that the dam's operation jeopardized the continued existence of both species. (Pikeminnow have since been extirpated from the river below the dam.) The FWS suggested a series of studies to determine the effects of temperature changes, to identify the needs of endangered species, and to develop other operational changes to meet those needs. This research became one of the most extensive sets of focused environmental studies in history, the Glen Canyon Environmental Studies.[16] The FWS issued a second draft biological opinion in 1987, a decade after the initial consultation. This time, the FWS recommended physical changes in dam operations and other steps to help the endangered fish, including a hatchery population of fish and establishment of a second spawning population of humpback chubs in Grand Canyon. In 1989, however, the secretary of interior decided to begin an environmental impact statement (EIS) on future dam operations, triggered by a proposal to upgrade the turbines to increase power production. Ironically, compliance with one federal law (NEPA) delayed further action to protect the endangered fish under another (the ESA). The FWS agreed not to finalize its new biological opinion pending completion of the EIS, which took another six years.

The draft EIS considered nine alternatives, each of which sought a different balance among the purposes for which the dam might be managed. The "no action alternative," which must be included in every EIS as a baseline against which other options can be weighed, would continue existing dam releases. A second option would maximize power production, with greater impacts to downstream resources and probably more harm to endangered fish. Four alternatives would reduce the daily fluctuations in dam releases to varying degrees, with a goal of reducing downstream environmental impact while still maintaining flexibility in power production. The remaining alternatives would eliminate daily fluctuations altogether, with steady flows throughout the year, on a monthly basis, or adjusted seasonally to mimic predam flows.

Hydroelectric power can be ramped up quickly to meet peak power demand (such as on hot summer days when everyone runs their air conditioners). The resulting daily water-level fluctuations, however, posed problems for boaters and fish alike. Reducing or eliminating daily fluctuations was one possible step toward reducing harm to the downstream environment. But the 1956 and the 1968 water project laws directed that the dams be operated to produce the "greatest practicable amount of power" consistent with the Law of the River. Water and power users argued that the Bureau of Reclamation could not even consider alternatives that would violate those statutes. An alternative cannot be "feasible" if it is prohibited by law.

Likewise, what happens if ESA compliance causes an irreconcilable conflict with the Law of the River? The Supreme Court held in the snail darter case described in chapter 5 that the ESA trumped ambiguous congressional appropriations bills for the Tellico Dam. The statutes approving Glen Canyon Dam and other Colorado River water projects, however, predated the ESA. In a case involving management of Lake Mead to protect endangered southwestern willow flycatchers (discussed in chapter 7), one federal court ruled that ESA consultation is required only for agency actions that remain within an agency's discretion. This meant that efforts to protect the flycatcher were limited to those consistent with existing legal mandates.[17]

In the face of these conflicts, the Bureau of Reclamation agreed to adopt new "interim operating criteria" to reduce the impacts of dam operations pending completion of the second round of scientific studies and the EIS. Before either of those steps was completed, however, Congress passed the Grand Canyon Protection Act (GCPA).[18] In the new law, Congress had the opportunity to resolve the existing statutory inconsistencies. Unfortunately, the new law could be used in an introductory philosophy text as an example of circular logic.

Section 1802 of the GCPA contains three distinct requirements, each of which seems to point in a different direction. The first instructs the secretary of interior to operate the dam "in such a manner as to protect, mitigate adverse impacts to, and improve the values for which Grand Canyon National Park and Glen Canyon National Recreation Area were established." This could be viewed as providing a broader ecosystem protection mandate than the ESA, because the Grand Canyon statute includes all of the values for which the parks were established. Although the ESA in theory requires protection of the ecosystems relied on by threatened and endangered species, as discussed in the previous chapter, the focus of implementation has been on the needs of individual species, and not on broader ecosystem goals.

The second provision of the GCPA, however, potentially reins in restoration efforts by commanding the secretary to implement the first provision "in a manner fully consistent with and subject to" the Colorado River Compact and other aspects of the Law of the River, including the 1956 and 1968 statutes. The third provision in the statute further blurs an already murky picture, providing that nothing in the Act "alters the purposes for which the Grand Canyon National Park and the Glen Canyon National Recreation Area were established," or affects the secretary's authority and responsibility to manage those protected areas for those purposes. The National Park Service Organic Act dictates that national parks be managed "in such manner and by such means as will leave them *unimpaired* for the enjoyment of future generations."[19] How can the secretary manage Grand Canyon National Park in an unimpaired condition for future generations while operating the Glen Canyon Dam in ways that fundamentally alter the Colorado River ecosystem?

Finally, Congress rejects any effort to resolve the conflict between the Law of the River and later-enacted environmental statutes by providing in section 1806 of the GCPA:

Nothing in this title is intended to affect in any way—

(1) the allocations of water secured to the Colorado Basin States by any compact, law, or decree; or
(2) any Federal environmental law, including the Endangered Species Act.

Congress did include in the GCPA one provision that belies the notion that the agencies were supposed to perform miracles, despite the dam's proximity to the Crossing of the Fathers. In the last provision of the act, Congress instructed the departments of Energy and Interior to "identify economically and technically feasible methods of replacing any power generation that is lost through adoption of long-term operational criteria for Glen Canyon Dam," including ways to increase power production at Hoover Dam.[20] Congress would not have included the power replacement provision unless it envisioned that revised dam operations might curtail power production to some degree. Read in this way, Congress did reorder priorities for operation of Glen Canyon Dam: water allocation and environmental restoration in a dead tie for first, and power generation second.

Because of these internally inconsistent provisions, the GCPA did not help the agencies resolve disputes over the competing alternatives for dam operations in the draft EIS. Eleven out of twelve federal agencies involved agreed on an option that would reduce but not eliminate daily fluctuations in flows from the dam, in an effort to balance water and power needs against restored habitat for native fish and other natural values within the Grand Canyon. The FWS held out for an alternative that would mimic natural seasonal flow patterns without any fluctuating daily flows, which would reduce power production but, the FWS believed, better restore the environment downstream from the dam.

To underscore its concerns and clearly to increase its leverage given that it was on the losing side of an eleven to one vote, the FWS restarted the ESA consultation process begun sixteen years earlier, and concluded again that proposed dam operations would jeopardize the continued existence of humpback chub and razorback sucker. In its draft reasonable and prudent alternatives statement, the FWS suggested a set of experiments in which seasonally adjusted steady flows would be conducted during low water years, which historically had occurred about half of the time since dam operation began. Dam operators would release high, steady flows during the spring, and low, steady flows in the summer to match the natural pattern of spring runoff followed by much lower summer flows. This was the beginning of an adaptive management approach in which, rather than choosing one fixed strategy for operating the dam, different alternatives could be tested and modified in an effort to learn as we go.

Ironically, however, the burden of proof Congress imposed in the ESA, and the complexity of the ecological issues involved, interfered with this proposed experiment. The Bureau of Reclamation argued that the FWS could not support its proposal based on the "best available scientific information" required by the law. In particular, the bureau asked, how did we know that the proposed flows would not help nonnative fish more than the natives, and thus do more harm than good? That was a fair question, to which we still do not know the answer. But if there was no way to experiment, and if we handcuff the agencies with this scientific burden of proof before any actions can be tried and studied, the law can impede rather than promote restoration efforts.

As a compromise, the feuding agencies agreed to allow steady flows, as well as shorter intense flows designed to carry sediment downstream to rebuild sandbars, backwaters, and other habitats that have been depleted because the dam blocks sediment flow downstream. The Fish and Wildlife Service issued its final biological opinion in December 1994 (now seventeen years after its initial request for consultation); the Interior Department released the final EIS the following spring. Yet Secretary of Interior Babbitt did not sign his Record of Decision for more than another year, in October 1996. Why the added delay after so much study, analysis, and debate, and after choosing an adaptive management approach that should have addressed the problem of scientific uncertainty? The answer lies in the Abe Lincoln principle.

The FWS proposed to conduct experimental flows to mimic the natural spring and summer flow regime during years when water was short. In those years, the Bureau of Reclamation ordinarily would release only the minimum amount of water needed to satisfy compact obligations. If one goal of a spring flood is to rebuild beaches and other structures that rely on sediment, it is better to do so in a dry year in which another heavy flow is not likely to follow, thus washing away the habitats that have just been formed. The upper basin states and power users, however, prefer to store water in dry years as a hedge against later shortages, and to release more water in wet years. Storing more water in wet years also reduces storage capacity in the event of another wet spring the next year, and increases the chance of unwanted "spills," which reduce power-generating capacity.

In balancing these competing goals, Secretary Babbitt gave a little bit to each side. He approved the kinds of experimental flows sought by the

FWS, but only in wet years when they would less likely interfere with water storage and power production. The degree to which the secretary made that decision on political or legal grounds depends on how you read the circular legal puzzle outlined above. Under one interpretation, the 1956 and 1968 statutes foreclosed any new flow regime that increased the chance of "spills." Under a different reading, the GCPA created an exception that allowed such spills in the name of restoration.

Neither the secretary nor the scientists on whom he relied knew for sure what effect experimental flows would have on efforts to restore the Grand Canyon ecosystem. Below the dam, an armada of research boats and an army of scientists waited to measure the effects of the experimental "flood" on virtually every aspect of the downstream ecosystem. One of the key characteristics they monitored in the physical environment was the movement of sand, an attribute that can affect almost every other natural or human value within the canyon. Even if we focus only on the resources affected most directly by changes in the movement of sediment through the system, questions about what changes in dam operations will be effective in restoring natural beaches and backwater habitats have been difficult to answer. Efforts to restore the flow of sand through the canyon provide our first example of the benefits of an adaptive management approach to restoration.

Revised Dam Releases to Restore Sediment Flows

If you ever built a sand castle on a beach, you have some experience to help understand the problem of sand in the Colorado River. You probably built your medieval masterpiece in the intertidal area, where the sand was wet enough to stick together. Your castle stood strong until the tide began to rise. Then the waves gradually began to erode the palace walls. You could keep it safe for prince and princess so long as you continued to reinforce the walls with at least as much sand as was washed away. Once more sand was exported than you were able to replace, the structure was doomed. Hopefully, you took a picture.

Although the Grand Canyon reach of the Colorado River obviously is not tidal, variations in flow make the analogy appropriate. As described in chapter 2, before the dams the river carried huge amounts of sediment every year. Most of the movement occurred during heavy spring flows (the high tides of the Colorado), but much of the material deposited during intermittent summer thunderstorms in the erosive canyon country, when low flows prevailed for many months at a time (low tide on the Colorado). Sand collected in the river channel in the low water summer months, and the rejuvenated river carried it both downstream and laterally to banks and side channels the following spring. As spring floods subsided, some sand deposited in the channel farther downstream, but much also settled along beaches and bars and backwaters, providing a source of replenishment to these constantly changing features, the sand castles of the Colorado.

Because Glen Canyon Dam trapped most of the sediment in the reservoir, the river no longer carried enough to replenish beaches and backwaters. At the same time, the relatively clear water now released from the dam eroded more sand from those structures and carried even more downstream. In not too many years, they began to wash away. With them went important habitats for native fish and beaches formerly used by boaters. Remaining beaches were overrun by tourists and choked with vegetation, which was no longer scoured away by the larger spring flows that occurred before the dam.[21]

This erosion also impaired cultural resources revered by Native American tribes. Traditional Southern Paiute lands along the river extend 600 miles from the Kaiparowits Plateau in Utah to Blythe, California. Southern Paiutes believe they have a responsibility to protect this land, its water, and its other resources. One Southern Paiute elder referred to the river and its tributaries as "the veins of the world." Grand Canyon (Piapaxa 'Uipi, or Big River Canyon) is the most special place. Plants and other resources in the canyon continue to be significant sources of food, medicine, and ceremony, but their value transcends economics.[22]

Native American peoples occupied the Grand Canyon for at least 11,000 years. They include the ancestral Puebloans, and later the ancestors of the present-day Hualapai, Havasupai, and Southern Paiutes. As a result, a large amount of Native American artifacts are buried in the sand that lines the river. National Park Service archaeologists identified 475 archaeological sites, 336 of which are in areas that might be affected by dam operations. With less sand in the river, and erosion over time, archaeological resources are exposed to weather and other impacts, placing them at risk of damage or permanent destruction.[23]

One main purpose of the 1996 controlled flood was to replenish bars and beaches in this stretch of the Colorado River, and to offset the erosion that placed archaeological and other cultural resources at risk. River planners designed flow patterns to store in the main river channel sand that had been delivered by summer storms in tributaries downstream from the dam (especially the Paria and Little Colorado rivers). Then, larger but shorter flows (beach habitat building flows) could move this stored material downriver and onto eroded bars and beaches. If this material stabilized before the ensuing high flows, perhaps habitats and campsites could be restored and cultural resources preserved. Although erosion will continue to eat away at those re-created or fortified features, hopefully they will remain long enough to allow more sand to accumulate in the river, to be transported by another spike flow later on.[24]

This plan depended on the assumption that at least enough sand enters the system over time to offset the amount carried away by the river under existing dam operations. The "mass balance" models used in the 1995 EIS predicted that tributaries below the dam deliver enough sand—and that enough of that material is stored in the main channel of the river—to rebuild the river's beaches and sandbars successfully. In these models, scientists predict how much sediment mass comes into the system from various sources over a period of time, where it goes, and how much is carried out of the system during the same period. If the equations "balance" properly, the total sand input must equal sand storage plus sand transport out of the system.[25]

The 1996 experimental flood did rebuild many beaches and bars in Marble and Grand canyons. Just like the tides that washed away your sand castle, however, the Colorado soon washed away much of this reconstruction. Within six months, almost half of the new beaches were gone, and most of the rest were significantly smaller than when first created.[26] Although the details of sand transport in the Colorado are extremely complex, the basic reason for this reversal of fortune is simple: less new sand was delivered and more of it was washed away than predicted by the models used in the EIS. Fortunately, the scientists who monitored the experiment took a whole lot of pictures, in a lot of different ways. In a classic example of adaptive management, those images helped them to improve their understanding of the dynamics of sand in the river through Grand Canyon, and *might* allow us to get better results in the future.

The models used in the 1995 EIS were limited by a lack of sufficient data. Before the dam was built, the Interior Department studied virtually every aspect of resource use and conditions in Glen Canyon, upstream of the dam site. But they did not do the same for downstream resources. (This was a decade before Congress passed NEPA, which would have required a comprehensive EIS.) Thus, very little information was available about predam sediment transport in the river. Even though some scientists predicted just ten years after the dam began operations that inadequate sand replenishment would cause erosion of downstream sand, additional sediment studies were discontinued until the Glen Canyon Environmental Studies program began a decade later. So scientists had very few years of sediment data before the 1995 EIS. The studies they had were done during the extremely high flow years of 1983–1985, and sediment flow characteristics vary with river flow. As a result, the models were based on data that did not properly reflect the highly variable nature of sediment movement in the river and its tributaries. It was the same general mistake as writing the Colorado River Compact based on just twenty years of data from wet years. The EIS predicted that four times as much sand was being stored in the river channel than scientists now believe occurs. Sand delivery to the river in Grand Canyon is only sixteen percent of what it was before the dam. As a result, less sand is available to rebuild beaches and bars when flood releases occur than predicted in the EIS.[27]

Inadequate and unrepresentative data also led to errors in predictions about how much sand the river would move, and how quickly. Earlier studies assumed that the size of sand grains did not matter, and scientists predicted that extremely fine sediment would move at the same rate as coarser sand. In fact, faster water carries heavier particles. During the high flow years when data were collected the sand deposited in the river channel was unusually coarse. As a result, the models calibrated based on those data did not properly predict the rate at which the river would move the finer sand needed to rebuild beaches and sandbars. Now we understand that most of that finer sand is transported all the way out of Grand Canyon within a few short months of normal flows from the dam. At the time of the experimental flows in March 1996, much less material was available in the river channel to rebuild our natural sand castles than expected.

Why, then, did the experiment restore beaches and bars at all, and why was the new sand washed away so quickly? Scientists took not only pictures, but aerial photographs of the beaches before, during, and after the experimental flood. Analysis of those photos revealed a surprising result. Beaches and bars in the river were re-created or reinforced not by sand carried from the river channel, as predicted in the EIS, but from lower-lying portions of the bars themselves. Rather than using sand from another part of the beach, we took sand from the back of the castle to reinforce the front walls in a futile race against the tide.

The results of the experimental flood also provided perspective on the full range of flow conditions on which restoration efforts should focus. Periodic floods are natural disturbances that reorganize material and energy within river systems. Scientists highlight these events as important to natural cycles,[28] and dramatic events such as floods, tornadoes, and volcanic eruptions are impressive to laypeople as well. But "normal" conditions, however dull, can be equally important in shaping ecosystems. The rapid depletion of fine sand from the river channel after it is added to the river from tributary storms, and the rapid erosion of newly created or reinforced sandbars after flood flows, are both explained more by low than by high river flows.

Before Glen Canyon Dam was built, spring floods through Grand Canyon were followed by a prolonged period of flows much lower than under current dam releases. Sand delivered during summer storms in the tributaries was stored in the channel when river levels dropped. The rising river then mobilized that material the following spring. As the river dropped once more, sand deposited on the downstream bars at curves and other places of natural deposition. Now, sand from a summer thunderstorm is carried out of the canyon within several months due to high summer dam releases, when more electricity is needed to run air conditioners. This sand is no longer available during an artificial beach building flow the following spring. Likewise, in the natural system a prolonged period of languid summer, fall, and winter flows followed the annual spring rebuilding and reshaping of beaches and bars. Rather than being washed away by routine dam releases, in the natural system newly formed beaches and bars remained in place during low flow periods.

The implications of this new understanding may be critical to the feasibility of restoring the Grand Canyon ecosystem, further demonstrating the wisdom of adaptive management. If there is not enough tributary sand stored in the channel, any chance to restore beaches and backwaters may erode along with the sand. If so, the only remaining option is to move sand from *above the dam* to the river below to supplement existing inputs from the tributaries. Ridiculed as a fantasy not long ago due to expense and logistical challenges, the idea of "busing" sand past the dam (or perhaps sluicing it through the dam's spillways) suggests restoration problems of a different kind. Although the issue has received relatively little study, available evidence suggests a buildup of mercury, selenium, and other contaminants in the sediment accumulating beneath Lake Powell. This presents a difficult choice between two options, neither of which is consistent with the statutory "nonimpairment" goal for Grand Canyon National Park. Either the river remains starved of the sand needed to replenish and reshape its bars and backwaters, or sediment from behind the dam contaminates the river below.[29]

There is still some hope for more successful restoration, however, based on lessons from the 1996 experimental flood. Rather than timing artificial beach building flows during the spring, when floods occurred in the natural system, scientists proposed two different ways to use new sand that enters the river during summer or fall storms. One approach is to reduce dam flows soon after those storms, to increase the amount of fine sand stored in the channel until an artificial flood flow can occur. This might reduce power generation. Alternatively, flood flows can be timed immediately after new sand enters the river, to move and deposit sand before it washes downstream. Both approaches require continuous monitoring of river conditions and flexibility in dam operations to respond to tributary storms quickly.

The first such experiment occurred in November 2004, nearly a decade after the 1996 experimental flood. The results are still being evaluated, but some scientists are becoming increasingly pessimistic about our ability to restore sand features in Grand Canyon simply by tinkering with existing dam operations. A report by the U.S. Geological Survey warned that "the possibility exists that no operational scenario will result in management objectives being achieved for restoring sandbars, simply because of the volume of water that must be released on an annual basis" given the severely depleted sand resources below the dam.[30] If so, the remaining alternatives may be increasingly expensive, with much greater impacts on dam and power plant operations. Significantly reducing flows to store sand during the summer through winter period would reduce power generation during the very months when power is in greatest demand.

The second option, scheduling flood flows during the fall or winter rather than the spring, would not bring us closer to the natural seasonal flows that occurred before the dams. That raises questions about the impacts of this new proposed beach-building plan on other river resources, especially on native species such as humpback chub, which evolved under conditions of torrential spring floods and much lower flows the rest of the year. Flows timed to restore beaches and backwater habitats might not be effective for, or perhaps might be detrimental to, other resources. An adaptive management approach can provide information about trade-offs between resources using different restoration strategies, but as more resources are considered, the choices become even more difficult. We explore that second level of choices and trade-offs next, in the context of efforts to recover humpback chub populations in the Grand Canyon.

Competing Strategies to Restore Endangered Fish Populations and Habitats

By the time Robert Rush Miller identified humpback chub as a separate species in 1942, they had probably already begun to decline. As discussed in chapter 4, little is known about long-term trends in humpback populations because of inadequate historical data and difficulties in data collection and analysis. Biologists are reasonably confident that the Grand Canyon population of humpbacks plummeted from more than 10,000 in the late 1970s and early 1980s to fewer than 3,000 by the early 1990s, although some recovery appeared evident by late 2005.[31] Only one successfully reproducing population of humpbacks remains between Lake Powell and Lake Mead, the large aggregation of fish at the confluence of the Colorado and Little Colorado rivers.

Efforts to recover chub populations are complicated because there are so many causes of impairment. Some adverse conditions predate the dam (especially nonnative fish), while others resulted directly from the dam (such as changes in flow patterns, physical habitat, temperature, water quality, and food supply). The health and fitness of the remaining chub may also be suffering from a new parasite, the Asian tapeworm, which was introduced to the United States accidentally along with grass carp from China and discovered in the Little Colorado River in 1990. Given all of those causes, it was not likely that a single experimental flood would miraculously rejuvenate the humpback population. But some evidence indicated that chub successfully reproduced and survived in the high water years of 1984 and 1985. And there was reason to believe that intense floods might favor native fish over introduced trout. But such a large flow experiment had never been tried in a major southwestern river.

If one goal of the 1996 flood was to improve humpback conditions and populations measurably, however, the results were disappointing. Scientists concluded that native fish populations were not affected significantly. Nonnatives were "moderately affected," but recovered quickly due to reproduction and reinvasion from tributaries.[32]

If another purpose of the flood was to learn more about the efficacy of a major restoration strategy in an adaptive management context, the results were sobering. Some researchers later suggested that floods would have to be significantly larger to affect nonnative fish populations measurably. The 1996 flood dislocated nonnative fish from the main stem to the mouths of tributaries, from which they rapidly reinvaded. A much larger flow might leave them with literally no place to swim. Biologist Richard Valdez suggested, based on studies in smaller southwestern rivers, that native species reclaim dominance after a flood 100 times larger than the mean natural flow of that river. If the same ratio is needed for the Colorado River, this would require an experimental flow forty times larger than the 1996 flood.[33] This raises the same dilemma as that posed for sediment restoration. Would any alternatives allow significant restoration of native fish populations absent such a radical change in the flow regime?

One possibility is to change the timing of experimental floods, but various canyon resources are likely to respond differently to a given set of flows. The 1996 experiment occurred in March and early April, to avoid adverse effects to young fish later in the year. Some scientists suggest that timing floods in May or June would be more consistent with predam flows. As explained further below, however, that timing would hurt parts of the "naturalized" ecosystem that continue to be valued by users. It could also disperse tamarisk seeds at a time when germination is more likely, thus hastening tamarisk invasion of the Grand Canyon.[34]

A second approach is to combine experimental flows with mechanical removal of nonnative fish—fish genocide reminiscent of the Flaming Gorge debacle, but this time to eliminate introduced species to benefit the natives. Now experts have vastly more experience with methods to remove fish selectively, and thus a much better chance of doing so without harming the native fish. The idea is to combine artificial suppression of introduced species with flows designed to aid native species as a sort of one-two punch. If native reproduction and survival is enhanced at times when nonnative populations are suppressed, a higher percentage of the young natives are likely to survive to reproductive age.[35]

The effectiveness of exotic fish removal to help native species remains experimental, the ultimate efficacy of which remains poorly understood, but also is complicated by another competing resource choice in the Grand Canyon. The removal program is designed to balance the goals of preserving the tailwater trout fishery at Lees Ferry while decreasing populations of trout and carp farther downstream, where large populations of nonnatives interfere with restoration of the humpback population. Initial removal efforts succeeded in the sense that large numbers of nonnative fish were removed, prompting the agencies to expand the experiment to a longer stretch of the river. Although initially there was no evidence that nonnative fish removal translated to increases in the population of humpback chub, this effort is one possible explanation for the subsequent recent increase in chub. That would suggest that more focus on nonnative fish eradication should continue.[36]

A third option is to change the temperature and other characteristics of the water sent downstream from the dam. Predam floodwaters were warmer, and scientists believe that releasing cold waters as part of experimental floods can shock young native fish.[37] Because reservoirs are stratified, with warmer water at the surface and much colder water in successive layers below, the depth from which water is drawn affects the temperature of water below the dam. Because water is released from Glen Canyon Dam at a single level, the temperature of the water released depends on how deep the reservoir is relative to the fixed intake structure. In most years, the temperature of water released from the dam is much colder and more constant than under natural conditions, especially during critical spawning and rearing months. In the predam river, low temperatures averaged 30°C below average summer highs. Water released from the dam is now virtually constant, varying by just 2°C over the course of the year.

Humpback chub and other native species reproduce in warmer water, and scientists believe that colder water from the dam suppresses natural reproduction, including both spawning and survival of juvenile fish. Even when chub spawn successfully in the Little Colorado River, juvenile fish experience a "thermal shock" when they swim down to the main stem, with water as much as 10–15°C colder than in the tributary. This hurts their ability to swim and makes them even more vulnerable to predators. During the recent prolonged drought, however, water in Lake Powell dropped to near the level of the intake structures, which caused warmer water to be released into the river below. Although that may have helped in the apparent contemporaneous recovery of humpback chub downstream, it also leaves a question about whether those increases were promoted more by nonnative fish eradication, by warmer temperatures, or both. The best strategy for future recovery efforts depends in part on the answer to that question.

One approach to the water temperature problem is to change the intake structures in dams so that water can be drawn from different reservoir depths. Dam managers mix waters from different levels to achieve an optimum temperature, like you do when you fiddle with the hot and cold taps on your faucet. The Bureau of Reclamation added a multilevel structure to the Flaming Gorge dam in 1978, allowing better temperature control for the water released downstream, although that modification was designed to help trout, not native fish.

In 1999, the Bureau of Reclamation proposed a relatively inexpensive way to use the existing intake structures on all eight turbines of Glen Canyon Dam to test alternative release temperatures. The design would have cost just $15 million, about ten times less than the traditional design used at Flaming Gorge and other dams. Because it was within the bureau's existing budget, it could have been accomplished quickly, without having to ask Congress for additional appropriations.[38] Due to adverse comments on the draft, however, the bureau did not proceed with the project. Ironically, the bureau's initial proposal to use existing intake structures would not have worked once Lake Powell reservoir levels dropped due to the ensuing drought.

Delay in the temperature control proposal stems in part from the same kinds of uncertainties that complicate the rest of the restoration program. Scientists do not fully understand the impact of different temperatures on the downstream ecosystem. A group of independent science advisors to the Grand Canyon Adaptive Management Program wrote, because of the complexities involved in ecological responses to different temperatures, that "expectation of precise and accurate quantitative predictions is neither likely nor appropriate." The science advisors warned that temperature is one of many factors that influence the relative fitness and survival of species (along with biotic relationships among species, flow regimes, and sediment input), and that it is not possible to evaluate the effects of temperature changes in isolation from other factors. As one example, warmer water will likely lead to more Asian tapeworms, which could further harm native fish.[39]

The biggest problem with efforts to help native fish by increasing water temperature is the prevalence and diversity of exotic fish species below the dam, including species that are desirable to some river users. Increasing river temperatures dramatically could harm trout, but smaller increases could actually improve their growth and abundance. This might delight fly fishers but harm native species if increased predation of juvenile chubs by trout more than offset any benefits to chub spawning and survival. In its 1999 proposal, Reclamation articulated its goal "to allow release temperatures to be controlled to improve conditions for endangered fish while at the same time protecting other important resources like the Lees Ferry trout fishery."[40] The Endangered Species Act mandates efforts to help the endangered endemic chub. Political pressure, economic forces, and a residue of Eurocentric fish bigotry favor the trout.

To complicate matters further, lurking in the wings are other introduced species such as catfish, carp, and bass, which thrive in warmer water. Warmer dam releases will definitely help those species, but the more difficult and more important question is whether catfish and carp will benefit more or less than native fish. Fisheries biologist Carl Walters warned that warmer flows from the dam could unleash what he called the "vampire in the basement." If a single catfish spawns successfully at the mouth of the Little Colorado River, he suggested, the 2,000 or so resulting offspring could consume the entire year's humpback spawn in five days.[41] Do native fish populations face greater risk from increased predation, or from the effects of cold water on reproduction and survival? The science advisors believed in 2003 that enough information was available to begin a series of experiments to test the impacts of different water temperatures on the downstream community. The Bureau of Reclamation declined this invitation, opting instead to initiate a new study, which still was not complete three years later and fourteen years after the FWS first called for temperature control experiments. Ironically, warmer water released from the dam during the drought might have constituted an unintentional experiment about the effects of warmer water on the downstream ecosystem.

Our failure to "cast lots" to decide whether it is more important to improve conditions for introduced trout or native chub might jeopardize efforts to achieve either result. But this is just one of many difficult choices managers of the Grand Canyon restoration program face. In chapter 4, we noted that there is no single, presumptively "correct" prior ecological condition to serve as a restoration goal. Ten thousand years ago, the climate and ecosystem of the Grand Canyon were quite different. Go back 140,000 years and the Colorado was dammed by massive lava flows, with natural reservoirs at least forty miles long.[42] Given that long history of change, is it presumptively correct to say that the predam ecosystem was better than what exists now? Although Glen Canyon Dam harmed some ecological and recreational resources, it improved conditions for others. Choosing among past and current resources presents a third level of difficult restoration decisions in the Grand Canyon Adaptive Management program.

Competing Resources in Grand Canyon Restoration

In 1938, Elzada Clover and Lois Jotter achieved two significant "firsts" in a single journey. With the help of pioneer river guide Norm Nevills, they became the first women to travel the 660 river miles of the Colorado from Green River, Utah, to Lake Mead. Along the way, Clover and Jotter completed the first systematic botanical survey of this little-explored stretch of river. The riparian ecosystem they studied, however, was not what we find today.[43]

Before Glen Canyon Dam, a very different river influenced riparian plant communities in Grand Canyon. Close to the river, high annual spring flows scoured any vegetation that might have taken root, leaving a typically barren margin of moist sand. The only plants in this zone were ephemeral grasses, herbs, and a few woody plants in protected areas. Above this elevation, the old high water zone hosted native trees and shrubs that benefited from the annual inundation of water and nutrients provided by spring floods. In this higher zone, floods brought sustenance but were not strong enough to uproot growing plants. Dense thickets grew in an arid zone that otherwise would support only scattered vegetation. Desert shrubs survived above this zone, where the river rarely if ever flooded. A riparian oasis existed in the Goldilocks zone, where there was neither too much nor too little water, but just the right amount. A newcomer lived on the block (tamarisk), but Clover and Jotter found only a few of them in scattered locations.

After the dam, the Grand Canyon riparian ecosystem changed quickly and dramatically. Because high flows were usually limited to those generated during maximum power plant capacity, trees in the old high water zone were slowly starved of the soil moisture and nutrients carried by natural spring floods. The zone of inundation shifted lower down and supported a new high water zone with thickets of vegetation closer to the river. By 1989, about 1,235 acres of new riparian habitat emerged due to the river's modified flow regime. The new, more stable flow regime also created conditions for riverside marshes and other wetland habitats that did not occur widely before the dam. Invasive tamarisk is a major component of the new riparian plant community, and as described in chapter 4, tamarisk provides a mixture of benefits and problems. It hosts large populations of insects, which provide food for a wide range of birds, including the largest population (although still small in numbers) of endangered southwestern willow flycatchers in Arizona. The new habitats support small mammals, reptiles, amphibians, and birds, including five to ten times more breeding birds than before the dam. All of these changes began something of an ecological chain reaction. The canyon now supports one of the highest densities of nesting peregrine falcons in North America, which prey on new or increased populations of violet-green swallows, swifts, and migratory waterfowl. Snake populations similarly increased along with higher populations of small rodents and lizards.[44]

The river ecosystem also changed in ways that are not easily characterized as either positive or negative. More light penetrates the much clearer water now released from the dam. Along with rich nutrients and oxygen in water released from the dam, this provides perfect conditions for the second largest recorded gross primary productivity (conversion of light to energy via photosynthesis) in North America. Large stands of filamentous algae *Cladophora* support increased populations of the amphipod *Gammarus lacustrius*, which was introduced from Asia in the 1930s as food for trout. Diatom populations increased 1,600-fold after the dam. Although dominated by only four species, diatoms are nutritious food for fish. These changes are good for trout, for people who chase them with artificial flies, and for equipment, guiding, and other businesses that depend on that quest. Other species also rely on the enhanced productivity of the new river ecosystem, such as the population of overwintering bald eagles at Nankoweap Creek. Spawning trout were first seen at Nankoweap in the mid-1970s. The first eagles were observed about a decade later, and eagle populations increased steadily thereafter, although no official monitoring of that population continues.[45]

These choices among competing ecological resources, however, tell only part of the story. On the river, many visitors value recreation over ecology, as observed by Ann Zwinger: "Three river parties unload into the cave, unlimber their frisbees for a fifteen-minute workout, and depart, having never left water's edge, having never seen an ant lion metropolis or a spider megalopolis, what six and eight legs can create, and what worlds lie between grains of sand." Demand for river trips through the canyon is so intense that the National Park Service recently prepared an EIS for a management plan to allocate river-days instead of hiring aquatic police to direct traffic.[46]

The usual starting point for a Grand Canyon river trip is Lees Ferry, fifteen miles below Glen Canyon Dam. At the boat launch, Lees Ferry is marked as "mile zero." People who know nothing about the Colorado River Compact must wonder what happened to the rest of the river upstream. River running in the Grand Canyon is one of the most coveted recreational adventures in the world. But dare not believe you can simply *go:* "Before launching a boat here, be sure you know what you are doing and where you are going. DO NOT go downstream without a permit and a park ranger permission," says one sign. An information board identifies who can rig boats where, who can camp when and where downriver, scheduled arrivals at Phantom Ranch and other destinations, and a daily launch calendar for all trips (including numbers of people and trip lengths), with just one or two "private" trips listed per day. Private launches must meet with a ranger for an equipment check, identification check, and one-hour orientation. Equipment needs include life jackets, first aid kits, repair kits, tool kits, extra parts, oars, paddles, an extra water purification system, a toilet system, straining screens, a firepan (open fires are prohibited), signal mirror, emergency panels, an air pump, ground-to-air radio (recommended), completed passenger list, and picture IDs for all passengers. These requirements are understandable given safety issues and public rescue costs, but so much for wilderness, solitude, and the spirit of adventure. So much for heading down "the great unknown." These are regulated trips down a regulated river, but the new flow patterns from the dam changed river running in several ways.

Before the dam, river runners coped with different flow levels and different challenges depending on the season. But absent flash floods, water levels were not likely to change quickly during a trip down canyon. All of that changed with the daily release fluctuations under the original "load following" power production schedules. Dam and power plant operators released more water through the turbines during the day, when peak power needs were high, and less water at night when baseline power production from other sources on the electrical grid sufficed to meet demand. River runners learned to "ride the wave" by knowing how long it would take a surge or ebb to reach a particular spot on the river, sometimes days after release from the dam. When boaters guessed wrong, they found themselves either beached by an unpredicted drop in water or riding a torrent of unexpected flow.[47] Daily flow fluctuations have been reduced in recent years as part of the modified flows designed to mitigate the impacts of dam operations to native fish (although some scheduled fluctuations are used now to suppress populations of trout).

The vastly reduced annual spring flows, however, may harm the river's recreational values more seriously in the long run. Most of the canyon's rapids were created by debris flows, the powerful slurries of water, mud, and boulders generated in steep tributary canyons during intense summer storms. The size of a rock that water can move increases with the square of the water's velocity, meaning that if the speed of a tributary quadruples in a flash flood, it can move a boulder sixteen times as large. Those fortunate enough to escape but still to witness such events report boulders the size of cars being swept into the main river channel. The resulting mass of rock and silt constricts the river channel below. This forces more water to pass a given spot in a shorter period of time. Debris flows are responsible for most of the big rapids in the canyon.[48]

Rapids, of course, are why the canyon is such a river running paradise. But rapids can be too hard to run, and certainly too dangerous. Before the dam, massive spring flows offset the effects of debris flows by moving deposited material downstream. Rapids were shaped and reshaped in a continuous but intermittent cycle, leaving a stretch of white water that was challenging but, to those with sufficient knowledge, skill, and the right equipment, navigable. Because of the dam, the river no longer experiences floods large enough to reshape new constrictions. If this trend continues, some boaters fear that the river eventually will be too dangerous to run, shutting down the lucrative river guiding industry. More intense flows than used in the 1996 experimental flood may be needed to reshape debris deposits in the river, although very short peaks may suffice to accomplish this task.[49] Some suggest that flows as high as 400,000 cubic feet per second may be needed to reshape Crystal Rapid. This is a water volume much larger than that tried in any of the experimental flows thus far.

To date, no one has answered these pivotal questions about the ends to which Grand Canyon restoration efforts should strive. Adaptive management is designed to assist us in making difficult choices about restoration strategies in the face of scientific uncertainty. Iterative experiments might provide more information about which methods will do the most good for which river resources, but they cannot substitute for some open process by which we make decisions about competing resources and values in restoration programs. We turn next to the processes currently used to make those choices in the Grand Canyon restoration program, and evaluate how well they are working.

Making Restoration Choices Democratically

Carl Sandburg wrote: "Every man sees himself in the Grand Canyon."[50] But every person sees something different in him or herself, and all want their personal values to be reflected in a restored canyon (and in other natural places), regardless of how incompatible those conflicting values may be.

Congress contributed to this dilemma by passing the loop of conflicting mandates in the statutes authorizing the Colorado River dams, in later environmental laws such as NEPA and the ESA, and in the GCPA. Perhaps the world was a bit simpler in 1956 and 1968, when Congress established a relatively clear hierarchy of uses that roughly paralleled the priorities set in the Law of the River—water first, power second, and environment at best third. By 1992, it was clear that all of those uses could not necessarily be met while still complying with newer laws designed to restore and protect the natural environment. Unfortunately, Congress was either unwilling or unable to make clear choices. Instead, it *delegated* those value decisions to the secretary of interior. The secretary's decisions, in turn, were to be guided by a collaborative process designed to bring together representatives of the various "stakeholder" groups whose uses and interests are affected by how we manage the dam and river.[51] In his 1996 Record of Decision (ROD) on the EIS on Glen Canyon Dam operations, Secretary Babbitt established the Glen Canyon Dam Adaptive Management Program (GCDAMP) to advise him and future secretaries on program implementation. The program includes a twenty-five–person committee known as the Adaptive Management Work Group (AMWG), consisting of representatives of the basin states, federal and state resource agencies, tribes, environmental groups, recreation interests, and power users.[52]

Having participated in several such processes, I was interested to observe the dynamics at an AMWG meeting in Phoenix in spring 2002. I wanted to see how well the AMWG functioned at making the kinds of difficult choices facing the Grand Canyon restoration program, so I attended several of the AMWG meetings over a two-year period. This meeting began with a tribute to a recently resigned chair, who was experiencing serious health problems. The committee reaction, which included a standing ovation, shows how well he was respected by all of the stakeholders. One benefit of collaborative efforts is to break down the personal barriers that often characterize highly polarized conflicts, in which the *people* on the other side of the debate are stereotyped, caricatured, or even demonized. That does not always mean that the underlying conflict is resolved, as was true in this meeting, but the parties at least discuss the issues on the merits rather than hiding in political trenches and firing off rhetorical bombs.

At this particular meeting, improved interpersonal relationships did not necessarily translate to agreement on the issues. Members of an ad hoc subcommittee had been conferring two or three times a month since 1999 (roughly three years) to develop a strategic plan with which to guide the AMWG's recommendations. The facilitator asked for a motion to adopt the subcommittee's unanimous recommendation to approve the proposed plan, but no one so moved! A long debate ensued over the fundamental purpose of the AMWG. Representatives of water and power interests argued that the sole purpose of the program was to evaluate how well the alternative selected in the 1995 EIS fulfilled the goals of the GCPA, and that any significant changes from that decision required more NEPA analysis. Environmental and recreation interests suggested a much broader program scope in which decisions in the EIS and ROD could be reconsidered if proven inadequate to meet the restoration goals of the statute. Ultimately, the AMWG agreed to compromise language that reads: "The Adaptive Management Program evaluates how well the preferred alternative of the EIS/ROD and other management actions meet the goals of the GCPA and the mix of resource benefits in the EIS/ROD." This motion passed easily because it resolved almost nothing. It identified what analysis will be conducted, but begged the question of what happens if actions outside the scope of the EIS fail to restore the canyon ecosystem. That issue still remains on the table. Neither side was willing to cast lots or find some other way to decide how far this program can go, and the degree to which impacts to other uses and values (such as water use and power generation capacity) might suffer as a price of restoring the Grand Canyon ecosystem.

This disagreement was not mere semantics. It went to the group's basic mission, somewhat like the choices faced by the Dominguez-Escalante expedition. Water and power interests believed the group had a single task, such as determining whether a new land route to California existed. Find out and report back to your superiors in Santa Fe (or in Washington, D.C.). Environmental and recreation groups believed they were on a much broader mission—to convert dam operations in ways necessary to restore the degraded resources and values in the Grand Canyon ecosystem.

The positions taken by the two sides were predictable as a matter of status quo politics. Any interest group that benefits from the current situation will use process to forestall change. Lawyers are familiar with this game, as are politicians and lobbyists. So long as obstacles exist, it is always easier to oppose than to propose, to prevent than to promote. In a large sense, the water and power interests "won" the 1995 EIS process. Although the secretary's decision sanctioned experiments to restore the aquatic ecosystem downstream from the dam, it allowed those flows only when enough water was stored to protect water and power interests. Daily fluctuations were reduced, with some cost to power production, but the basic electricity generation regime remains intact. So water and power users gain by confining the debate to the specific issues and options considered in the 1995 EIS, interpreted as narrowly as possible. Any significant change or new initiative requires referral to the Technical Working Group and then the full AMWG, and sometimes multiple referrals back and forth, with additional guidance from the science advisors and other sources, and sometimes compliance with NEPA, ESA, or other legal requirements. The process ensures that all values and interests are aired fully and openly, and if possible accommodated, before making major changes in dam and river management. That every significant change requires approval through this cumbersome process, however, raises questions about whether we can *afford* the delay inherent in all of this discussion before time runs out for the chub and other species.

To be fair to the hard work and accomplishments of this group, this is just one isolated example of process breakdown in a long program, although there are other such examples. Since that meeting the AMWG has reached consensus on a number of important recommendations, some of which resulted in experimental flows and other changes to dam and river operations. For example, the AMWG agreed to support the modified flow regime designed to mobilize sediment after intense summer storms and then to maintain a period of low, stable flows to allow accumulated sediment to restore beaches and sandbars in the canyon. Power interests accepted some generation losses to accommodate that experiment. Likewise, AMWG agreed to experimental fish suppression flows, and to mechanical removal of trout and other nonnative fish near the Little Colorado River junction (not an easy concession for sport fishing interests).

More broadly, the AMWG ultimately succeeded in adopting a strategic plan for the GCDAMP. Some aspects of the plan provide useful guidance regarding some of the more difficult choices facing the restoration program. For example, the plan asserts that construction of Glen Canyon Dam and the introduction of nonnative species "irreversibly changed the Colorado River ecosystem," an apparent agreement that complete restoration to predam conditions is either not possible or infeasible. But the plan also advocates management changes designed to "attempt to return ecosystem patterns and processes to their range of natural variability." So the stakeholders agreed on a compromise philosophy in which the program will seek to restore natural processes and ecosystem functions, but not necessarily full predam ecological conditions.

The AMWG also appears to have made some difficult choices between competing ecosystem resources. One program goal articulated in the plan is to maintain or attain viable populations of native fish, and to prevent adverse modification to their critical habitat. The plan also seeks to maintain a naturally reproducing population of rainbow trout, but only "to the extent practicable and consistent with the maintenance of viable populations of native fish." The AMWG thus urges that the program continue to maintain both a recreational trout fishery and viable populations of native fish. But in apparent recognition of overriding ESA mandates, the plan recognizes that if one must prevail over the other, it will be the native species.

In other respects, moreover, the strategic plan continues to beg important questions, reiterating the basic contradiction in the GCPA: "These recommendations must recognize the environmental commitments of the Glen Canyon Dam Environmental Impact Statement and Record of Decision, and comply with the Grand Canyon Protection Act. The Adaptive Management Program must also remain in compliance with the Law of the River and relevant environmental statutes, regulations, and policies."

One particularly ambiguous goal in the plan is to "[m]aintain power production capacity and energy generation, and increase where feasible and advisable, within the framework of the Adaptive Management ecosystem goals." Where actions might benefit some resources to the detriment of others, the plan establishes a hierarchy in which actions that help all resources will be tried first, actions that help some resources but are neutral to others second, and "as a last resort, actions that minimize negative impacts on other resources, will be pursued consistent with the Glen Canyon Dam Environmental Impact Statement and the Record of Decision." Humpback chub seem to have won out over rainbow trout, but not necessarily over power production.

A 1999 National Academy of Sciences review panel noted that the GCDAMP seeks to maximize many goals simultaneously, with few clear decisions about which resources have priority in the face of an impasse. The panel criticized the program's entirely pluralistic approach, which assumes falsely that all uses and interests can be accommodated equally, rather than having to make some difficult choices. Others who have worked in the program similarly warn that no single set of management prescriptions can achieve the full range of values sought by all of the involved interest groups, and that the broader public must make some key choices because "values, not science, underlie the choice."[53]

Possible solutions to this decision-making impasse appear to point in opposite directions. We might delegate more flexibility to scientists and managers who plan and implement restoration efforts, or we might provide for more open, inclusive decision making. In fact, the apparent contradiction reflects a recognition of the relative strengths and appropriate roles for different parties in the restoration process.

Adaptive management requires iterative, real world experiments. Most scientists, however, do not design experiments under the watchful eyes of a large stakeholder committee that must decide how each experiment will affect their interests before it can proceed. When years of discussion occur between iterations, and when experiments may not proceed if they affect other uses such as water and power, the science part of adaptive management bogs down. A decade and a half after Congress passed the GCPA, and a decade after Secretary Babbitt launched the GCDAMP, the program has managed only two high flow experiments (the beach habitat building flow in March 1996, and the November 2004 experimental high flows), and only two other significant test flows (a low summer steady flow test in 2000 and experimental fluctuating nonnative fish suppression flows from 2003 to 2005). Such delays present serious barriers to learning, especially when natural experiments must await uncertain environmental conditions, such as a particularly wet or dry year. The environmental window within which to conduct the experiment might open and close before a stakeholder or other process is completed.

Some analysts argue that collaborative stakeholder involvement is essential to adaptive management, which seeks to integrate scientific and social learning to "prevent policy gridlock," and that social learning is as important to restoration as is physical and biological science. But merging those goals into a single interactive process serves neither goal well, and fundamentally misconstrues the concept of adaptive management. Walters and Holling, some of the scientific pioneers of adaptive management, did favor more interaction between scientists and policymakers. They did not, however, propose to involve a collaborative stakeholder panel in every aspect of the scientific program design in ways that render it too slow to respond to rapidly changing environmental conditions and to take advantage of time-limited experimental opportunities. Likewise, in *Compass and Gyroscope*, Kai Lee argued that adaptive management must be *accompanied by* social learning, through political processes and conflict management, so we can take advantage of new scientific knowledge. But he viewed the two as distinct processes, with social learning the desired outcome of increased scientific understanding, and vice versa. Lee referred to adaptive management as our *compass*, "a way to gauge directions when sailing beyond the maps," and democratic resolution of conflicts as our *gyroscope*, "a way to maintain our bearing through turbulent seas."[54]

The Grand Canyon restoration program merges the two into a single instrument that is ineffective for both tasks. The compass needle is slow to respond to the constantly changing directions of the canyon and its ecosystem, meaning that the adaptive management program is bogged down in stakeholder politics. This allows less scientific learning than would be possible if scientists were able to operate with more flexibility in designing and testing strategies as variable environmental conditions occur. But it is the gyroscope that is really lacking in the Grand Canyon program. Restoration efforts are hampered by the futile assumption that all uses and values of the Grand Canyon ecosystem can be maximized simultaneously.

To date, no one has been willing to "cast the lots" to choose among the various competing uses and values at stake in the Grand Canyon restoration process. We still want to cast flies for trophy cold-water trout while restoring populations of warm-water natives. We value the new marshes and habitats created by the new high water riparian zone, and the birds and other species they support, but still want open, vegetation-free beaches to support the masses of river runners. We want to use dramatic flood flows to move sand downstream, and to reshape debris flows to keep the rapids navigable, but are not willing to sacrifice cheap power to do so.

After a full decade of effort, it seems clear that the existing AMWG process is not suited to making these hard choices. None of the interest group representatives are likely to concede that their interests can be eliminated or even significantly curtailed to accommodate a common vision for canyon restoration. As some program scientists and officials have observed, consensus is not "likely in the future because of mutually exclusive objectives." Expressing frustration about the "mismatch" between the relevant time frames for science and decisions, they complained that "the science cannot be forced and the management decisions cannot be delayed until all the science is complete."[55]

One possible solution to this paradox is to distinguish between the kinds of choices discussed above. Where experimentation is needed to learn more about which restoration strategies will be most effective, or to determine the effects of different management options on various river and canyon resources before adopting longer term management changes, perhaps we need to free the scientists and managers from many of the rigors of the existing process. As suggested in another context in chapter 5, for adaptive management purposes the environmental statutes and regulations designed to promote prevention and mitigation approaches to environmental protection may do more to impede than to facilitate environmental restoration. Even where restoration experiments will cause temporary impacts to some current river uses, such as power generation, more flexibility is warranted so that longer range decisions can be better informed. Stakeholder groups such as the AMWG can provide guidance to restoration program scientists and managers about the kinds of issues to study and the kinds of risks that are acceptable in formulating and conducting adaptive management experiments, but free them to do so without constant micromanagement. Limited exceptions to NEPA and other procedural requirements might be warranted to facilitate a more flexible form of adaptive management.

Restoration actions that involve more fundamental choices among different resources and their uses, however, cannot and should not be delegated to scientists, bureaucrats, or any other narrowly defined set of interests or individuals. Those decisions must be made openly, with attention to all legitimate interests and values. Collaborative processes such as the AMWG are one legitimate way to make such decisions, but only if the group actually makes those choices. It is difficult for interest group representatives to let go of resources they value the most. In many cases, it is simply naïve to expect them to do so.

Another viable solution is for Congress (or in different contexts, other elected officials) to make clearer choices than are reflected in the laws governing management of the Colorado River. That prospect is surprisingly frightening to many, who fear that their "side" will lose in any particular political climate and depending on who controls Congress, the White House, or any other seat of political authority. But the same is true for any important issue of public policy. The onus is on advocates for particular values to educate themselves, their elected officials, and the public at large to achieve the depth of understanding necessary to make wise choices. More important, those who favor environmental restoration but fear that elected political leaders will make poor choices will lose just as surely under approaches that lead to gridlock. Maybe it is even better to lose a political battle or two over particular programs and places, and to see whether the public is really willing to accept those environmental losses, than to leave the public with the false impression that all resources and values can be maximized simultaneously. Perhaps it is a choice between a quick death and slow suffocation due to endless processes in which no real choices are made while species and other key resources continue to decline.

Returning to Abe Lincoln's famous maxim, good government does not mean a futile effort to please all the people all the time, for in doing so we risk pleasing nobody. Before we can restore the aquatic ecosystem of the Grand Canyon, we need to restore democracy, and the admittedly difficult choices that sometimes entails. In the case of the Colorado River, the choices get even more complex as we focus on larger sections of the river, and accordingly, even more tradeoffs between competing resource uses and values.

Expanding Restoration Program Focus

To date, the choices at stake in the Grand Canyon restoration program reflect a relatively narrow focus. Although dam management decisions affect large numbers of species and resources, the principal emphasis has been on efforts to restore populations of endangered humpback chub, and the impact of those efforts on other canyon resources. The AMWG strategic plan alludes to possible efforts to reintroduce extirpated species like Colorado River pikeminnow "as feasible and advisable," but that goal is entirely aspirational thus far. Are we recovering a single species, or restoring an entire ecosystem? Geographically, the program is limited to the reach of the Colorado between Lake Powell and Lake Mead, and laterally to the elevation inundated by peak dam releases. That scope largely ignores the relationships between river management within Grand Canyon and similar restoration efforts upstream and down, and the interactions between different portions of the Colorado River ecosystem.[56]

Downstream, a much broader collaborative river restoration program seeks to address twenty-seven threatened, endangered, or otherwise jeopardized species. This program involves not just the river itself, but the broad riparian zone along its length. That breadth of attention, in turn, affects environmental resources and programs all the way to the Sea of Cortéz, and to the Salton Sea in Southern California. However complex the science and the value choices in the Grand Canyon program, imagine how daunting the choices become when the focus expands to the river below it. We explore those issues in the next chapter.

CHAPTER SEVEN

Ownership of Unownable Things: Property Rights and Environmental Restoration at the Water's Edge

In his science fiction fable *Kirinyaga*, novelist Mike Resnick described why the Kikuyu tribe of Kenya considered their god Ngai more powerful than the god of the Europeans:

> Eons ago, when the Europeans were evil and their god decided to punish them, he caused it to rain for forty days and forty nights and covered the earth with water—and because of this, the Europeans think that their god is more powerful than Ngai.
>
> Certainly it is no small accomplishment to cover the Earth with water—but when the Kikuyu heard the story of Noah from the European missionaries, it did nothing to convince us that the god of the Europeans is more powerful than Ngai.
>
> Ngai knows that water is the source of all life, and so when He wishes to punish us, He does not cover our lands with it. Instead, He inhales deeply, and sucks the moisture from the air and the soil. Our rivers dry up, our crops fail, and our cattle and goats die of thirst.[1]

Communities in the arid Southwest fear both of these gods. Although dams provide electricity, recreation, and other services, fear of flood and drought was the most compelling original reason for the massive series of dams and other engineered changes along the lower Colorado River. That construction changed the physical, hydrological, and biological properties and processes of the river below the Grand

Canyon even more than in the upper and middle river reaches, and programs to address those ecosystem impacts confront the same kinds of science and policy issues as discussed in chapters 5 and 6.

But ecosystem restoration in the lower Colorado faces challenges of a different kind. Dams and flood control structures along the lower river accompanied and promoted much more extensive development within the floodplain and riparian habitats than occurred upriver. This development right up to the water's edge presents a fundamental conflict between private property rights and ecosystem restoration and protection. Floodplain and riparian habitat restoration will be impeded significantly if we continue to accept extensive development in those zones, and to maintain the publicly subsidized structures designed to protect that development from the ebb and flow of the river. We also must decide the degree to which restoration goals should be more modest in highly modified environments than in less severely altered areas.

This chapter begins with a description of the dams and other engineered changes designed to promote development, flood control, water, and power needs along the lower river, and the serious impacts that development caused to species and ecosystems. It then explores the multi-species conservation program designed to allow that development to continue in the face of Endangered Species Act (ESA) legal requirements. We then discuss the extreme difficulty of restoring the lower river environment through a painstaking process of re-creating individual patches of habitat when the basic flow processes of the river remain so significantly modified. A more effective and efficient restoration strategy would return the river's flow patterns to something closer to the natural hydrological regime, which would allow natural regeneration of floodplain and riparian habitats along much wider stretches of the lower river. That, however, would present more serious conflicts with the development that occurs throughout the lower river corridor. The chapter closes with an evaluation of whether a broader restoration approach requires us to rethink the nature of property rights in the critical ecological transition zones between land and water.

River Restoration Goals for the Radically Altered Lower River

Wholesale modification of the Colorado River was not just an accidental side effect of programs designed to achieve other benefits. Although we did not envision all of the ecological changes that would occur, we *wanted* to change the river's basic flow patterns and behaviors. The

Bureau of Reclamation's 1945 general plan to develop water resources in the basin, aptly named "A Natural Menace Becomes a National Resource," explained: "Yesterday the Colorado River was a natural menace. Unharnessed it tore through deserts, flooded fields, and ravaged villages. It drained the water from the mountains and the plains, rushed it through sun-baked thirsty lands, and dumped it into the Pacific Ocean—a treasure lost forever. Man was on the defensive. He sat helplessly by to watch the Colorado River waste itself or attempted in vain to halt its destruction."[2]

The lower Colorado River (figure 7.1) flooded its banks regularly. That created conditions suitable for the vast riparian forests, marshes, and other habitats that provided refuge for countless species amid the sparseness of the surrounding desert. Annual spring floods spread over the river's broad floodplain, bringing moisture, nutrients, and seeds to replenish the diverse native plant communities, and to support an equally impressive range of birds and other species. Before the dams, settlers reported a major flood roughly once a decade. Lieutenant George Wheeler, who led a U.S. government survey of the lower river in 1871, estimated a maximum flow past Yuma of 400,000 cubic feet per second, more than five times larger than the spring average. Native American tribes learned to live with the god of floods, moving villages and fields to match the river's flux. European settlers preferred more permanent communities. Violent floods in 1891 overran the levees designed to protect Yuma, destroying virtually the entire existing town.[3]

Not all of the flooding, though, was natural. Legendary in the lore of the lower Colorado River is the 1905 "flood," in which virtually the entire river broke through the temporary headgates of the Alamo Canal, the water supply for the newly developing agricultural empires in California's Imperial Valley and Baja California's Mexicali Valley. During the sixteen months in which the river spilled into the Imperial Valley, about 30,000 acres of arable lands were inundated, damaging or destroying farms, homes, roads, and railroads and causing millions of dollars of damages. It took two years for workers to stem the flow of the river as it surged into the Salton Sink, creating what is now known as the Salton Sea (discussed further in chapter 8). Other floods continued to interfere with development in the Imperial Valley and along the lower Colorado River as people moved into the region to farm in the early 1900s, and to retire as the 20th century progressed. More so than water storage and power generation, flood prevention was the main rationale for the Hoover Dam and other projects to follow.[4]

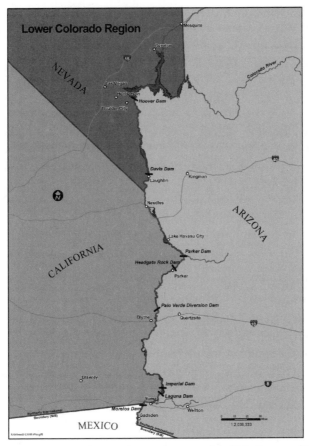

Figure 7.1. Lower Colorado river basin. *Courtesy of the U.S. Bureau of Reclamation*

But water shortages also influenced Colorado River development deeply. Shortly after the 1891 flood, the basin experienced a nearly decade-long drought, with flows of just over 9 million acre-feet (maf) in 1902. New dams stored water in wet years to prepare for the dry, and facilitated diversions to growing irrigation districts and cities in California, Arizona, and Mexico. Laguna was the first dam built on the lower river—indeed, the first ever by the newly created federal Reclamation Service (later the Bureau of Reclamation). That low structure was designed not for storage, but to divert water into the Yuma Main Canal for irrigation. The dam filled with sediment within weeks of completion, and its storage capacity shrunk from 85,000 to 1,000 acre-feet. Construction of large storage and flood control dams on the Salt and Gila rivers began shortly thereafter, however, and over time water diversions reduced the Gila to a trickle by the time it reached the Colorado.[5]

The hydrology and ecology of the lower river changed most dramatically after completion of Hoover Dam in 1935, with its capacity to store twice the river's average annual flow. Downstream flows plummeted as the dam filled, and most of what remained was diverted for irrigation. Over the next two decades the bureau built dams of various sizes and with different purposes, including Parker Dam (completed 1938), Imperial Dam (1938), Headgate Rock Dam (1942), Morales Dam (1950), Davis Dam (1950), and Palo Verde Dam (1957). Combined, those structures inundated lands along nearly half of the river from the Grand Canyon to Mexico.

The dams alone, however, did not eliminate flood damage, especially as the steady influx of settlers tilled farms and built communities closer and closer to the water's edge. If anything, dams changed the river channel and flows in ways that increased flooding in Needles, California, Laughlin, Nevada, and other riverside communities. To reduce property damage and to increase the speed with which water could be channeled into diversions, the Bureau of Reclamation dredged the river into deeper, straighter channels; armored its banks with rip-rap and sometimes even concrete; and confined it between dikes and levees. Those projects affected hundreds of miles of what remained of the river between the reservoirs. This lowered groundwater levels adjacent to the river; dried out adjacent backwaters, wetlands, and riparian areas; and severed the river from its natural floodplain. In places, dredging and channelization narrowed a meandering river channel once two miles wide into a confined ditch 150 yards across. As two biologists observed: "The lower river had become a water delivery canal."[6]

Ironically, nearly $2 billion in flood control efforts (not including the cost of the dams) did not prevent $200 million in property damage during the floods of the early 1980s. With full faith in modern engineering and little respect for the power of nature, society continued to build within the floodplain. When the river overflowed its newly engineered banks, even more development lay in its path.

Dams, diversions, development, and flood control structures fundamentally transformed the ecology and hydrology of the lower Colorado River. As was true upstream, most biologists believe that native fish in the lower basin declined severely even before the dams due to a combination of overharvest, entrainment and stranding in canals and other water diversions, and predation and competition from introduced species such as carp, catfish, and bass. Dams and channelization "sealed the fate" of native species by creating conditions that favored introduced species and enabled them to outcompete and more easily prey on the remaining natives. Five of the nine freshwater fish native to the lower Colorado River were extirpated, including humpback chub, Sonoran topminnow, desert pupfish, woundfin, and Colorado pikeminnow. Only small numbers of bonytails and razorback suckers remain, and both are endangered.[7]

Reservoirs and river channelization also affected birds and other species that relied on the hundreds of thousands of acres of marshes and riparian forests that lined the river. Much of this habitat was drowned beneath the new reservoirs. Between the artificial lakes, many areas of native habitat succumbed to development; other areas degraded due to declining groundwater and elimination of annual spring flooding; and in much of the rest invasive salt cedar replaced native forests of cotton-wood, willow, and mesquite.[8]

It is legitimate to ask what restoration goals are appropriate for such a highly modified system. The 9 million people who visit Lake Mead and millions of others who happily race their boats, camp, and live along other parts of the lower river do not seem to think that anything is "wrong" with the modified river.[9] Just as American society ceded Man-hattan Island to human development, why not simply accept that the lower Colorado River is a casualty of civilization in the Southwest? And as discussed more extensively below, perhaps anything close to full restoration of the lower Colorado simply is not possible given how much change has occurred.

Congress has spoken with multiple voices on this key issue. In the Clean Water Act, Congress rejected the idea that *any* of the "waters of the United States" can simply be sacrificed as a development zone. Instead, it set a national objective to "restore and maintain the chemical, physical, and biological integrity of the Nation's waters." What Con-gress meant by "chemical, physical, and biological integrity" is a more difficult question. Although the Clean Water Act is commonly under-stood as a law designed to regulate the release of chemicals and other pollutants into rivers and other waters, Congress defined "pollution" more broadly to include *any* human alteration of the chemical, physical, and biological integrity of the nation's waters.[10] This definition could cover virtually all of the existing human changes to the environment of the lower Colorado River. Physical modifications designed to prevent flooding, such as armoring banks, channelization, and levees, radically alter the physical integrity and continuity of the river. Widespread human introduction of nonnative species changes the biological integrity of the river in equally fundamental ways, prompting some scientists to refer to that phenomenon as "biological pollution."[11] Reservoirs transform the fundamental nature of the aquatic ecosystem, and substantial depletion of in-stream flows modifies the river's hydrological integrity. Did Congress intend the river to be restored by eliminating all of those changes, regardless of how beneficial they are to the American economy?

There are some indications that Congress did intend the concept of restoration to extend that far when it adopted the comprehensive restoration goal in 1972. The Senate Committee Report on the legislation expressed an intent to restore aquatic ecosystems as closely as possible to their natural state: "Maintenance of such integrity requires that *any changes* in the environment resulting in a physical, chemical or biological change in a pristine water body *be of a temporary nature*, such that by natural processes, within a few hours, days or weeks, the aquatic ecosystem will return to a state *functionally identical to the original*." The House Committee similarly explained: "The word 'integrity' as used is intended to convey a concept that refers to a condition in which the *natural structure and function of ecosystems* is maintained." Aquatic ecologists define biological integrity in much the same way as the House Committee, as "the ability of an aquatic ecosystem to support and maintain a balanced, integrated, adaptive community of organisms having a species composition, diversity, and functional organization comparable to that of the natural habitats of a region."[12]

To date, the Clean Water Act has not been used as a major legal tool for Colorado River restoration, although perhaps it should be, as it has been for large aquatic ecosystems such as the Chesapeake Bay, the San Francisco Bay delta, and the Everglades.[13] Those other large aquatic ecosystem restoration efforts were prompted initially by more traditional concerns about chemical pollutants. But all evolved into efforts that focus more broadly on a range of chemical, physical, and biological impairments, because eliminating individual sources of harm did not suffice in restoring the integrity of whole ecosystems. For the Colorado River, the statutory focus for program design has largely been the ESA. Just as the Clean Water Act became the focus of restoration in other systems in which chemical water quality was the initial concern, the ESA became the main focus in the Colorado River because endangered fish and other species were the driving source of concern. That statutory focus, however, may inappropriately limit restoration scope and goals because of various ways in which the ESA has been implemented and interpreted.

The ESA mandates recovery programs for all threatened and endangered species, just as the Clean Water Act requires restoration of the integrity of all waters of the United States. By requiring the U.S. Fish and Wildlife Service (FWS) to designate certain areas of "critical habitat" for those species, however, Congress envisioned that species need not be restored to all of their previous habitats. Rather, portions of habitat should be restored adequately, and in sufficient areas and locations, to ensure species recovery. That more targeted approach is taken by the large consortium of agencies that collaborated to design the effort known by the cumbersome name Lower Colorado River Multi-Species Conservation Program (MSCP), which seeks to address the needs of the large number of imperiled species from Lake Mead to the Mexican border. But even that more selective effort is no simple task. As discussed next, the MSCP reflects even more trade-offs and compromises than needed in the upper and middle parts of the river, and falls short of the species and ecosystem recovery goals established in federal law.

Multi-Species ESA Implementation in the Lower River Corridor

"London Bridge is falling down." But not any more. Entrepreneur Robert P. McCulloch purchased the original London Bridge for $2.5 million, shipped it across the Atlantic, and reassembled it as a tourist attraction to promote his housing developments at Lake Havasu, the reservoir created by the Parker Dam.[14] The bridge comes complete with its own adjacent "English Village" and a "Windsor Beach." Despite the costs and logistical challenges, like the Dewey Bridge far upstream, it was possible to "restore" this item of human engineering.

A different children's rhyme may be appropriate to describe efforts to restore the more intricate piece of natural engineering called the lower Colorado River: "And all the King's horses and all the King's men, couldn't put Humpty together again." Some scientists suggest a sense of futility about whether successful species restoration here is even possible: "The future is grim for native fish in the Lower Colorado River. Remnant native fish communities continue to decline, except for small refugium populations. Their fate has been sealed by the dependence on the river by 30 million water users in the United States and Mexico. Societies' dependence upon water makes native fish recovery economically and politically unlikely, and perhaps impossible." Other researchers are more optimistic based on studies that demonstrate successful reintroduction of native fish to parts of the upper river.[15]

Major environmental programs came much later to the lower river than to areas upstream. Although the FWS identified endangered species in the lower river as early as 1967, initial planning for the MSCP did not even begin until 1995, when the FWS formed a conservation partnership with the Bureau of Reclamation, other federal and state agencies from Arizona, California, and Nevada, and Native American tribes. Environmental groups participated initially, but withdrew when other participants declined to address related environmental problems in Mexico, as discussed in chapter 8. The ensuing decade saw a flurry of biological assessments, biological opinions, draft conservation plans, scientific reviews, and multiple revisions. The participants published the final plans and environmental reviews in December 2004, and signed the formal legal documents to implement the effort in April 2005.[16]

The MSCP is notable both for what it does and does not purport to do. Like its sister programs upriver, the MSCP considers the impacts of a large number of individual actions that may further harm various imperiled species, and seeks to redress those cumulative harms together. The covered area extends from the upper end of Lake Mead to the southern end of the U.S.-Mexico border, which runs north-south for approximately twenty miles where the southwestern corner of Arizona meets the Mexican states of Sonora and Baja California. The program also covers the river's historical floodplain. Rather than focusing on one species or even a group of related species, the MSCP addresses twenty-seven threatened, endangered, or otherwise imperiled species selected from a candidate list of 149 "special status" species that might occur in the region. To address so many species, each of which faces different but overlapping threats and each of which has particular recovery needs, the MSCP employs a strategy of habitat restoration and protection, along with selected efforts to augment some species populations.

The MSCP is dubbed a "conservation" rather than a "recovery" program, which is not just a semantic difference. The program's environmental goal is not full species recovery, as envisioned in the ESA, but to "conserve habitat and *work toward* the recovery of threatened and endangered species." Likewise, the program seeks to "reduce the likelihood of additional species being listed," which is less than a firm commitment to preventing new listings. Perhaps this reflects the highly disturbed nature of the lower river and its habitats compared to some areas in the upper river, and the accompanying difficulty of full species recovery. ESA section 4(f) requires the secretary of interior to develop and implement recovery plans for each listed species. Although the FWS has issued recovery plans for southwestern willow flycatcher and other species, the MSCP is the main implementing program that will dictate how much recovery actually occurs.[17]

In many ways, the species protection recovery goals and the design of the MSCP are even more subservient to ongoing development interests than the Recovery Implementation Program–Recovery Action Program (RIP/RAP) in the upper river. The MSCP, in fact, provides far more legal certainty to those responsible for the ongoing harm than it does to the protected species. The stated program goal is to "accommodate present water diversions and power production and *optimize opportunities* for future water and power development, to the extent consistent with the law." It does so by granting "incidental take authorizations"— special permissions that provide a long list of water users and power generators virtually complete immunity against ESA violations for a half century, and for a wide range of actions that might put the covered species at risk. In return, all the covered parties need do is to pay a preestablished amount of money to support the program, with a virtual guarantee of no additional costs or other requirements through the fifty-year period. This is quite a deal, but one that requires an understanding of the law to appreciate fully.

In addition to the agency consultation requirements discussed in earlier chapters, the ESA makes it illegal for "any person," meaning both governments and private parties, to "take" endangered species. Congress defined "take" very broadly to mean "harass, harm, pursue, hunt, shoot, wound, kill, trap, capture, or collect" individuals of those species. Although words such as "harass" or "pursue" or "hunt" require conduct directed intentionally at members of the listed species, terms such as "harm" or "kill" are much broader. They might, for example, include construction and operation of a water diversion in which endangered fish are stranded and killed. The FWS interpreted the word "harm" by regulation to include "significant habitat modification or degradation where it actually kills or injures wildlife by significantly impairing essential behavioral patterns, including breeding, feeding, or sheltering," a reading later upheld by the U.S. Supreme Court.[18] This might include water withdrawals that eliminate backwater breeding habitats, or levees that block the movement of endangered species to and from riparian habitats.

Congress provided a relief valve, however, which allows the secretary of interior to permit takings that are "incidental to" an otherwise lawful activity. An incidental take permit must ensure that the person covered will "minimize and mitigate the impacts of the taking" as much as possible, and that the taking will not "appreciably reduce the likelihood of the survival and recovery" of the affected species. As part of the MSCP, the FWS issued a single incidental take permit to forty-one agencies, boards, power companies, irrigation companies and districts, and cities and towns in Arizona, California, and Nevada. But the permit also allows the permittees themselves to issue "Certificates of Inclusion" to any *other* landowner, water rights owner, developer, farmer, or "other private and public entities" to take species, although only for the kinds of actions included in the covered activities. The list of possible parties who may be covered by this single permit, then, is essentially indeterminate.[19]

The permit also provides virtually complete certainty regarding the compliance obligations of the covered parties for the full fifty-year term of the permit. No matter how much more the program ultimately costs than currently projected, the permittees need not pay a penny more. MSCP designers predict that the whole program will cost $626,180,000 (in 2003 dollars) over the full permit term. All nonfederal permittees combined are responsible for no more than half of that amount (about $314 million), or no more than half of the projected program costs in any given year, with the federal government responsible for the other half. That's just over $6 million a year spread over three states (with California permittees paying half and parties in Arizona and Nevada twenty-five percent each) to satisfy the parties' ESA obligations fully. And the federal agencies are responsible for actually implementing the conservation measures identified in the program.[20]

As with the upper Colorado River restoration program, it is not clear how costs can be determined so precisely, and with such certainty of meeting the program's biological goals. The cost estimates depend on assumptions about things such as the amount of restoration on public as opposed to private lands (which must be purchased), the price of any land that must be bought or leased, the revegetation cost for an acre of land with particular species of plants, contingency funds in case projects fail, and costs for planning, design, and long-term management.[21] Those assumptions may be perfectly reasonable for planning purposes, with an understanding that the crystal ball may be fuzzy and costs might change. It seems inappropriate, however, to confer legal compliance assurances on the basis of so many unconfirmed assumptions, however plausible they may be based on current knowledge.

Consider, for example, how much it will cost to revegetate an acre of riparian habitat successfully. Estimated habitat creation costs range widely from $4,600 to $30,500 per acre, depending on the species of plants and the kind of land being used. This does not include land acquisition costs, which vary dramatically depending on whether restoration occurs on federal, tribal, or private lands. Restoration costs per acre will also vary depending on soil conditions, water needs, and other factors. But because actual restoration sites have not yet been selected, only educated guesses are possible for those variables. Existing revegetation cost estimates are based on relatively small plots, in which individual cuttings were planted by hand. But one of the first site planning documents for the program notes that individual planting methods "do not translate well to large-scale habitat creation," and "may prove too costly and labor intensive" for the scale of restoration required by the MSCP. It may cost less to use mechanized planting methods, but those techniques have been tried only recently, and we do not know how successful they will be at a large scale.[22]

The MSCP permittees also obtained a firm guarantee under something called the FWS "no surprises rule" that the costs of compliance cannot increase over the permit term, no matter how much the parties may have misjudged the effectiveness of the proposed program, and no matter how badly the covered species might need additional help. The no surprises rule provides incentives for parties to enter into conservation agreements such as the MSCP without fear that requirements will change over time. Its legality has been challenged by environmental groups elsewhere in the country, but no court has yet decided that issue.[23]

Another key difference between the MSCP and the RIP/RAP programs is that "effectiveness" will be judged not by reference to the health of the affected species or the size of their populations, as defined by the species recovery goals. Instead, success will be measured entirely based on implementation of the specific conservation measures identified in the plan, whether or not they work. Additional measures may be taken, but only at federal taxpayer expense (or by voluntary donations). The permit prohibits the FWS from imposing any additional compensation in the form of land or water, or any restrictions whatsoever on the use of land, water, or natural resources to protect or recover the listed species. Thus, *all of the risk* of uncertainty in a highly uncertain program falls on the species or on the federal government.

In fact, even if additional measures or other changes are needed, the program includes an astonishing example of role reversal. As explained earlier, the National Environmental Policy Act (NEPA) and the ESA require that persons who engage in activities that might jeopardize endangered species evaluate all feasible alternatives to minimize impacts to those species (or under NEPA, to other environmental resources). Here, if additional measures are needed to protect the species, the federal agencies are required to "adopt those actions or measures that will have the least effect upon the Permittees and the respective constituents served by the Permittees."[24] Rather than requiring those who harm threatened and endangered species to minimize the impacts of their actions, this contingency requires those who seek to protect the species to minimize the impacts of conservation measures on those who cause the harm!

The MSCP permit is also remarkably broad in the potentially damaging new development impacts and activities it allows. The permit authorizes the incidental take of twenty-seven species that are currently listed as threatened or endangered, or that *might* be listed in the future. This unusual prospective approach is in itself legally suspect. And again, the permit insulates the protected parties by specifying that if a court rejects this approach, the FWS will accept the program as is for purposes of a new permit application, without any additional changes or requirements. From this perspective, the MSCP's inclusion of so many species is actually more of a liability than a benefit to the environment, because the permit insulates so many parties from culpability for the consequences of their actions to so many species.

The single-price permit is also expansive because it covers virtually every existing and future activity needed to support water diversions and power generation and transmission in the lower Colorado River, including all existing water diversions, future changes in the place from which water is withdrawn, additional future withdrawals, related operation and maintenance activities, and operation and maintenance of electrical generation and transmission facilities. This includes a major proposed change in the diversion point for more than 1.5 maf of water, roughly ten percent of the river's average natural flow. Water that currently flows into the All American Canal, close to the Mexican border, will be diverted instead from Parker Dam for conveyance to cities in Southern California and Arizona. *Where* water is diverted from the river is important ecologically, because upstream withdrawals reduce flows to a much longer portion of the river below.

The water transfer from Southern California agriculture to growing coastal cities serves a legitimate purpose. For many years, California has been using almost 1 maf more than its 4.4 maf apportionment under the compact. As part of an intricate set of agreements to help California curb this excess thirst, cities will pay farmers to use water more efficiently, and the cities will use some of the saved water.[25] But the change will deplete even more water from the lower river corridor, with potentially significant environmental impacts. Groundwater tables, for example, may drop even more than in the past, potentially drying out more riparian habitat and severing more backwaters from the river.

How could the FWS make a sweeping determination that such a broad scope of actions would not "appreciably reduce the likelihood of the survival and recovery" of all twenty-seven species covered by the permit? The long history of recovery efforts in the upper river and the Grand Canyon proves how difficult it is to determine the impacts of specific activities and recovery efforts on particular species given the complexity of the environment and the available data. After decades of study, answers to questions about which actions and conditions cause the most harm, and which restoration efforts will succeed and to what degree, remain elusive and uncertain. Yet in the biological opinion supporting the MSCP incidental take permit, the FWS proclaimed "that this level of anticipated take is not likely to result in jeopardy to the covered species, or destruction or adverse modification of critical habitat."[26]

In part, this risk allocation decision is explained by some legal sleight of hand in which the FWS and the Bureau of Reclamation narrowed the scope of what was considered in the incidental take permit and in the MSCP. The FWS combined its analysis about incidental take by nonfederal parties and its evaluation of the actions and decisions of the federal agencies into one decision, arguing that the standard for mitigation of impacts is more rigorous for incidental take permits than it is for interagency consultation.[27] Although the FWS did not fully explain this conclusion, presumably it is because the standard in the section 10 incidental take permit is mitigation "to the maximum extent practicable," while under section 7 the agency may approve actions based on "reasonable and prudent alternatives" that are "necessary or appropriate to minimize such impacts," without the "maximum extent practicable" qualifier. Because the FWS used the stricter of the two standards for both decisions, it reasoned, it is actually doing more to protect the species.

But neither of these two provisions of the ESA promotes full species recovery much less broader ecosystem restoration. The incidental take provision requires the FWS to find that the taking will not "appreciably reduce" the chance of species survival and recovery. This fits the first strategy of environmental law discussed in chapter 1, mitigation (stop the bleeding). Although it is generally illegal to take endangered species, the law allows nonfederal parties to cause some harm to accommodate otherwise lawful activities, so long as the impacts are mitigated as much as possible, and do not "appreciably reduce" the likelihood of species

survival and recovery. This tempers the harshness of an otherwise abso-
lutist law, allowing the economy to function so long as species survival
and affirmative recovery do not suffer. The incidental take provision,
however, is not the part of the law that promotes species recovery
(restoring good health). The interagency consultation process (section
7) of the ESA uses a similar mitigation standard. The FWS must find
that any federal agency action is "not likely to jeopardize the continued
existence" of any threatened or endangered species, or cause destruction
or adverse modification of critical habitat for those species. The FWS
regulations define "jeopardy" as "an action that reasonably would be
expected, directly or indirectly, to reduce appreciably the likelihood of
both the survival and recovery of a listed species in the wild by reducing
the reproduction, numbers, or distribution of that species."[28] The main
goal of section 7 consultation is to prevent further declines that might
hinder species survival or recovery. Again, it is not to promote affirmative species recovery.

The MSCP approach to the lower river, then, will not accomplish
the more affirmative duties Congress assigned to the FWS and other
federal agencies to promote species and ecosystem recovery. The fundamental goal of the ESA is to ensure that not just individual species
but the "ecosystems upon which endangered species and threatened
species depend may be conserved." Congress defined "conserve" to
mean "the use of all methods and procedures which are necessary" to
allow species to be "delisted," meaning that they have self-sustaining
populations in healthy ecosystems. The ESA requires the FWS to
issue and implement recovery plans for each threatened and endangered species, with specific management actions to achieve recovery
and objective, measurable criteria to determine when that has
occurred. A different provision within section 7 (aside from interagency consultation) requires federal agencies to use their authorities to
support species recovery efforts. The law prohibits nonfederal parties
from causing more harm to threatened and endangered species while
engaging in otherwise lawful activities, and it requires the federal government to work toward species recovery.

Perhaps more important in terms of the openness of restoration
decisions, MSCP documents obscure some very key choices about the
scope of restoration efforts. The Bureau of Reclamation's public MSCP
Web site boasts that the program is a "coordinated, comprehensive,
long-term multi-agency effort to conserve and work towards the recovery
of endangered species, and protect and maintain wildlife habitat on the
lower Colorado River."[29] By couching its decisions in "incidental take"

terms, however, and by separating federal agency obligations to minimize harm from their affirmative recovery duties, the FWS quietly
determined that the focus of the MSCP should be mitigation rather
than restoration. Perhaps the lower river is too far gone for full successful restoration. Perhaps water, power, development, and other economic
uses of the lower river are too important to sacrifice to restore species
and their ecosystems. But Congress seems to have decided otherwise, in
both the ESA and the Clean Water Act. Any decision to reverse this
choice with respect to several hundred miles of one of the great rivers
on the continent should be considered in a far more transparent and
open process.

Even with respect to the narrower "no jeopardy" finding, however,
the FWS and the Bureau of Reclamation defined the MSCP as narrowly as possible, in a way that obscures an equally important value
judgment. The ESA requires consultation for *any* federal agency
actions that might jeopardize the continued existence of threatened or
endangered species. But the FWS narrowed this requirement for all
ESA consultations by regulation to address only "actions in which
there is discretionary Federal involvement or control," meaning
actions other than those mandated by some existing law.[30] The Bureau
of Reclamation argues that it is bound by the Colorado River Compact, the treaty with Mexico, federal statutes governing management
of the river, and the Supreme Court decree in *Arizona v. California*.
Therefore, all of those aspects of dam management and operation are
beyond the consultation requirements of the ESA. Apparently, razorback suckers lack the panache of snail darters. Unlike the laws approving the Tellico Dam in the snail darter case, the bureau asserts that the
statutes governing operation of Colorado River dams trump the plain
requirements of the ESA.

The Bureau of Reclamation and the FWS made this argument in a
lawsuit brought by environmental groups challenging a decision in the
1990s to refill Lake Mead to higher levels, which inundated willows and
cottonwoods that provide nesting habitat for southwestern willow flycatchers.[31] The bureau argued that it lacks discretion to spill water from
Lake Mead to protect flycatchers because it may only manage the reservoir for river regulation, navigation, flood control, irrigation and other
water supply uses, and power generation. The FWS adopted mitigation
measures requiring that lost flycatcher habitat in Lake Mead be replaced
elsewhere in the lower Colorado River corridor. The court found that
this alternative mitigation was reasonable and complied fully with the
ESA, and therefore did not decide whether consultation was required

only for discretionary agency actions. (The next chapter discusses a later case in which a federal court did rule on this important issue.)

Most of the actions that virtually eliminated the native species of the lower Colorado River occurred long ago. Yet Congress clearly identified restoration as an important goal of the ESA, just as it did in the Clean Water Act for aquatic ecosystems. Presumably, Congress enacted a special law to protect *and restore* threatened and endangered species and their habitats because existing laws, regulations, policies, and practices led steadily to extinction. By interpreting the statute in a way that renders past practices immune to ESA scrutiny, the federal agencies have sided instead with the status quo.

The impacts of this narrow interpretation of the scope of the MSCP are dramatic. Over the next fifty years, the program proposes to "design and create" approximately 8,000 acres of habitat suitable for use by the affected species, including 5,940 acres of cottonwood-willow forests, 1,320 acres of honey mesquite forests, 512 acres of marsh, and 360 acres of backwaters. The program will also spend money to protect existing habitat, and "augment populations" of some species, which means stocking of fish spawned in hatcheries. Notably absent from the lower river program are any efforts to modify dam operations to benefit native species or their habitats. Presumably, this is because the dams are taken as a given, as are the water deliveries to the lower basin states and to Mexico mandated by the Law of the River, all of which are therefore defined as beyond the scope of ESA analysis. The program also lacks any significant efforts to address exotic fish species, although it does include some efforts to eradicate tamarisk where native vegetation will be established, and to control brown-headed cowbirds in nesting areas for southwestern willow flycatchers.

More important, by defining the program to include only the effects of new, discretionary changes to river operations, the scope of required habitat protection and restoration is minuscule compared to the dramatic cumulative changes to the ecosystem that have occurred over the past century. The MSCP relies on computer models to simulate how the river, groundwater, backwaters, and riparian zones will respond to changes in operation of dams, flood control structures, diversions, hydropower, and other related facilities. Given the uncertainties inherent in computer modeling over such a long period of time, and the complexity of the affected environment, predicted losses are remarkably precise: 2,132 acres of cottonwood-willow habitat, 590 acres of honey mesquite forest, 243 acres of marsh, and 399 acres of backwaters.[32] Some of that uncertainty is offset because, in most cases, the program seeks to create two to three times as much habitat by acre as the predicted losses. (Fewer acres of backwaters will be created than will be lost [360 acres to replace 399]). This is a typical strategy in environmental mitigation, to adopt a greater than one-to-one "mitigation ratio" to account for multiple uncertainties—how many acres of current habitat will be lost, how successfully new habitat can be created or restored, and the degree to which those habitats provide the same ecological benefits as natural habitats that are lost or degraded.

The bottom line is that, at best, 8,000 acres of new habitat will be restored along some 500 miles of the lower Colorado River ecosystem. It is difficult to believe that this is sufficient to compensate ecologically for all that has been lost along the lower river. The reservoirs alone inundated more than 210,000 acres of riparian habitats, spanning about half the length of the river from the Grand Canyon to Mexico. (I was tempted to round that number down to 200,000 acres, but the rounding error would exceed the total acreage to be restored.) Hundreds of miles of additional river are channelized and bound within levees. At least another 300,000 to 350,000 acres of riparian habitat along the lower Colorado River have been lost to agriculture and development, and much of what remains is covered with monocultures of invasive tamarisk or mixtures of tamarisk and native plants. The FWS now believes that only about 23,000 acres of native vegetation remain along the entire lower river.[33] All of that past harm was not considered because it does not fall within the rubric of new, discretionary actions.

So although billed as a "comprehensive plan to conserve, monitor, and manage populations and habitat of covered species,"[34] the MSCP is really just a program to *mitigate* the impacts of any incremental harm caused by proposed future changes in river operations. Although the MSCP might contribute somewhat to overall species recovery, that is not its real goal, and it does not constitute comprehensive environmental restoration.

Even the success of those mitigation efforts, however, depends on the degree to which newly created habitats will be used by the affected species of fish, birds, and other wildlife. Efforts to restore degraded wetlands and other ecosystems around the country have had very mixed success, and attempts to "create" habitats from scratch (as opposed to restoring degraded native habitats) are particularly difficult.[35] Whether you believe in the god of floods or the god of drought, successfully creating new habitats that approximate natural ecological conditions requires us to play God ourselves. We next consider the difficulties restoration specialists face in returning native species to the degraded habitats of the lower Colorado River.

The Efficacy of Habitat Restoration Along the Lower Colorado River

On the road north from Blythe to Needles, California, is a turnoff to the Blythe Intaglios—giant figures carved by an ancient people on the desert floor by scraping the surface rocks down to bare soil. The figures are formed by the contrast between the rocks and the soil. California historical landmark number 101 reads: "Times of origin and meaning of these giant figures, the largest 167 feet long, smallest 95 feet, remains a mystery. There are three figures, two of animals and a coiled serpent and some interesting lines." These figures persisted through centuries of wind and rain, but now they must be protected from vandalism with locked fences and signs. Yet outside the fences visitors have made modern replicas of the originals. Even here, we tamper with the originals and try to create newer versions in their place.

Up and down this stretch of the river I visited small plots of riparian revegetation like the Pratt Lease (described in chapter 1), ecological gardens that try to re-create small portions of the cottonwood-willow or mesquite forests that once lined the lower river. Everywhere the story was the same. It was difficult to tell which plots would succeed, which would fail, and why.

I first toured restoration sites at the Akahav Tribal Preserve, a part of the Colorado River Indian Tribe Reservation, with ecological coordinator Shannon Traub. Traub was a recent graduate of Indiana University with a degree in environmental science and conservation biology, who found this out-of-the-way position through E-jobs. Workers were preparing a new area to plant with mesquite, cottonwood, and willows, approximately seventeen feet apart. Planted in straight rows, the area looked like a tree farm. Eight other sites had been planted so far, for a total of 250 acres. Although the older sites looked far more natural, tamarisk continued to invade, crowding out the young mesquite. The strategy to remove the intruders involved chainsaws and a tamarisk-specific herbicide called "Garlon 4," which is painted on their trunks. When I asked whether birds use the newly vegetated areas, Traub said they believe so, but no formal surveys had been conducted.

The Akahav restoration project also includes two backwaters, which are regulated by gates to allow water in when the river is high and to hold water in when the river is low. I was confused, though, when Traub told me that these backwaters are stocked with bass, trout, and carp for recreational fishing. A Bureau of Reclamation official touted this site as part of their aquatic restoration efforts. Traub explained that nonnative fish enter from the river anyway, and that the tribes prefer to use the area as a recreational fishery. On the banks opposite this backwater sit new, expensive-looking homes.

Just north of Yuma Crossing I visited the "31-mile site," a small revegetation patch used as mitigation when the bureau realigned the river in 1997 and 1998 to keep water away from the levee. This project does not really constitute restoration of areas harmed in the past, because the new habitat is being created only to offset new harm. At most, it prevents even greater losses in total habitat, assuming the revegetation succeeds. From a distance, the site did not seem to be doing well. I saw cottonwoods with yellowing leaves, very sparse in some places. As I walked closer past a locked gate, rows of cottonwoods and willows with drip irrigation hoses seemed to be doing better. The site was protected by a "No Trespassing" sign that was shot to pieces. But immediately adjacent to the fenced restoration site sat a dense thicket of tamarisk, which made me wonder who was the native and who was the trespasser. A program monitoring report indicated reasonably good success at this site, although high soil salinity required careful selection of planting locations and irrigation to leach salts from the soils before planting.[36]

Farther upriver I visited the Cibola National Wildlife Refuge, established by the FWS in 1964 in cooperation with the Bureau of Reclamation, again as mitigation for efforts to stabilize the river against flooding. "Stabilize" is a euphemism. The bureau dry-cut a new, deeper river channel to replace the natural maze of shallower, shifting channels that meandered across a broad floodplain. This kind of mitigation explains most of the national wildlife refuge acreage along the lower river. As I approached the visitor center, a hand-written sign announced the daily counts: 5,925 Canada geese, 532 snow geese, 677 sandhill cranes, and 1,034 ducks. Not bad, but it illustrates what is true for much of the national wildlife refuge system—managed factories for waterfowl that are so popular with hunters. Although that serves a legitimate purpose, the Cibola refuge was not initially designed to restore native habitat.

That is beginning to change, at least in some areas. Refuge manager Mike Hawes described several restoration sites within the refuge. One thirty-five–acre plot was planted in spring 1999 with one-gallon pots of willow, cottonwood, and mesquite that two years later were thirty feet high. Although bird use was low, more was expected every year. Another site was planted across the river in the late 1980s. But because the area is dry, with no chance of flooding to moisten the soil, and because there is little understory vegetation, no flycatchers use the area. The refuge is basically farming trees. When I later toured the refuge by car, I saw that even these restoration sites represent a tiny percentage of the overall acreage.

Open water habitats in the Cibola refuge illustrate a similar dichotomy. Cibola Lake is managed for sport fishing, like the backwaters in the Aka-hav Preserve. Three small "finger lakes" were dredged for native fish, but at the time had not yet been stocked. One pond next to the levee is used successfully as a grow-out pond for hatched razorbacks and bony-tails, which are reared and released into the river every year. Because this is all managed hydrology, the refuge pumps water from the river to the ponds. The electric bill at the time was $30,000 a year.

My amateur touring and informal discussions provided only anec-dotal snapshots of a complex undertaking. But they underscore the sig-nificant, multiple uncertainties involved in the centerpiece of the MSCP—efforts to create new riparian, marsh, and backwater habitats to replace some of what has been lost or impaired, and which provide usable habitat for the target species.

So after viewing these selected sites, I visited Dr. Bert Anderson in the small office of the Revegetation and Wildlife Management Center in Blythe. Anderson coauthored the principal survey study about changes to the ecosystem of the lower Colorado River in 1988, chaired one of the scientific peer reviews of an early draft of the MSCP, and has been involved in revegetation projects for three decades. He is a study in contrasts, and unlike understandably cautious agency scien-tists, is not afraid to express his opinions. He noted that Lake Powell and other reservoirs have radically altered the hydrology and ecology of the watershed. Yet he stated confidently that tamarisk is not as bad as people think (as discussed in chapter 4), and actually provides good habitat for some species, such as cactus mouse, Lucy's warbler, blue grosbeak, yellow-breasted chat, and white-winged doves. But it also produces more biomass than anything else, rendering whole habitats unsuitable for native species, reducing biodiversity, and disrupting the food chain.[37]

The main purpose of my visit, though, was not to debate the virtues and vices of salt cedar, but to learn what makes revegetation efforts suc-ceed or fail. His answer was not surprising. He has been evaluating the issue for more than twenty years and still does not know, in spite of about 100 project-years of experience with maybe 20,000 trees. Possible sources of failure include variable propagule quality, beaver and insect damage, salinity and other soil characteristics, soil moisture, and depth to groundwater. Yet because managers wanted to accomplish most existing revegetation projects quickly and inexpensively, data for those efforts are missing, making retrospective analysis difficult. Multiple conditions probably determine success at different sites, and conditions vary even within small plots.

A large percentage of the region is suitable for mesquites, Anderson said, verifying Mary Austin's instincts of a century earlier: "The Mesquite is God's best thought in all this desertness." But Anderson indicated that it is more difficult to predict whether cottonwood revegetation will succeed, even though he has the most data for those species. Cottonwoods often grow well initially, but then develop a yellowish cast (like what I saw at the 31-mile site), meaning their tips are dead, and the trees decline in their fourth or fifth year. The trees may hang on, but are worthless as habitat. So new sites may look impressive, but their long-term viability is more difficult to assess. Other replanting efforts have failed altogether due to high soil salinity or other factors.[38]

One major uncertainty involves the features that must be present to ensure that target species will use new habitats. Southwestern willow flycatchers, for example, do not breed in all stands of native trees. Nesting occurs in stands that have moist soils or are located very close to open water, and in areas with particular amounts of cover, understory vegetation, and other characteristics. Native trees survived at the revegetation sites I visited in Cibola National Wildlife Refuge and at the Pratt Lease. Yet later surveys identified bird use by many species, but not by southwestern willow flycatchers. And although MSCP designers hope that revegetated patches will provide habitat for multiple species, habitats created for one species may not work for others, and specific habitat needs are not well known for all program species. A year after the MSCP was adopted, consultants submitted a "species profile report" because "clear and thorough knowledge of habitat requirements must be gained to successfully implement any restoration plan." Although the report identified some habitat characteristics based on a review of currently available literature, it also identified gaps that must be filled with additional research. The report included detailed information for only nine out of the twenty-seven species in the program, including razorback sucker, flannelmouth sucker, bonytail, California black rail, Yuma clapper rail, least bittern, southwestern willow flycatcher, western yellow-billed cuckoo, and Colorado River cotton rat.[39] The program's habitat-based goals for success, which form the basis for ESA guarantees to the covered permittees, were established without full understanding of the characteristics necessary to provide good quality or even acceptable habitat for all of the species.

Even if all habitat needs were known for all of the program species, it is far more difficult to assemble a particular ecological habitat than it is to put a bridge back together to look the same as when it spanned the Thames. The MSCP Habitat Conservation Plan acknowledges that "experience in development and creation of [these species'] habitats and management of their populations is limited," and each restoration site poses unique problems that might impede success. The program also assumes that successful revegetation in small plots can be reproduced for much larger areas. This may not be the case, especially where revegetation requires favorable soil conditions, which can vary greatly even within a contiguous site. Most existing revegetation sites are smaller than seven acres, but MSCP plans involve plots of hundreds of acres, for which there is no similar experience.[40]

Similar problems confront efforts to create backwaters along the lower Colorado River. Program goals are modest relative to what has been lost—just 360 acres compared to the many thousands of acres of backwaters that were destroyed or impaired. This target is based on hydrological models that predict an additional loss of 399 acres of backwaters due to future changes in river flows. Although this small amount of restored acreage would not compensate for past losses, new backwaters could provide important rearing habitat for razorback suckers, flannelmouth suckers, and bonytail, increasing their chance of survival to a size that will enable them to escape predation and compete with nonnative fish in the main river.[41]

The strategy of creating new backwater habitat in the lower river is based in part on the work of fishery biologists who began to rear native fish in off-stream habitats to avoid the inevitable loss of young fish to nonnative predators when they are released too young into the main channel or connected habitats. Off-stream habitats consisted of excavated but isolated backwaters and oxbow lakes designed to resemble those once found on the pristine Colorado River floodplain. Native fish are allowed to spawn in these habitats, where young are protected until large enough to survive in the main channel. Prototypes have succeeded in promoting razorback recruitment to adulthood at places such as Havasu National Wildlife Refuge and Beal Lake, Arizona. Based on these initial successes, these researchers advocate a comprehensive recovery program in which larger numbers of native fish are translocated among a more diverse combination of predator-free, off-channel habitats; the main river channel; natural backwaters; and artificial reservoirs. They call this mixed set of artificial and natural habitats "channel plus connectives." The translocation proposal is designed in part to avoid some of the problems of hatchery culture. By spawning native fish directly in protected backwater habitats, much larger numbers of native fish from different subpopulations will contribute to the gene pool than in hatcheries, which use fewer parents.[42]

As with riparian habitats, however, creating new backwaters that provide suitable native fish habitat is not easy. Between 2000 and 2002, the FWS and Ducks Unlimited created four new ponds totaling thirty-five acres at the Imperial National Wildlife Refuge north of Yuma. The ponds were connected to the main channel in two places to allow fresh water in and drainage out, treated with rotenone to eliminate nonnative fish, and stocked with 658 razorback sucker fingerlings. Within several months, only an estimated twenty-five razorbacks survived, compared to as many as 17,000 nonnative sunfish. Scientists suspect two significant causes: nonnatives are persistent and can reinvade from the main channel; and inadequate connections with the main channel caused low oxygen levels (hypoxia) in the water, which may have killed the razorback fingerlings. Unfortunately, those two problems may have conflicting solutions. Adding connections to improve water flow and thereby increase oxygen levels provides more pathways for nonnative fish to reinvade. To avoid this problem, the FWS plans to reconstruct the ponds using fresh groundwater rather than river water, and to simulate conditions at more successful backwater ponds created at the Cibola National Wildlife Refuge. The hypoxia problem is exacerbated because the ponds are adjacent to farm fields. If the ponds are filled to high enough water levels, adjacent fields will flood, killing trees or crops. When water levels are kept low enough to avoid that problem, aquatic plants take over the pond. Decaying plants consume too much oxygen, making the bottom water hypoxic. Lower water levels also allow birds to prey too easily on the young native fish. It is difficult to re-create habitats that existed under the volatile conditions of the native river, while still protecting adjacent investments in farms and homes.[43]

This brings us to what may be the real problem. The MSCP will not create habitats that approximate nature under natural conditions. At best, it envisions simulated habitats. Only facsimiles are possible because the river's flow patterns and ecological conditions have changed so dramatically. Because the river no longer overflows its banks, irrigation will be used to leach salts from the soils, and artificial flooding will replace natural spring inundation. People will plant trees—by hand or with machines—because the annual floods no longer bear the silt and seed that once promoted natural germination throughout large swaths of riverside. Because the river is channeled within deep banks and between high levees, and because of the prevalence of nonnative species, created "backwater" habitats will really be off-channel artificial ponds. The biologists who pioneered the use of isolated backwater habitats acknowledge that the scheme reflects a compromise to reconnect the river to an artificial facsimile of its natural affiliated marshes, backwaters, and floodplains.[44]

All of this reminds me of "Betty's Kitchen" (described in chapter 1), on lands inundated by the 1983 floods along the lower river. That site provides a sharp contrast to the painstaking, expensive, and uncertain acre-by-acre habitat creation envisioned in the MSCP. The 1983 floods destroyed homes and other structures all along the lower river, an unfortunate cost of our compulsion to build to the water's edge. But the floods also allowed nature to reclaim its own habitat. Restoration required no intensive planting or soil preparation. Yet the site teemed with seep willow, cottonwood, honey mesquite, and quailbush, most of which sprang naturally from dormant seeds or new ones provided by the river. I wondered whether this kind of natural revegetation would accomplish more, with less time and money and with more certainty, than the planned MSCP. A more comprehensive and potentially effective approach may be to restore flow conditions to approximate the natural hydrology that characterized the lower river. But this highlights the difficulty of restoring natural river and riparian habitats in a region that has supported so much investment, often on private property, in reliance on engineered flood control.

Reestablishing native riparian habitat through individually planted plots is a long, expensive, and painstaking effort. Only limited funds are available, and despite the success of restoration in small patches, only a tiny fraction of the river corridor has been restored. Even the MSCP goals of 8,000 acres represent a small portion of the habitat that has been lost. Although it is difficult to estimate how much acreage will be needed to achieve species recovery, clearly we are nowhere near that goal.

Far more acreage would regenerate naturally if we restored river flows to something closer to natural conditions, and if the river were freed from its engineered prisons. Almost 2,300 acres of cottonwood-willow forest regenerated naturally during the floods of the mid-1980s, three times what is envisioned for the entire fifty-year MSCP. But most of that habitat dried out or was replaced by tamarisk in the dry years that followed.[45] Maintaining such habitats would require repeat floods every several years. Yet unlike the effort to revise flows from Flaming Gorge, Glen Canyon, and other dams upriver, no such changes are being considered as part of the MSCP. Just as the FWS was reluctant to fill new fish-rearing ponds with enough water because of potential impacts to privately owned farms adjacent to Imperial National Wildlife Refuge, simulating a natural flow regime along the lower Colorado would conflict with currently developed property.

The Upper Colorado River Endangered Fish Recovery Program includes some efforts to breach the dikes that sever the main channel of the river from adjacent floodplains and wetlands, and thereby to recreate inundated floodplain, backwater, and wetland habitats. If sufficient habitat of this kind is restored, razorback and other native fish larvae may be able to survive to adulthood in their natural habitats. Then perhaps the species can begin to reproduce and recover without the perpetual aid of hatcheries and protected, artificial rearing habitat. Private property owners, however, will likely resist these efforts along more heavily developed portions of the river. Despite laudable intentions, to date only three discrete habitat restoration sites have been improved within the Ouray National Wildlife Refuge on the Green River near Vernal, Utah (on federal land), and two on the Colorado River near Grand Junction. One was on private conservation lands owned by Grand Valley Audubon, and the other was in a state wildlife area. Another 1,200 acres of floodplain and wetland habitat have been purchased along the Green, Gunnison, and Colorado rivers; but funding for additional acquisitions is limited.[46]

So although natural restoration might be feasible on federal or even tribal lands farther south, the river cannot overflow its banks selectively, and increased flows likely would affect private property. Properties at risk would include casinos that rim the waterfront in Laughlin; riverfront condos and resorts at Lake Havasu; miles of luxury homes, trailer parks, and campsites that line the banks of the river in many other areas; and thousands of acres of farmlands in the river's floodplain. Any effort to change the pattern of dam releases, and even to approximate the river's seasonal floods, would force a retreat from the water's edge, not only in areas targeted for riparian restoration, but in other areas as well. Landowners might protest that such efforts would constitute an unconstitutional "taking" of private property without just compensation.

But how much of this "property" within the natural floodplain of the Colorado River can legitimately be called "private"? Most of this development was made possible by Hoover Dam and its downstream sisters, and is maintained by dam operations, levees, channelization, and other efforts designed and built with public funds. It is ironic that private property rights impede ecological restoration in areas where development was made possible only through public investments. More important, although legal plats show property boundaries in rigid lines, the natural river respects no such formal boundaries. This suggests a closer look at the nature of property, and how it affects environmental restoration goals and methods.

Private Property and Restoration Along the Lower River

Novelist Barbara Kingsolver pondered our western compulsion to own not only property, but nature itself, in her collection of essays *High Tide in Tucson*. Kingsolver's own attempts to "make the desert bloom" behind her desert Tucson home failed as nature intruded on her western notions of what a garden must be. Abandoning those futile efforts brought respite rather than failure: "Life is easier since I abdicated the throne. What a relief, to relinquish *ownership of unownable things*." Kingsolver was hardly the first American writer to comment on the limits of property ownership. Ralph Waldo Emerson described neighbors who owned fields and woodlands, "but none of them owns the landscape." His friend Thoreau likewise advised that "the landscape is not owned." More recently, William Kittredge wrote: "The truth is, we never owned all the land and water. We don't even own very much of them, privately. And we don't own anything absolutely or forever. As our society grows more and more complex and interwoven, our entitlement becomes less and less absolute, more and more likely to be legally diminished. Our rights to property will never take precedence over the needs of society. . . . Ownership of property has always been a privilege granted by society, and revocable." Although these could be dismissed as the musings of poets and novelists, in fact they have a deep tradition in the history of property law.[47]

Anglo-American property law is based in part on the premise that private land ownership safeguards individual liberty and maximizes societal welfare. John Locke, the English political philosopher and intellectual forebear to the authors of the Declaration of Independence and the U.S. Constitution, wrote: "God gave the World to Men in common; but since he gave it them for their benefit . . . it cannot be supposed he meant it should always remain common and uncultivated." According to Locke, individuals have a right to "own" property if they combine land resources with their own labor and capital to generate wealth. By adding value to the land, they are entitled to ownership as a natural right so long as similar opportunities are available to others. The right to retain profits provides an incentive for landowners to maximize the economic value of each parcel, and the aggregate values of property maximize social welfare. Individual property rights also guard against an intrusive central authority.[48]

Under this absolute "ownership" theory, however, landowners may maximize profits by imposing costs on other landowners or on society as a whole, with a classic example being pollution. According to the famous chronicler of English law, William Blackstone, land ownership is relatively "absolute," but is subject to common law principles designed to prevent one owner from causing harm to others. But land also contains a hybrid of private and public and ecological values, not all of which are maximized by private ownership. Ecologists, economists, and legal scholars have identified a range of "ecosystem services" that are not measured by private markets. Riparian wetlands, for example, buffer floodwaters, filter pollutants, cycle nutrients and other materials through the ecosystem, and provide breeding and spawning grounds and other habitats for fish and wildlife. Although many of these services benefit human economies, their ecological "value" is not limited to humanity alone.[49]

Anglo-American property law also relies on fixed boundaries and stability of title, without which landowners could not have the certainty of return necessary to invest labor and capital. Land can usually be defined by fixed boundaries. Aquatic ecosystems, however, are not always amenable to such rigid delineation. Large lakes and coastal waters change with the tides. Rivers and streams expand and contract seasonally and annually through a shifting mosaic of floodplains and side channels. Other aquatic ecosystems, from wetlands to ephemeral streams, vary in time and place. Thus, although it is desirable for purposes of property law and the human economy to delineate fixed boundaries between land and water, from an ecological perspective such boundaries are elusive. The U.S. Supreme Court recognized this problem in the context of regulating discharges under the Clean Water Act:

> In determining the limits of its power to regulate discharges under the Act, the Corps must necessarily choose some point at which water ends and land begins. Our common experience tells us that this is often no easy task: the transition from water to solid ground is not necessarily or even typically an abrupt one. Rather, between open waters and dry land may lie shallows, marshes, mudflats, swamps, bogs—in short, a huge array

of areas that are not wholly aquatic but nevertheless fall far short of being dry land. Where on this continuum to find the limit of "waters" is far from obvious.[50]

In areas of fluctuating or variable shorelines, legal methods exist to set boundaries between competing property owners. It is much more difficult to define such a dividing line between private and public rights in aquatic ecosystems. For most purposes, water bodies and the water they hold are considered "public," although we dole out rights to use (but not to "own") water if those rights are consistent with the public welfare.[51] Adjacent land is considered "ownable," whether by private parties or by a government, and therefore open to development. That distinction, however, artificially assumes a sharp dividing line between land and water.

The idea that some parts of the world cannot be owned privately has deep roots. Under Roman law, resources such as the air, the water, the ocean, and the shores of the seas could not be owned by individuals, but instead were held for the common benefit of the public at large. Many European countries accepted this concept during the Middle Ages, and England adopted a variation known as the public trust doctrine. The public trust doctrine restricted the ability of the Crown to grant tideland resources to private individuals, and imposed on the government a duty to manage and protect those resources for common purposes of commerce, navigation, and fishing, just as a financial trustee manages and protects assets for a minor child.[52]

American judges expanded the public trust doctrine to meet the needs of a nation with changing public values and a different geography than England's. State courts expanded the doctrine to include areas such as nonnavigable tributaries to navigable waters, all waters usable for public recreation, dry sand beaches, and state parks. Beginning in the 1970s, scholars argued for revitalization of the public trust doctrine to address water pollution, loss of species and habitat, and other forms of environmental harm. Some courts agreed, most notably in a case in which the California Supreme Court curtailed water diversions by Los Angeles from tributaries to Mono Lake because declining lake levels jeopardized birds and other wildlife.[53]

The public trust doctrine, however, retains the anthropocentric focus of property law in which "trust assets" are held by the government for the common benefit of human users. Perhaps the bigger problem is that the ecological values inherent in aquatic ecosystems are not amenable to *either* private or public ownership. This concept of "nonownership" has an equally long legal history, but has not received the same scholarly or judicial attention outside the arena of wildlife law. Some scholars read the original Roman law to mean that some common resources cannot be owned *at all*. Private individuals cannot "own" wildlife even if wild animals reside on their land. Individuals may own domesticated animals reduced to human control and wild animals reduced to physical possession through hunting or capture. No one can "own" a species, however, or even a population of wild animals. A rancher might own domesticated horses but not the wild mustangs grazing on her land.[54]

The U.S. Supreme Court clung for many years to the notion that states owned wildlife in trust for their people, but gradually abandoned this concept. In *Missouri v. Holland*, Justice Oliver Wendell Holmes questioned the idea that state "ownership" of birds that migrate across state lines could impair federal regulatory power: "To put the claim of the State upon title is to lean upon a slender reed. Wild birds are not in the possession of anyone; and possession is the beginning of ownership. The whole foundation of the State's rights is the presence within their jurisdiction of birds that yesterday had not arrived, tomorrow may be in another State and in a week a thousand miles away." In later cases the Supreme Court referred to the ownership concept as a "legal fiction" or "fantasy," and ultimately ruled that state authority to regulate wildlife is grounded in sovereign authority to protect common resources and the common welfare. It is not based on ownership.[55]

The nonownership principle conforms to a growing realization that nonhuman components of the natural world are not merely resources for human use and consumption, but have intrinsic value. Just as the law evolved in the 19th century to reject the idea that people could own slaves, law in the 20th century changed to conform with society's growing ethical rejection of human dominion over all other living species. At least since the early 1970s, some scholars began to propose legal rights for nonhuman species. The idea that wildlife cannot be owned also makes sense in light of the realization that species provide ecosystem services beyond those measured in the market economy.[56]

So what does this have to do with restoration of the Colorado River? No one claims ownership of razorback suckers or Yuma clapper rails. But private property rights at the water's edge limit the government's ability to restore the natural relationship between land and water. If inundation of private property constitutes an unconstitutional taking of property, modified dam flows that even periodically inundate riparian habitats or backwaters might be prohibited. Or, it might require the government to compensate landowners financially for any harm, which would cause restoration costs to skyrocket. But the degree to which Colorado River restoration efforts constitute takings of private property turns on the nature and extent of private property rights in this transition zone between land and water. Landowners have built heavily in seasonally inundated areas that are as much a part of the river's aquatic ecosystem as the main channel of the river. What claims do private property owners lay to resources that benefit society and the environment rather than individual owners? Can private parties "own" ecological resources in what was the natural floodplain of the Colorado River?

Scientific understanding of the place of individual wildlife populations and species in a broader ecological context continues to evolve, and societal values shift accordingly. Discrete wildlife populations and species are connected to the ecosystems of which they are component parts. Ecologists define an "ecosystem" as an assemblage of species and their related chemical, physical, and biological interactions with each other and the physical habitats in which they exist. It makes little sense to say that humans cannot own wildlife without accepting the corollary that they cannot own the other ecosystem components, values, and services of which those populations and species are component parts. The difficult part is determining how far that proposition extends at the water's edge, and how it can be accepted without rejecting the important idea of private land ownership, which is such a critical part of our political heritage.[57]

One way to resolve this conflict is to accept that landowners have rights against other people, but not against the land or water, or against species and ecosystems. The fact that a person cannot own the aquatic ecosystem, and accordingly has no right to destroy or degrade those resources significantly, does not mean that she has no rights to use or profit from them. A landowner may have the exclusive right of access to wildlife on her property, and to charge others to view or lawfully hunt that wildlife, without owning the animals. A landowner along the lower Colorado River may have the right to build a wharf and to exclude others from using that dock. That does not, however, necessarily include a right to cause severe erosion that degrades the ecological productivity and the resources of the river below. In the case of wetlands, instead of drawing an artificial and ecologically meaningless line between land and water, one analyst urged a boundary defined by rights and uses: "If a wetland is privately owned but happens to also be critical habitat for a public trust natural resource such as a migratory duck, what rights belong to the land owner and what rights to the public as a whole?"[58]

Rather than deciding that property ownership in aquatic transition zones is an all-or-nothing proposition, with the private party owning all on one side of the artificially defined boundary (the "dry" side), and the government owning all on the other (the "wet" side), a "nonownership" doctrine suggests a sliding scale in which the relative degree of ownership rights shifts along with the temporal and geographic transition from water to land. The realization that values, services, and components of aquatic ecosystems cannot be owned does not render private title in the land-water transition zone meaningless. It does, however, limit the degree to which owners have a property right to develop in ways that would infringe on aquatic ecosystem values. The landowner maintains the right to use property for the traditional purposes for which land is put, so long as the aquatic ecosystem values that cannot be owned are not substantially impaired. Of course, that suggests that someone—scientists, judges, politicians, or agency officials—must decide what constitutes "substantial impairment."

Likewise, the nonownership concept helps to clarify the degree to which government actions to protect or restore the aquatic environment constitute a "taking" of private property. When governments impose restrictions on the use and development of private land to protect aquatic environmental resources, landowners often assert constitutional takings claims. To the extent that resources are not subject to private ownership, by definition they cannot be "taken" by a government acting in its capacity as legal guardian of those resources. One cannot assert a "taking" with respect to property that is inherently "unownable," and that does not meet the threshold justification for property rights. This result has been upheld in several recent takings cases. In Oregon, a court rejected a takings claim for denying beach development where Oregon law did not recognize an absolute property right.[59]

To be very clear, I am not arguing that we should intentionally flood valuable homes and other structures, or that we should casually dismantle levees or change river flows in ways that would cause such harm. But when individuals build right up to the water's edge, supported by billions of dollars of public flood control structures and in some cases federal flood insurance, property rights should not hamstring efforts to restore public or intrinsic ecological values that are impaired, in part, by that very development. Public dollars are being pledged to an expensive, painstaking, and possibly futile program to restore habitats along the lower Colorado River. But those efforts ignore the need to restore natural hydrological and ecological processes that might promote a healthy riparian and floodplain ecosystem.

Some farmers along the Green River complain that restoration flows from Flaming Gorge Dam flood their crops. The Bureau of Reclamation acknowledges that this is true, but only for relatively short periods of time.[60] Where landowners use resources in the river's natural floodplain, some of their property "rights" must be balanced against common public rights in public resources. Restoration of the backwaters and riparian zones in the lower river corridor similarly would benefit from efforts to restore some of the higher spring floods that can naturally revegetate and replenish the riparian ecosystem. Although such flows should be designed in ways that minimize property damage to the maximum extent possible, they should not be rejected merely because some impacts would occur to private property within the river's natural floodplain. That policy allows private individuals "ownership of unownable things."

There is, however, another major reason why the Bureau of Reclamation and the basin states have avoided restoration flows in the lower river. Any water used to restore riparian ecosystems in the United States, and that could not be diverted to the All American Canal just north of the Mexican border, would increase flows to Mexico beyond the minimum requirements of the 1944 treaty, something the United States has long shunned. Ironically, however, extensive development does not occur within the river's natural floodplain in Mexico. Increased flows at the international border would naturally restore much more native habitat than is envisioned under the MSCP, without flooding as many homes and farms. But MSCP managers elected to sever the program at the artificial geopolitical boundary, rather than treating the lower Colorado River as an integrated, international ecosystem. We explore those issues in the next chapter.

CHAPTER EIGHT

An Elusive and Indefinable Boundary: Restoration and Political Borders

From my motel in Yuma, Arizona, a few miles from where the Colorado River reaches the Mexican border, I took an early morning run through fields already filled with farm workers picking produce. I was here to learn more about conditions along the lower river as it exits Arizona and courses through Mexico to its terminus in the Sea of Cortéz. On small dirt paths I crossed over a lined canal, the banks of which seem to serve as levees as well, protecting the valuable fields from a river that is usually tamed but still unpredictable. I continued to the river through surprisingly "natural" sloughs with herons and egrets and other birds, but the vegetation was dominated by tamarisk. The river still seemed to have quite a bit of water, but surged ahead through straight, channelized, and armored banks.

Later that day, I headed south with Karl Flessa and his students from the University of Arizona. As we passed even larger fields, I wondered how much acreage is irrigated with Colorado River water. In the 1860 report of his exploration northward from the delta, Lieutenant Joseph C. Ives described this area south of Yuma as a bleak greasewood desert. "The barrenness of the [Yuma] region, the intense heat of its summer climate, and its loneliness and isolation have caused it to be regarded as the Botany Bay of military stations."[1] I doubt he would recognize this as the same place in the bright greens of today.

In Mexico, a much smaller percentage of land was irrigated, and the fields were not nearly as lush, compared to the industrial farming to the north. We crossed the Colorado River, turned onto a small dirt road and drove down to the river. Once again, I was surprised at how natural it looked, with snowy egrets, great egrets, and great blue herons. Flessa pointed to the relatively new bridge and explained that the 1983 floods washed out an older bridge.

These snapshot views, a few facts, a few observations, were something of an illusion—what John Steinbeck, chronicling his own journey to the Sea of Cortéz, called "this myth of permanent objective reality," where observed facts do not portray a holistic truth. Steinbeck described a technique to preserve fish and to record relevant data but observed how the raw numbers belied what lay beneath: "There you have recorded a reality which cannot be assailed—probably the least important reality concerning either the fish or yourself." The idea of illusion recurs in descriptions of the lower Colorado River and delta. In describing the surrounding desert, John Van Dyke wrote: "This is the land of illusions and thin air. The vision is so clear at times that the truth itself is deceptive." Phillip Fradkin also described the delta as a "land full of illusions." And recounting his effort to navigate the lower river in late 1857 and early 1858, Lieutenant Ives lamented that "no dependence can be placed on appearance."[2]

An egret here and a heron there belied what was missing, but the truth was not much farther south. Dry, sparsely vegetated desert gradually changed to broad expanses of mudflats, which from the road appeared devoid of vegetation. I wondered out loud how different the area would look if it were still subject to the huge annual floods that characterized the area before the dams. Flessa agreed that the mudflats likely would vegetate, with seeds carried by upstream waters germinating on newly moistened soils.

We crossed the international border into Mexico with only minimal delay, but what significance does this artificial boundary have for the river? By the time the river reaches the border, roughly nine out of every ten gallons have been diverted to cities outside the basin, consumed by crops or other human uses, or evaporated into the air, due to decisions reached almost entirely in the United States. These depletions have perhaps even more serious impacts on the river and its associated ecosystems as it travels its last few miles in Mexico. Yet there is no restoration effort in Mexico comparable to the Recovery Implementation Program–Recovery Action Program or Glen Canyon Dam Adaptive Management Program. And as mentioned in the last chapter, Lower Colorado River Multi-Species Conservation Program participants declined to consider impacts in Mexico. Any such effort would depend on increased water flows from north of the border, and especially on at least a partial return to the flood regime that occurred before the dams.

In the last chapter, I argued that restoration of the Colorado River and other aquatic ecosystems is impeded by the failure of the legal regime to recognize fully the elusive ecological and hydrological boundary between land and water. Instead, it is based on the illusion of a fixed demarcation between dry land and wet water. The law, in short, does not properly reflect natural reality. This chapter examines a similar mismatch between legal and institutional arrangements and ecological realities along the U.S.-Mexico and other geopolitical borders. The Colorado River and its ecosystems are continuous, but due to economic, political, and other factors, for many purposes we treat the U.S. and Mexican components of the river as separate entities. The United States agreed by treaty to protect some of Mexico's share of the river in both quantity and quality, but mainly for economic uses. No similar arrangements ensure that actions within the United States safeguard the ecosystem in Mexico, and as a result, transboundary restoration efforts are essentially nonexistent.

We begin by examining the degree to which the Colorado River in Mexico has been damaged through the cumulative effects of dams, water withdrawals, and other changes north of the border. We then explore the connections and tradeoffs between water use and management in the United States and the health of the ecosystem in Mexico, and in the Salton Sea in California.

The End of the River

Whatever water, silt, and nutrients reach the lower river and delta (figure 8.1) today support some plants, birds, and other species. But this pales compared to what predated the dam, as portrayed by Aldo Leopold in *The Green Lagoons*:

> A verdant wall of mesquite and willow separated the channel from the thorny desert beyond. At each bend we saw egrets standing in the pools ahead, each white statue matched by its white reflection. Fleets of cormorants drove their black prows in quest of skittering mullets; avocets, willets, and yellowlegs dozed one-legged on the bars; mallards, widgeons, and teal sprang skyward in alarm. As the birds took the air, they accumulated in a small cloud ahead, there to settle, or to break back to our rear. When a troop of egrets settled on a far green willow, they look like a premature snowstorm.[3]

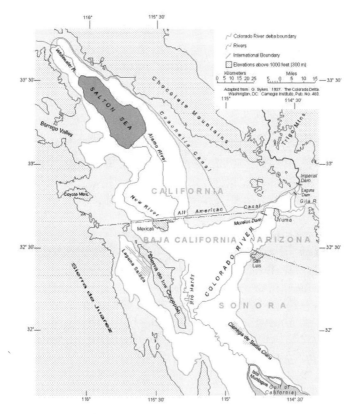

Figure 8.1. The Colorado River Delta region. *Courtesy of the Pacific Institute*

Before the dams, the delta supported one of the world's great desert estuaries, in which nearly 2 million acres of riparian and tidal wetlands hosted a vast diversity of plants, birds, and other wildlife. Ives described "immense numbers of wild fowl," and "innumerable flocks of pelicans, curlews, plovers, and ducks of different varieties." Early trapper James O. Pattie described a densely vegetated river bottom below Yuma six to twelve miles wide, filled with birds and other wildlife. Most of that was gone by the 1970s when Phillip Fradkin wrote the first edition of *A River No More*. But the high water years of the 1980s and smaller floods in the 1990s brought the modest recovery I witnessed on my way south. Now scientists estimate that the delta supports 150,000 acres of wetlands, but those areas remain at risk unless floods recur periodically to rejuvenate them. Mexico designated the northern part of the Sea of

Cortéz and the southern part of the delta as an International Biosphere Reserve.[4]

Wetland losses and other changes in the river dramatically altered the ecosystem of the delta, the estuary, and the upper Sea of Cortéz. Because sediment and other material that once reached the lower river is now trapped behind dams, the delta is eroding rather than expanding. Tamarisk, arrowweed, iodinebush, and other exotic plants supplanted native cottonwood/willow communities, and spring floods no longer supply new seeds and conditions that allow native plant germination. Reduced freshwater flow changed the delta's aquatic environment by increasing the salinity of water in the estuary. Estuaries experience a tug-of-war between freshwater from the river and saltwater brought in by the tides; and the Sea of Cortéz has some of the most dramatic tides in the world. Ives described bore tides extending several feet high from bank to bank, with "a deep, booming sound, like the noise of a distant waterfall." Before the dams, salinity at any point in the river changed with the daily tides, the seasons, and the annual flow variations of the river. Now, at any place and time salinity is much higher than it otherwise would be. Circulation patterns in the estuary changed as well. Because freshwater is not as dense as seawater, large freshwater flows used to flow at the surface while denser seawater entered the river at the bottom, causing a circulation pattern extending thirty-seven miles from the river's mouth. That mixing area is now much smaller.[5]

These changes in flow and salinity affect marine species in various ways, not all of which we understand well. Some species evolved to lay eggs in lower salinity estuarine conditions that no longer exist in the same areas. Gulf shrimp populations have plummeted, although scientists documented some recovery in years when more freshwater reached the sea. Various gulf fish species are also declining, and although commercial fishing and other factors contribute to those problems, some scientists suspect that changing salinity in the estuary plays a role as well. The Mexican government has listed as endangered one species of marine fish (the totoaba) and one small marine mammal (the vaquita porpoise), both of which are endemic to the delta.[6]

Karl Flessa's main work in the delta documents the impacts of reduced river flows over a much longer period of time. Flessa is a paleoecologist, a scientist who reconstructs past environments and ecosystems. Sometimes that involves obscure clues, such as the remains of dead animals. He sent me a t-shirt that reads "Centro de Estudios de

Almejas Muertas," which means "Center for the Study of Dead Clams."[7] (I wear this when I travel to Latin America because it prompts double takes.) The punch line is "Putting the dead to work since 1992."

On the shores of the estuary we found literally acres of dead clams mounded in huge piles, reminding me of the folk song: "And think of my happy condition, surrounded by acres of clams."[8] We used them in any way possible, once as spoons to eat canned peaches. Flessa and colleagues estimate that there are more than *two trillion* shells stored in large beach deposits known as "cheniers." Earlier cheniers formed when sediment was diverted from the delta during times in geological history when the river spilled into the Salton Sink, discussed below. More recent cheniers formed when dams robbed the river of sediment. Just as clear, "hungry" water scours sediment from the Colorado River in Grand Canyon (chapter 6), sediment-starved river water and daily tides erode existing sediment and expose the shells of clams that lived up to 1,000 years ago. By dating shells and comparing them with live clams, Flessa's group learned that the species composition shifted dramatically since the dams were built. The Colorado delta clam (*Mulinia coloradoensis*), a species endemic to the region, once dominated with ninety-one percent of the shells. A smaller clam species, *Chione*, comprised only four percent of those in the cheniers. Now, the opposite balance predominates, with eighty-seven percent *Chione* and only four percent *Mulinia*, which is found in only a few small locations and is at risk of extinction.

But how do we know that reduced freshwater flows have anything to do with this change? Here Flessa's work is most clever, at least to this layperson. Freshwater and saltwater have different ratios of oxygen isotopes, each of which has a different atomic weight (^{16}O and ^{18}O). By drilling holes in clam shells and testing for the isotope ratio of the calcium carbonate powder, Flessa can calculate the water conditions in which the specimen grew. *Mulinia* thrived in waters with lower salinity than exist today. Predam salinity ranged from thirty-two to thirty-five percent, but has now increased to thirty-five to forty-five percent. *Mulinia* may have not tolerated the increased salinity that resulted once freshwater flows declined, or large freshwater inflows may have triggered *Mulinia* growth or spawning.

Although I know that dramatic changes in species composition are usually environmentally significant, I still asked Flessa what difference it makes that the delta clam population shifted from *Mulinia* to *Chione*. He

looked surprised at the question, but offered some possibilities in defer-
ence to my lawyer's quest for relevance compared to his scientific satis-
faction with improved understanding. (There is that culture clash
again.) *Chione* shells are much thicker than those of *Mulinia*, making
them much harder for foraging birds to pierce. The reasons for declines
in totoaba and vaquita populations are confounded by commercial fishing losses and other causes, but *Mulinia* are not collected for food.
Reduced freshwater flow is the most probable cause of that decline,
making it a better indicator than the totoaba or vaquita of how dams
changed the delta's ecology.

Probably more ecologically significant than the change in species,
however, is the dramatic drop in clam density Flessa and colleagues
documented. The shell record in cheniers suggests an estimated standing population of 6 billion clams in the delta's subtidal and intertidal
zones over the past millennium, at densities of between twenty-five
and fifty clams per square meter. Average clam density in the region is
now only three clams per square meter, up to ninety-four percent
lower than before the dams, perhaps due to the reduction in nutrients
that reach the gulf. Fewer clams probably nourish far fewer birds and
other species.[9]

This dramatic harm to the lower delta and the upper Sea of Cortéz
begins not at the U.S.-Mexico border, but more than 1,000 river miles
north at the Grand Ditch, where water is first diverted from the river. The
change continues downriver with each gallon taken for consumptive use,
each acre-foot diverted out of the basin, and with each molecule evaporated or lost in transportation or storage. The sum of this impact is defined
by a single number: 1.5. One and one-half million acre-feet (maf) in most
years is what the United States promised to Mexico in the 1944 treaty.
Most of that is diverted within Mexico for agricultural and urban uses, and
none is dedicated to environmental restoration. To restore the river in
Mexico and the delta below, we need to rethink the fundamental decisions
that produced those circumstances, decisions that were based on political
rather than ecological or hydrological boundaries.[10] In the next section, we
explore the legal and political reasons why so little Colorado River water
crosses the border in most years.

Allocating the Colorado River Across the U.S-Mexico Border

Acres of Clams is a traditional American folk song. For contrast, try those
masters of hydrological wisdom, the Grateful Dead, from their aptly
named "Mexicali Blues":

Is there anything a man don't stand to lose,

when the Devil wants to take it all away? . . .

You just might find yourself out there on horseback in the dark,

Just riding and running across those desert sands.

With respect to the lower Colorado River, who is the devil and who is left in the dark, riding and running across the desert sands? Who are the winners and who are the losers? An obvious loser is the environment along the lower river and the delta. Another is the people of Mexico, who receive about one-tenth of the water that used to flow across the border, including the Cucupa Indians (literally "People of the River"), who have used the river and its resources for nearly 1,000 years.

The most direct explanation for the dramatic reduction in the flow of the Colorado at the international boundary was the persistent competition for water in the United States beginning at the turn of the 20th century. As discussed earlier, lower basin states wanted to secure water for ongoing growth, while the upper basin wanted to preserve the right to do so into the future. This quest to store as much water as possible for use within the United States matched the prevailing Progressive Era philosophy that water allowed to reach the sea wasted a precious resource. In his December 1901 State of the Union address, President Theodore Roosevelt said: "The western half of the United States would sustain a population greater than that of our whole country today if the waters that now run to waste were saved and used for irrigation."[11] Dams and diversions built over the following six decades hoarded water north of the border rather than allowing it to "run to waste" into the Sea of Cortéz.

Sheer nationalism—and at times blatant racism—also explain the stingy approach the United States adopted toward Mexico in allocating the Colorado River. Although this story is told wonderfully elsewhere,[12] a brief history puts the environmental fate of the lower river in context.

At roughly the same time as settlers started to divert water from the lower Colorado River to California's Imperial Valley, similar development proceeded in the Mexicali Valley of Baja California. During the 1870s, Mexico issued land grants that concentrated almost all arable land in the Mexican portion of the delta in a single landowner, Guillermo Andrade. Andrade sold 800,000 acres of his vast holdings to the Colorado River

Land Company, incorporated and controlled by Harrison Otis and Harry Chandler, owners of the *Los Angeles Times*. Meanwhile, developers sought to divert water to Imperial Valley through an improved natural conduit known as the Alamo Canal, which passed through Mexico. In return for permission to do so, Mexico required that half of the diverted water be used within Mexico. Because it had the capital necessary to irrigate lands in Mexicali, however, the U.S.-owned Colorado River Land Company obtained most of those rights.[13]

Imperial Valley agricultural interests, however, did not like this arrangement. To prevent further flooding due to the unstable nature of the Alamo Canal, and to end the need to share half of its water with Mexico, California pushed for construction of the All-American Canal (so named because it traversed only U.S. soil) as part of the Boulder Canyon Project Act (the law that authorized Hoover Dam construction). Even more provocative were racist assertions by representatives of Arizona that providing water to Mexico threatened the United States with an invasion by Asian laborers hired by the Colorado River Land Company to till fields in Baja. Mexico resented U.S. dominance of the river and the infrastructure through which it flowed, and later expropriated lands held by the Colorado River Land Company and pushed for a treaty that would guarantee her some reasonable share of the river.[14]

The U.S. government initially supported its policy of hoarding as much water as possible with what one Mexican analyst described as that "ominous thesis" known as the Harmon Doctrine. Proclaimed in 1898 by Attorney General A.G. Harmon in connection with the dispute over the Rio Grande, the United States claimed absolute sovereignty over all water originating within its boundaries, and accordingly, no legal obligation to leave any for its downstream neighbor. As described by one contemporary source: "[T]he rules of international law imposed upon the United States no duty to deny to its inhabitants the use of the water of that part of the Rio Grande lying wholly in the United States, although such use resulted in reducing the volume of water in the river below the point where it ceased to be entirely within the United States, the supposition of the existence of such a duty being inconsistent with the sovereign jurisdiction of the United States over the national domain."[15]

Mexico naturally rejected the Harmon Doctrine, arguing for rights to at least enough water to support existing irrigation in Mexico. To bolster its claim, Mexico raced to irrigate more acres in Baja and Sonora, just as its northern neighbors rushed to develop as much U.S. land as possible. It was a race to scarcity. Under international law, Mexico was entitled to apportionment of the river according to principles of equity that consider a wider range of interests, including the needs of its native peoples and its environment, and that prevent one country from causing significant harm to a downstream nation. Most international treaties allocating water between two or more nations reflect the idea that an upstream country should not cause significant harm to downstream neighbors.[16]

The United States ignored Mexico's rights to Colorado River water during the Colorado River Compact negotiations. As discussed in chapter 5, the U.S. State Department rejected Mexico's request for a seat at the negotiating table. Commission chair Herbert Hoover asserted that Mexico had no legal rights to the river, but the compact included a placeholder for a later treaty allocation to Mexico, and Congress authorized negotiations with Mexico over the Colorado in 1927.[17] But by allocating a minimum of 16 maf to the seven U.S. basin states twenty years before that treaty would be negotiated, the compact effectively limited the amount of water the United States later would be willing to deliver to Mexico.

Eventually, prompted by continued pressure from Mexico and a desire to further its Good Neighbor Policy during World War II, and by linking concessions on the Colorado to allocation of the Rio Grande, the United States negotiated a water treaty with Mexico in 1944. But with only 1.5 maf guaranteed, Mexico received less than half of the 3.6 maf it initially requested. Most of this water already was being used for irrigation within Mexico, mainly via diversion at the Morelos Dam just south of the border. Except in high flow years in which more than the treaty minimum reaches the border, not much is left for the ecosystem or for the Cucupa and others who use it between the Morelos diversion and return flows from the Mexicali Valley, by which time it is badly polluted.[18]

In negotiating, signing, and ratifying the treaty with Mexico, the United States abandoned the Harmon Doctrine as an official policy or legal strategy.[19] Intentionally or not, however, and certainly not explicitly, the United States would resurrect that "ominous thesis" at least two more times. The first resurrection led to more negotiations and at least a temporary resolution of conflicts between the two nations over high levels of salinity in water the United States delivered at the border. Some of the ways in which the salinity issue has been addressed illustrate that it is possible to engage in river management programs on a watershed scale, in ways that transcend geopolitical boundaries. The second resurrection continues to this day, and impedes our ability to restore the

lower river and its delta. We address the salinity issue in the next section, and move to broader ecosystem restoration issues thereafter.

Resolving the U.S.-Mexico Dispute over Salinity

On a sixty-acre site outside of Yuma, adjacent to the Colorado River, sits
a massive desalination plant. The Bureau of Reclamation built this $260
million facility as the last defense against water too salty to meet U.S.
obligations to Mexico. For the past three decades, treaty amendments
have required the United States to deliver to Mexico water that meets
acceptable quality standards as well as minimum amounts. Saltier water
would damage or ruin the crops for which most of Mexico's treaty share
of Colorado River water is used. The salinity control program designed
to meet these requirements has its own lessons for restoration efforts
throughout the basin. But some of the methods by which the United
States meets the negotiated salinity requirements are pieces in an intricate puzzle that will affect future efforts to restore other aspects of the
lower river and delta ecosystems.

I visited the desalination plant to learn more about these relationships, and was met at the gate by a Yuma County sheriff's officer. "Post-
9/11 security?" I asked. "No," he explained. A disgruntled employee
threatened the plant before then, so they posted a guard but extended
the precaution after 9/11. It is not clear why any terrorist would view
this facility as a likely target, because it has not operated beyond pilot
capacity since it was completed in 1992. Most of the time it is idle,
although it costs U.S. taxpayers $1.5 million a year to maintain at "ready
reserve" status.[20]

A Bureau of Reclamation program manager gave me a tour of the
plant, which is an impressive work of engineering. The control room is
ringed by computers, which reminded me of the control panel for a
nuclear power plant. In the operating area are stacks of reverse osmosis
pipes about a foot in diameter, with membranes and spacers wrapped
tightly around the cylinder to increase the total surface area of the filters.
When the plant is shut down, as it was when I visited and still is several
years later, the reverse osmosis elements need to be dried and stored with
biocide. The bureau tests the equipment every year or two, but the equipment has limited design life and must be replaced periodically.

Why so much money and so much equipment for a desalination plant
that sits in "ready reserve" status? Happily enough, the answer lies in part
in the relative success of the only environmental protection or restoration
program in the Colorado River basin that can, in any sense, be called a

whole watershed program. The salinity program addresses sources of salt from the river's headwaters to the Mexican border, and is designed to solve salinity problems on both sides of the international boundary. At least some parts of the salinity program also recognize the relationship between land use and aquatic ecosystem health rather than artificially separating water and land management at the water's edge. The salinity control program is far from perfect, but those aspects of the program separate it from much of the fragmentation that characterizes other efforts to restore the Colorado River and its interrelated ecosystems.[21]

Throughout much of its geologic history, inland seas covered large portions of the Colorado River watershed. Saline marine sediments deposited over millions of years, leaving soils and subsoils with high concentrations of soluble salts. Due to natural erosion of these materials into the river and its tributaries, as well as salt leaching by subsurface flows and saline discharges from natural springs and seeps, the predevelopment river was saltier than most rivers or streams, but usually not too salty for most human uses. Salinity levels were particularly high during annual low-flow periods, when little water was available for dilution. The Colorado River historically carried salt loads of between 200 and 1,000 milligrams per liter (mg/L) (or parts per million [ppm]) of total dissolved solids, depending on flow conditions and other factors. Hohokam Indian farmers experienced problems with soil salinization, and early in the 20th century Godfrey Sykes anticipated the salinity problem that would be caused by rising groundwater tables, agriculture, and soil leaching.[22]

The U.S. Environmental Protection Agency estimated in 1971 that human development more than doubled the amount of salt in the Colorado River system. Inefficient irrigation of arid lands overlying saline formations causes excess water (water not used by crops or evaporated or transpired from the soil or plants) to seep down through saline soils and groundwater into the Colorado River or its tributaries. Land use changes in the watershed, such as grazing, road construction, and development, can increase erosion of saline soils, especially during heavy storms and flash floods. Consumptive water uses in the basin, transbasin water diversions, and evaporation from reservoirs reduce the volume of water for dilution, causing higher concentrations of salts. Smaller but more concentrated discharges of salt come from municipal and industrial wastewater discharges and abandoned or operating oil, gas, and mining wells.[23]

High salinity in Colorado River water can harm agricultural, municipal, and industrial water users on both sides of the international border.

Although crops have different salt tolerances, saline water reduces yields of most major crops irrigated with Colorado River water. Over time, soils irrigated with saline water can concentrate salts until irrigated agriculture is no longer possible. This can force farmers to remove land from production, shift to lower value but more salt-tolerant crops, or spend money for tile drains or other controls. Highly saline water used for municipal, commercial, and industrial purposes damages household appliances and car radiators, clothing and textiles, and water and waste-water pipes and facilities. A 1988 study estimated annual economic damages due to salinity in Colorado River water at between $311 million and $831 million per year (in 1986 dollars), not including economic damages to farmers in Mexico. The Bureau of Reclamation assumes that economic damages due to Colorado River salinity are approximately $750 million per year in the United States, and could exceed $1.5 billion per year if not controlled.[24]

Until 1961, the salinity of water delivered to Mexico was about the same as that received by U.S. users of water from Imperial Dam. This changed abruptly when the United States began to dump return flows from the Wellton-Mohawk Irrigation and Drainage District (WMIDD) in southern Arizona into the river south of Yuma. The salinity of water delivered to the Morelos Dam ranged between 1,340 and 2,500 ppm, due to the discharge of extremely saline irrigation return flows (roughly 6,000 ppm) from WMIDD. Farmers in the WMIDD had been flooding saline soils repeatedly to grow crops. Irrigation water grew saltier with each repumping until crop damage ensued. Congress authorized the Bureau of Reclamation to bail out WMIDD farmers by drilling deep wells to pump briny groundwater, which the bureau then piped to the Colorado River upstream of Mexico's diversion. Water law expert David Getches referred to this "water" as a "crop-killing brew," and environmental historian Evan Ward charged that it "immediately touched off an ecological crisis."[25]

To add insult to injury, the United States claimed that this wastewater counted toward its annual treaty delivery to Mexico. It argued that nothing in the 1944 treaty specified minimum water quality standards, and that the treaty allowed the United States to use water from "any and all sources," including irrigation return flows. The bureau also argued that Mexico could address its own salinity problems by installing drains and adopting other control techniques, as were U.S. farmers. At the time, Glen Canyon Dam was nearly finished and ready to fill. If the United States could meet part of its delivery obligations with polluted groundwater from Arizona, more water could be used to fill the massive new bathtub upstream. Mexico understandably disagreed, filed formal protests, and threatened legal action in the World Court. Although the treaty did not specify numeric water quality requirements, Mexico claimed that both parties understood that the water was to be used for irrigation and urban uses, and must be fit for those purposes.

The salinity debacle could be characterized as the "first resurrection" of the Harmon Doctrine. With respect to water allocation sixty years earlier, Attorney General Harmon believed that a country retains complete control over all water and other resources originating within its borders. Therefore, he reasoned, it bears no obligation to share those resources with another nation. This position artificially separated the water from the river. It is one thing to say we need not share our natural resources with another country. It is another to claim that we can cause significant harm to a downstream neighbor, its people, one of its critical natural resources, and its environment, by dewatering a transboundary river. The situation in 1961 was conceptually no different in light of what we knew at the time. By withholding water of acceptable quality and replacing it with saline brines pumped from an aquifer polluted by U.S. farmers, we injured the very resource we had agreed to protect in the 1944 treaty. Both manifestations of the Harmon Doctrine assumed that the United States bore no responsibility for the effects of its actions across the border. Both exalted the artificial boundary between the two nations over the integrity of a river that knew no such boundary.

Over the ensuing decade and a half, the Kennedy and Nixon administrations again rejected this doctrine by which the United States could export environmental harm to Mexico, and agreed to two amendments to the U.S. treaty obligation, although neither president acted out of complete altruism. Just as the Roosevelt administration promoted its Good Neighbor Policy for building alliances during World War II, U.S. foreign policy during the Cold War was premised on maintaining good relations with our neighbors. Beginning in 1974, the United States agreed to deliver at least 1.36 maf of water with an average annual salinity not significantly higher than that provided to U.S. farmers in the Imperial Valley. To meet those treaty obligations, and to provide better quality water to agricultural and urban users domestically, the United States adopted the Colorado River Basin Salinity Control Act.[26]

Salinity in the Colorado River increases gradually as it flows downstream, and the resulting harm affects users on both sides of the border. The problem is basinwide in both origins and impacts; it does not respect artificial political boundaries. In a remarkable example of interstate cooperation with respect to a river so fraught with conflict, in 1972 the states agreed to adopt, and the U.S. Environmental Protection Agency approved, binding water quality standards to guide salinity control throughout the basin: 723 ppm below Hoover Dam; 747 ppm below Parker Dam; and 879 ppm at Imperial Dam. To meet these standards, the salinity program embraces a basinwide approach, although only for this single issue. The full range of parties responsible for salinity control is involved in a single program, including at least six federal agencies and all seven basin states under the auspices of the Colorado River Basin Salinity Control Forum. The program sets basinwide targets for reducing salt inputs into the river, considers all potential sources and solutions, and seeks a combination of solutions within the basin to reach the established targets.[27]

Initially, the United States simply shunted saline brines from WMIDD through a drainage channel extension that allowed wastewater to be discharged either above or below Morales Dam, at Mexico's option, and to make up the balance of Mexico's treaty water from other sources (primarily higher quality groundwater). These measures reduced salinity somewhat, but did nothing to actually remove salt from the system, and did not satisfy Mexico's claim to water of approximately equal quality to that diverted at Imperial Dam. To achieve the latter goal, Congress authorized massive federal expenditures similar to those used to build dams and other water projects. Congress authorized the Bureau of Reclamation to build the desalination plant I visited in Yuma, as well as diversion structures and a large bypass drain, called the Main Outlet Drain Extension (MODE) Canal, to carry reject waters from the desalination plant to the Santa Clara Slough in Mexico. Congress also paid to line portions of the Coachella Canal, which diverts water from the All-American Canal to farms in California's Coachella Valley. Canal lining saves water that used to seep into the ground. The United States used that water, along with water from new groundwater wells, to supply replacement water to Mexico. This was business as usual for the bureau. Big money to build big projects to solve big problems.

But in the Salinity Control Act, Congress also adopted other methods advocated by environmentalists as a preferred solution to the problem,[28] based on a rethinking of how and where it is appropriate to farm in the desert Southwest. The act included programs to reduce salinity in irrigation drainage by improving water use efficiency and by taking certain lands out of production, first in the WMIDD and later throughout the basin. This was not business as usual for the bureau, whose reputation was built on expanding water supplies to irrigate as many acres as possible. Now, it would to try to do more with less water, and to reduce or eliminate irrigation of lands with highly saline soils. As discussed in more detail in chapter 9, this is an example of how environmental restoration can involve a reconsideration of economic strategies as well as direct, physical approaches.

The new salinity program improved irrigation efficiency through subsidies in which the federal government paid part of the costs for farmers to line ditches and canals, change field layouts and level fields to reduce runoff, and install drip irrigation and automatic sprinkler systems to replace inefficient surface flooding of fields. If less excess water is applied, less runs off into surface waters along with salt and other contaminants, and less percolates through the saline soil layers below.

Addressing salinity through improved irrigation efficiency was a win-win solution that maintained the partnerships among the bureau, the U.S. Department of Agriculture, and farmers, who received modern irrigation equipment and other improvements. In addition to improving water quality in the river, irrigation efficiency helps farmers by protecting their crops from increasingly saline soils and water supplies. WMIDD irrigation efficiency improved from fifty-six percent (meaning that fifty-six percent of water applied is used by crops, and forty-four percent is lost to runoff, infiltration, or evaporation) to an estimated peak of seventy-seven percent in 1985. However, although the bureau reports that all "permanent" measures implemented by WMIDD are still in use, irrigation efficiency dropped back to as low as fifty-nine percent in some years. WMIDD officials believe that some of this drop in efficiency relates to changes in crops rather than less efficient irrigation.[29]

The effort to end irrigation on highly saline soils, on the other hand, was not well received and reflected even more of a shift from past reclamation philosophy. It did, however, make a lot of sense. Irrigation with federal reclamation water is already heavily subsidized, and many have argued persuasively elsewhere that water shortages in the West, in the Colorado River basin and elsewhere, could be resolved by requiring end users to pay market prices for their water. It makes even less sense to provide federal subsidies to farm poor, saline soils, and then to add more subsidies to pay farmers to fix the resulting environmental harm.

Perhaps the most important way in which the salinity control program evolved from "business as usual" to innovation was its transformation from a public works to a market-based effort. The original program funded salinity control projects identified by the bureau. Although these efforts succeeded in reducing salinity, they were expensive and inflexible because they required congressional approval of every specific project. Several studies concluded that this system resulted in missed opportunities for more cost-effective salinity control. In 1995, Congress amended the statute to provide for an open, competitive program under which any party—public, private, or mixed—may bid for salinity control funding. The bureau and the Salinity Control Forum issue annual requests for proposals and select the most cost-effective salinity control projects (measured in dollars per ton of salt reduced) for funding each year. So far, this competitive process has resulted in dramatic improvements in the cost-effectiveness of salinity control. The bureau initially expected the average cost-effectiveness of controls to be fifty dollars per ton of salt removed. Four years into the new program, selected projects averaged twenty-six dollars per ton, less than half of the average cost-effectiveness of projects under the old program. Project bids remained low for several years, and although they rose to forty-six dollars a ton in 2005, they remain below those in the prior program.[30]

The Upper Colorado River Endangered Species Recovery Program (chapter 5) allows a participant to pay a small fee to essentially buy out of its responsibility to restore critical habitat or to prevent harm to threatened or endangered species. As a reward, the participant then is authorized to withdraw heavily subsidized federal reclamation water without regard to environmental impacts. This transfers to the federal and state governments the responsibility to implement species protection efforts and to bear the risk that those efforts will fail. In the competitive salinity program, by contrast, participants receive federal subsidies only if they design and implement effective salinity controls. Each proposal must identify the salinity control methods to be used; a project management plan and schedule (how, when, and by whom); annual projected salt load reductions supported by documented methods; the expected project life; the proposed costs, payment method, and schedule; an evaluation of environmental impacts and proposed mitigation measures; and a risk analysis characterizing the likelihood of project success. Rather than allowing someone to pay a fixed amount to avoid their environmental responsibility, salinity program dollars are used to pay people to *assume* responsibility for the problem.

It would not be so easy to apply a competitive bidding approach to endangered species protection and restoration. Although a ton of salt is a ton of salt, restored habitat varies greatly in quality and other attributes, and restored habitat might be far more valuable in one area than in another. With a little creativity, however, competitive bidding could improve some aspects of restoration efforts along the river. In the upper river, efforts to restore backwater habitats are limited by available public land. Private landowners could bid for program funds to purchase conservation easements. With a viable way to account for the relative value of different habitat locations and qualities, more habitat might be protected through a market process than with public spending. In the lower river multi-species program, available program dollars could be used for a competitive process in which participants bid for the most cost-effective projects to restore flycatcher habitat, measured in dollars per acre of successfully restored land. Virtually everywhere along the river, restoration efforts would profit from keeping more water in the river for a longer distance. Federal program dollars could be used to "buy back" water rights with a competitive bidding approach. Those users who profit the least from existing water allocations would submit the lowest bids. Those water purchases would involve the least economically beneficial uses, and maximize the amount of water returned to the environment.

Depending on where you sit, the Colorado River Basin Salinity Control Program could be viewed as a major success or just an innovative way to continue to feed farmers from the public trough now that the era of big dams is over. Federal dollars are clearly being used more cost-effectively than in the past, to maintain salinity levels in the river below target levels, and to help meet treaty obligations to Mexico. Unlike other Colorado River restoration programs, this one addresses salinity on a whole watershed basis. To purists who believe in the "polluter pays" principle, however, the bureau now subsidizes farmers to correct environmental problems that it subsidized them to create in the first place.

Either way, existing salinity controls have only deferred the day of reckoning. As the upper basin uses more of its compact share, less water will be available to dilute salinity to acceptable levels. When that happens, the bureau likely will have to operate the desalination plant at Yuma. Hypersaline water now diverted from the river through the MODE Canal will be run through the plant, and treated water will be discharged into the main river channel upstream from Morelos Dam. That will trigger another suite of difficult restoration decisions that cannot be answered on the basis of political rather than ecological and hydrological boundaries. It will also raise in another context the question of what we seek to restore, and to what ends. Next, we look at the relationship between water use and salinity control in the United States, and ecosystem restoration in Mexico.

Water, Salinity, and Ecosystem Restoration in Mexico

Backpacker and writer Colin Fletcher quoted a curious definition of "river" from the *Oxford English Dictionary* as "the boundary between life and death."[31] This definition is surely true for the Colorado River, where shifting boundaries between wet and dry define where life is rich and where it is sparse.

Although the mudflats I passed on my way to the delta are now dry and bare, robbed of spring flood flows that once reseeded and replenished them, on the east side of the delta lies a vast wetland complex that is much wetter than it was before we took steps to mitigate the salinity crisis. In 1977, the United States began to divert about 130,000 acre-feet a year of saline agricultural drainage water into the MODE Canal as part of its temporary solution to the salinity problem. This water drains into the Cienega de Santa Clara and two smaller adjacent wetlands (El Doctor and El Indio). The El Doctor and El Indio wetlands occur naturally due to artesian springs, but have grown due to agricultural drainage water from farms in Mexico. The much larger wetland in the Cienega is typically described as artificial because most of its water now comes from agricultural drainage. But the reality is somewhat more complex.

The Cienega de Santa Clara was at times part of an active arm of the Colorado River before the river shifted westward early in the 20th century, leaving a largely dry depression where the river once flowed. Before we tamed the river, its boundaries were elusive. Because flows varied so much from year to year and season to season, and because of the region's flat terrain, the river reached the gulf (or as explained in chapter 7, in some years the Salton Sink) through an ever-shifting set of channels portrayed by Aldo Leopold: "On the map the Delta was bisected by the river, but in fact the river was nowhere and everywhere, for he could not decide which of a hundred green lagoons offered the most pleasant and least speedy path to the Gulf. So he traveled them all, and so did we. He divided and rejoined, he twisted and turned, he meandered in awesome jungles, he all but ran in circles, he dallied with lovely groves, he got lost and was glad of it, and so were we." Godfrey Sykes described how the river broke from one channel to another, some of which connected to the Santa Clara Slough and the gulf through a more easterly route. Although modern analysts assert that the river simply shifted westward to its current main channel, according to Sykes water diversions to the Imperial Valley helped the process. Eastward movement of the river jeopardized water diversions into the Imperial Canal. To prevent this unwanted shift in flows, and in part to control flooding, the Volcano Lake Levee was built in 1908, aided by a $1 million U.S. government appropriation, even though the levee was located in Mexico. At least in part, then, the river's westward shift was part of a deliberate effort to divert water from the delta to the Imperial Valley.[32]

Water taken from the Santa Clara depression with one hand, however, was returned with another more than a half century later. Before waters were diverted through the MODE Canal, the wetland covered about 490 acres, fed by artesian springs and agricultural drainage from Mexican fields. MODE Canal water expanded the wetlands to up to 50,000 acres. The sudden influx of water transformed the Santa Clara depression into the largest cattail marsh in the Sonoran Desert, densely vegetated with cattails as well as common reed, bulrush, and other species. Most of the Cienega de Santa Clara suffers from the pollution expected in a wetland whose water source is largely agricultural return flows. Selenium levels in MODE Canal water are two and a half times higher than in water diverted from Imperial Dam, and elevated selenium has been measured in sediments, fish, and plants taken from the Cienega wetlands. Nevertheless, the Cienega supports a remarkable amount of wildlife, including more than 6,000 endangered Yuma clapper rails, the largest remaining population in the world. The Cienega also provides habitat for endangered desert pupfish and thousands of resident and migratory waterfowl, and constitutes an important feeding area within the Pacific flyway. So although the wetlands created by MODE Canal drainage are in a sense artificial, they lie in an area through which parts of the river flowed naturally during some time periods. Whether we label these wetlands "natural" or "artificial," they support large amounts of wildlife in a region otherwise depleted of those resources.[33]

What will happen to this newfound ecological bounty once the United States needs the water now diverted through the MODE Canal to meet its treaty obligation to deliver 1.5 maf of acceptable quality water a year to Mexico? The Bureau of Reclamation will begin to operate the Yuma desalination plant, and most of the treated water will flow into the main channel of the Colorado. If this drainage water is not replaced by another source, the vast wetlands complex in the Cienega de Santa Clara will shrink, with significant consequences for the plants and wildlife that rely on it. As the scale of restoration efforts grows, so does the potential that a solution to one problem will create or exacerbate problems elsewhere.

This also raises another difficult set of legal and political issues regarding the extent to which the United States is responsible for the consequences of its actions on the environments of other countries. There is little doubt that the Bureau of Reclamation will have to prepare an environmental impact statement to begin full operation of the Yuma desalination plant. This decision is a "major federal action" that will "significantly affect" the human environment. Likewise, if all impacts of operating the plant occurred in the United States, this decision would trigger consultation under section 7 of the Endangered Species Act (ESA), because it might jeopardize the continued existence of threatened or endangered species and destroy or adversely modify critical habitat for those species. But to what extent must impacts in Mexico be addressed under those U.S. laws?

From an ecological perspective, this issue seems trivial. The ecosystem does not end at the U.S-Mexico border, and the environmental impacts of actions upstream do not disappear at that imaginary boundary. Legally, however, it is not clear whether or how these two U.S. environmental laws apply to this problem.

Federal courts have reached inconsistent conclusions on whether NEPA applies to extraterritorial actions, or to actions within the United States that may significantly affect the environment of other countries, and to date, the U.S. Supreme Court has not chosen to resolve that dispute. Parts of the law limit goals to protecting environmental quality for U.S. citizens. One statutory goal is to "fulfill the social, economic, and other requirements of present and future generations of Americans." Later in the statute Congress asserted that federal agencies are responsible to "assure for all Americans safe, healthful, productive, and esthetically and culturally pleasing surroundings." In the provision that requires environmental impact statements, however, Congress chose broader terms. It required impact statements for "actions significantly affecting the quality of the *human environment*," and repeatedly required agencies to evaluate "any adverse environmental impacts" and impacts on "man's environment." Congress admonished agencies to "recognize the worldwide and long-range character of environmental problems." None of this language seems to limit an EIS to actions and impacts within the United States.[34]

Similarly, federal courts have not resolved whether the ESA applies to extraterritorial impacts of federal actions on threatened and endangered species—either species listed in the United States that also exist in other countries (such as the Yuma clapper rail and desert pupfish), or species listed by other nations but that may be harmed by actions in the United States (including the totoaba and the vaquita porpoise in the Sea of Cortéz). In one case, the environmental group Defenders of Wildlife challenged the U.S. Fish and Wildlife Service (FWS) regulations that expressly excluded extraterritorial actions from the section 7 consultation obligations. Defenders cited as examples U.S. agency funding decisions for dams and other projects that would threaten endangered species such as elephants in Sri Lanka and crocodiles in Egypt. The U.S.

Court of Appeals for the District of Columbia Circuit agreed with Defenders of Wildlife, ruling that the ESA does apply to U.S. agency actions with such international impacts on threatened or endangered species. The U.S. Supreme Court reversed this ruling, but on "other grounds." In a heavily split decision, the court held that the plaintiff lacked legal standing to challenge these regulations because it could not show direct and immediate impacts on its members from the challenged actions, and because the courts could not fully redress plaintiffs' injuries because those projects might be built even without U.S. funds.[35] Extraterritorial applicability of the ESA was left for another day.

Defenders of Wildlife, along with other U.S. and Mexican environmental groups, raised this issue anew in the context of the lower Colorado River. In this case, the groups challenged the Bureau of Reclamation's and the FWS's failure to consult fully regarding the effects of lower river operations on the totoaba, the vaquita, the desert pupfish, the Yuma clapper rail, and the southwestern willow flycatcher. A federal district court judge rejected these claims, but not because he found the ESA inapplicable to extraterritorial impacts. In fact, although the bureau wanted to limit its consultation responsibility to the lower river north of the border, the FWS specifically directed the bureau to address impacts to species in Mexico as well.

The bureau argued, however, and the court agreed, that Reclamation's consultation duty was constrained for the same reasons as the Lake Mead flycatcher habitat case discussed in chapter 7. Under FWS regulations, agencies must consult only on "discretionary" actions. Here, the only way to prevent additional harm to threatened and endangered species would be to send more water across the border. This, the bureau argued, is outside of its discretion because the Law of the River prevents it from reallocating any more water to Mexico than required by the 1944 treaty and later developments, including the Supreme Court's decision in *Arizona v. California*. Judge Robinson put it succinctly: "[I]t seems unlikely that any case will present facts that more clearly make any agency's actions nondiscretionary than this one: a Supreme Court injunction, an international treaty, federal statutes, and contracts between the government and water users that account for every acre foot of lower Colorado River water."[36]

This ruling brings us full circle to where we started: 1.5 maf per year. And it underscores the conflict between stability and change discussed in the opening chapter. More than a half century ago the United States decided to limit its legal obligations to Mexico to save as much water as possible for use in the United States. At the time, little was known about endangered species. There was no NEPA, no ESA, and no Defenders of Wildlife. But water contracts, investments, and other decisions were made in reliance on that agreement, affecting some of the largest agricultural operations anywhere in the world. Those investments and decisions led to an environmental restoration dilemma of equal proportions.

About 200 miles northwest of the Cienega de Santa Clara is the Salton Sea, another ecosystem whose natural or artificial status is subject to some dispute. The Salton Sea as we know it today was "created" when a hastily constructed and unprotected diversion to the Alamo Canal overflowed in 1905, shunting the entire flow of the Colorado River for a period of two years into the Salton Sink, a closed basin north of the rapidly developing Imperial Valley farming communities. In addition to the flooding and property damage described in chapter 7, the legacy of this incident is a large, saline inland sea, which by the 1950s became a destination for boaters, campers, and tourists. Many scientists also consider the sea to be a crucial habitat in the Pacific flyway in a state that has lost the vast majority of its wetlands. The sea supports almost 400 species of birds and large populations of fish. These include endangered species, but also exotic species such as African tilapia (*Tilapia mossambica*). Irrigators introduced this fish to the region to control weeds in irrigation ditches, but some fish escaped accidentally into the Salton Sea and proliferated until it dominates the fishery today. Tilapia provide food for hundreds of thousands of birds, and support a major fishery. The Sonny Bono Salton Sea National Wildlife Refuge sits at the southern end of the sea.[37]

Like the Cienega de Santa Clara, the status of the Salton Sea as a "natural" ecosystem is a matter of opinion. Although human actions caused its most recent rejuvenation, the Colorado River naturally diverted into the Salton basin numerous times over recent geologic history, resulting in a periodic inland lake (Lake Cahuilla) that gradually dried up as the river shifted back to its channel to the Sea of Cortéz. Scientists dispute the frequency of those events. Geologic evidence appears to show that the river diverted into the Salton basin several times in the past 1,000 years, and that Lake Cahuilla occupied the basin as recently as 300 years ago. The river flowed into the basin at least eight times between 1824 and 1904, but evaporated shortly thereafter. Some claim that the basin contained water more often than not over time, but others scoff at this idea as unsupported by available evidence, and believe it is propaganda designed to support Salton Sea restoration efforts. The Salton Sink was part of the Colorado River watershed just like the Santa Clara depression, but both went through periods of birth and death as the river changed course over time. Anyone who wants to restore what is *really* natural would advocate that we tear down the levees, restore full springtime flows, and allow the Colorado to range at will across its natural floodplains, shifting its path through an elusive and indefinable boundary.

Also like the Cienega de Santa Clara, the ecosystem that evolved in the Salton Sea because of human water diversions depends on continued artificial manipulation to sustain its health. The sea is now maintained by irrigation return flows from the Imperial and Coachella valleys, as well as agricultural drainage and industrial and municipal wastewater from Mexico. The City of Mexicali discharges wastewater into the Alamo and New rivers in Mexico, which flow northward into the Salton Sea. Inflow to the Salton Sea is already salty, and evaporation increases salinity levels even more. As a result, the sea is twenty-five to thirty percent saltier than the ocean, a trend that will continue even if existing inflow levels persist. Deteriorating conditions have caused massive fish kills and bird kills, including the deaths of 150,000 eared grebes in 1992 (about three to five percent of the North American population), nearly 8 million fish in 1999, and 1,400 endangered brown pelicans. Suspected causes include diseases and low oxygen levels and toxins in the water due to algal blooms.[38]

Unless remedial measures are taken, this situation will worsen because of the complex set of agreements negotiated to bring California within its 4.4 maf compact allocation. As described in chapter 7, in one strategy to achieve those reductions Southern California cities will pay farmers to conserve water, some of which will be conveyed to the growing cities. This kind of market-based water transfer has worked in the past in California, and serves several desirable goals. Although improving overall water efficiency, trading shifts scarce water to growing urban areas where it is needed. However, this program will reduce inflow into the Salton Sea from about 1.3 maf to perhaps as low as 800,000 acre-feet. That will cause the sea to shrink and salinity to double within twenty to twenty five years, beyond levels at which fish can survive.[39]

Unlike the Cienega de Santa Clara and the Colorado River delta, however, there appears to be no doubt that extensive efforts will be taken to restore and preserve the Salton Sea. The federal and state governments are planning major restoration efforts, at considerable expense. Congress passed the Salton Sea Restoration Act in 1998, directing the secretary of interior to evaluate options that will serve several goals for the sea: continued use as a reservoir for irrigation drainage; reduced and stabilized salinity levels; stabilized surface elevations; long-term, healthy fish and wildlife populations and habitats; and enhanced potential for recreation and economic development.[40]

These goals are curious from the perspective of environmental restoration. Even if the sea is "natural" because it existed periodically throughout recent geological history, it was fed by overflows from the Colorado River, not by agricultural, industrial, and municipal wastewater. The sea was anything but "stabilized" in either elevation or salinity. It came and went in cycles, perhaps supporting wildlife for periods of time, but later suffering a slow death until the river arrived once more. Any fish and wildlife it supported were native, not an aquaculture of fish imported from Africa. On the other hand, before human development birds and other wildlife populations that relied on a receding Lake Cahuilla could move to other habitats along the Pacific flyway that have now been destroyed. The best argument to protect and maintain the Salton Sea is that so little remains of similar habitats in the region. But it is difficult to compare this program to efforts to restore the natural hydrology and ecology of the Colorado River.

The Salton Sea Authority studied several options to meet the goals in the Salton Sea Restoration Act. Some of those options have implications for the delta, and all ring of the artificial engineering suggested by the statutory goals. Engineers tested an array of desalination techniques, ranging from evaporation ponds to bizarre-looking devices reminiscent of snowmaking machines to spray seawater into the air to evaporate water and separate out the salt. A second set of proposals involves pumping saline water away from the sea. One possibility is to discharge that water to the delta or the Sea of Cortéz. Although that would increase flows to the lower river's wetlands and other areas, it would carry high levels of salinity and other pollutants. The third set of options would produce the most artificial habitat of all. Managers would build a huge set of dikes from the east shore to the west, dividing the sea into two parts. Highly saline water would be pumped from one part of the sea to the other, leaving a relatively low salinity arm to support fish, wildlife, and recreation, and a briny dead zone as an effective waste receptacle for the rest. (Ironically, this is what we have done *unintentionally* to Great Salt Lake in Utah by building a railroad causeway across the lake, with serious adverse effects to the ecosystem.) Salton Sea diking would cost an estimated $730 million to build (more than the entire estimated costs of the 50-year multi-species conservation program for the lower river) and another $10 million a year to operate. Other options, however, ranged in cost from $1 to $4 billion for various desalination methods, to *tens of billions* of dollars to pump Salton Sea water elsewhere.[41]

Spending so much time, effort, and money to "restore" the Salton Sea can be justified to support replacement habitat for all that has been lost throughout the rest of the region. And despite the almost zoolike qualities of the proposed result, none of the programs in the Colorado River watershed will restore the environment to an entirely unimpaired state, free from any anthropogenic influences. Still, it seems difficult to explain why the Salton Sea program moves forward steadily although no similar effort is being made to restore the entire lower river and delta region in Mexico. One explanation is that the various restoration "programs" are fragmented into so many parts (upper river, Grand Canyon, lower river, Salton Sea), which gives us no single forum in which to compare the ecological merits of money spent and water diverted to one part of the basin versus another. Decisions are made according to political rather than natural boundaries. For many years, analysts have urged that a single entity be formed to address water and environmental issues throughout the entire Colorado River basin, but to no avail.[42]

A more likely explanation is that Salton Sea restoration would keep water in the United States. The United States can hide behind the 1944 treaty and the limiting ESA regulations to justify its refusal to consider the impacts of its actions on species and other resources across the border. In reality, it is continuing to cling to every last drop of Colorado River water it can, without regard to impacts in Mexico. It is fair to characterize this position as the second resurrection of the Harmon Doctrine. Even though we now know that limiting water deliveries at the border jeopardizes endangered species and other components of downstream ecosystems, the United States insists that it bears no legal or ethical obligation to correct the problem. Just as the United States gradually abandoned the original Harmon Doctrine in signing the 1944 treaty, and just as it abandoned the first resurrection of the doctrine when it resolved the salinity crisis of the 1960s and early 1970s, it is time to reject this new iteration of the doctrine. The United States should rethink the water treaty with Mexico and devote more water to restoring essential components of what is an internationally shared ecosystem.

Restoring Water Flows and Colorado River Ecosystems in Mexico

On a gray, cloudy spring day, I watched a media stunt in front of the federal building in Salt Lake City. This was the kick-off event for the "Sustainable Water Project Tour," billed as an "1,800-mile restoration journey for dried-up rivers in the Southwest."[43] The organizers later wound down the Colorado River watershed through New Mexico, Arizona, Nevada, and California, seeking "water conservation donations" for the tour's then-empty tanker truck, the contents of which would be ceremonially emptied into the Colorado River at the Mexican border. A sign on the truck read "Revive the Colorado: An Ecosystem, Not a Plumbing System." All along the river the groups presented letters asking water agencies to voluntarily give back one percent of their Colorado River water allocations to help replenish the delta. Hence the campaign slogan: "one percent for the Delta."

I arrived early and watched a group of about a dozen grow to maybe three dozen people, with an interesting mix of young neohippies and silver-haired, conservatively dressed supporters. With two official government types and two security guards standing back by the building, and a half dozen bicycle police circling, at first the security detail almost outnumbered the demonstrators. The effort, however, was endorsed by 120 organizations, which campaign leaders said represent 8 million people collectively.

What the event lacked in numbers was offset by enthusiasm. Several people danced around with a large red Chinese dragon which, we were told, was the "River Dragon, which came all the way from China to help save the delta." Later someone arrived in a cow costume bearing a sign: "#1 water polluter in the U.S." Other signs read: "Re-Water the Delta," "Save the Grand Canyon," "Stop the Waste," "Let it Flow," "Protect Rainbow Bridge," "Free the River, Free the Soul," "Save the Vaquita Porpoise," "Living Rivers, Not Dead Fish," "WWW.Drainit.org," "Rivers for Life," "Restoration, Not Reclamation," "Subsidized Cattle Kill Rivers!," and my personal favorite, "Hug a Humpback Chub." (It may be difficult to wrap your arms around that idea.) After a series of speeches, the rally ended with activist John Weisheit pouring the first water contribution from an orange bucket into a hole in the top of the tank truck, to a cheering crowd and chants of "one percent for the Delta" and "one percent now."

Media stunt or not, the proposal to dedicate more water to restore the lower river and delta in Mexico is quite serious. It is supported by several respected scientists, and legal experts have identified various ways to transfer more water to the delta via water markets and other nonregulatory means.[44] The idea developed as researchers tracked natural wetland restoration along the lower river following the flood flows of the 1980s and early 1990s. When runoff from heavy winter snows reached Lake Powell, Lake Mead, and other reservoirs that were already near capacity, the Bureau of Reclamation had no choice but to send more water downstream than required by the treaty, for the first time since it was signed in 1944. Those "accidental" flows restored approximately 150,000 acres of the river's natural floodplain and associated wetlands. Although nowhere near the extent of the original acreage of nearly 2 million acres, the positive effects were significant. Periodic inundation is needed, though, to maintain those renewed marshes and riparian habitats.

Scientists and organizations promoting water for the delta are optimistic because relatively little water was needed to restore significant portions of an ecosystem that had previously been proclaimed "dead." By studying satellite images of vegetation and other evidence before, during, and after periods of increased flows, and by estimating vegetation density and calculating plant needs and evapotranspiration rates, researchers determined that flows of between 3,500 and 7,000 cubic feet per second suffice to inundate the floodplain. Vegetation in the area uses an estimated 304,000 acre-feet of water a year, some of which is provided by groundwater. Based on this analysis, the scientists estimate that periodic flood flows of about 260,000 acre-feet are sufficient to inundate the delta between existing levees. Those flows could occur without losses to other users during prolonged wet periods when upstream reservoirs are full. In drier periods, water would have to be reallocated from other sources. Flows of just 32,000 acre-feet would suffice to maintain habitat on an annual basis—a trivial amount compared to the millions of acre-feet that flowed down river even in the heaviest predam droughts. Annually, that represents about one percent of the average flow of the Colorado River.[45]

The strategic opportunity presented by this restoration proposal is underscored by comparing habitat values in the delta to those remaining in the United States. Scientists believe that a restored delta can support at least 117,000 resident and migratory birds, more than twice as many as in the entire lower river corridor in the United States. Even at 150,000 out of 2 million predam acres, the revegetated floodplain in Mexico is almost twice as large as in the United States.[46] Only the artificial geopolitical boundary appears to stand in the way of using such a small percentage of water to restore so much high quality habitat.

The one percent proposal came to a head during negotiations within the United States over how to divide any surplus waters in the basin. Recall that the Colorado River Compact allocates 7.5 maf each to the upper and lower basins, but then authorizes states in the lower basin to use an additional 1 maf of "surplus" water when it is available. This begs the question of *when* there is a surplus under the somewhat complex set of competing factors contained in the compact, the Supreme Court decree, and the 1956 and 1968 water project statutes. In debating these "surplus guidelines," upper basin states preferred more surplus water to be stored in Lake Powell as a hedge against drought, to allow them to meet their annual delivery obligations even during low runoff years. Lower basin states urged a lighter trigger to send more water downstream from Lake Mead for use as surplus. Because the 1968 statute requires the Department of Interior to maintain Lake Powell and Lake Mead at approximately equal levels, releasing more water from Mead also requires larger releases from Powell. The Interior Department viewed this dispute as leverage to prompt California to develop a plan to stay within its allocation under the Supreme Court decree when surpluses were not available (the so-called "4.4 Plan"). Ultimately, Interior issued "Interim Surplus Guidelines" that tied release decisions to reservoir levels in Lake Mead (the numeric details of which are of interest only to masochistic Law of the River junkies), conditioned on California's adoption of the 4.4 Plan.

Amidst this new round of poker over how to divide the river within the United States, Defenders of Wildlife and other environmental groups threw in a wild card. Having lost their argument that the lower river and delta in Mexico should be considered as part of the multispecies conservation program (chapter 7), they now argued that if surplus water was available and needed to protect and restore endangered species and their habitats across the border, at least some "surplus" water should be dedicated to those goals. However, the Interior Department simply dealt them out of the hand and declined to consider that possibility, arguing once again that it lacked any discretion to dedicate more water to Mexico than required by the 1944 treaty.

Ironically, shortly after environmental groups launched the "one percent for the delta" campaign, the Colorado River basin moved from a period of excess to one of severe drought. As discussed earlier, the first five years of the new millennium saw the lowest consecutive flows on record. Just a few years after Interior and the basin states developed the Interim Surplus Guidelines, they were negotiating instead about drought management guidelines for the basin. The upper and lower basins again fought over when water should be released from the dams, but this time the focus was on Lake Powell rather than Lake Mead. The upper basin argued that the secretary of interior should not release more than the bare minimum amounts of water necessary to meet compact and treaty obligations during periods of drought, to ensure that Lake Powell does not drop too low. The lower basin states want surplus flows to continue as long as possible, arguing that enough water is stored in Lake Powell to accommodate periodic droughts, and that wetter years are certain to follow. Either way, efforts to supply dedicated water to the delta will be more difficult, but also more critical, during prolonged droughts.

Even before the drought, none of the water agencies along the river took the requested one percent donations seriously, and certainly none complied. Among other objections, water users questioned how they could be assured that Mexico would actually use additional water to restore the river and the delta, rather than diverting even more for agricultural and urban uses. Mexico has demonstrated some commitment to the ecology of the delta by designating the region an International Biosphere Reserve, but that step might be largely symbolic unless it was actually willing to dedicate water to restoration. Mexico has limited capacity to use more river water for agriculture, because Morelos Dam is merely a diversion dam through which water is shunted into canals to the Mexicali Valley and other destinations in Baja and Sonora. Morelos Dam has only 20,000 acre-feet of surplus capacity. So if more water reaches the border, from a purely practical perspective Mexico might have no choice but to let it flow downstream, as it did in the 1980s and 1990s.[47] But it also might use increased flows to recharge depleted groundwater supplies, a strategy that would require no additional surface storage capacity. The United States could condition additional flows on their use for restoration. If Mexico diverts more than treaty water for nonrestoration uses, the spigot remains entirely within U.S. control. But such a heavy-handed, paternalistic approach likely would offend political sensibilities within Mexico, especially with relations already strained over U.S. economic and immigration policies and other bilateral issues. A more cooperative approach to binational management of the river would probably be more effective in ensuring that water released in the United States would be used for environmental restoration in Mexico.

A more important reason, however, explains why the basin states and others north of the border are reluctant to concede even the relatively small amounts of water needed to restore wetlands and riparian areas in Mexico. Once again it relates to the tension between certainty and change. Although scientists can estimate the amount of water needed to restore wetlands and riparian vegetation in the floodplain, based on current knowledge they cannot similarly quantify flows needed to restore marine species such as the totoaba and vaquita porpoise. Shrimp populations in the Sea of Cortéz recovered following the flood flows of the

1980s, but there is no evidence of similar recovery by other marine species. Not enough is known about the marine and estuarine ecology of the region to reach conclusions about restoration needs for those resources.[48]

Similarly, several other restoration issues have not been addressed by existing research, certainly not to the degree they have been in the portions of the river in the United States. Successful ecological restoration may require replenishment of sediment and nutrient as well as water flows to the lower river and the delta. No one has evaluated seriously the long-term implications of restoring the region with water the quality of which is so significantly different from what it was before the dams. Scientists also have little information about other species that live in the region, including fish, reptiles, mammals, and birds, or about the relationships between those populations and their cousins in the United States. Arid lands botanist Edward Glenn said, "This region can still be described as a scientific 'blank spot' on the map of North America, deserving much more study to inform those who make decisions about its future."[49]

In fairness, the United States has not entirely ignored scientific or environmental problems in the Mexican portion of the river corridor. But to date the only serious response has been the familiar call for more studies. The United States and Mexico have signed several agreements to increase cooperation, to engage in pilot projects, and to coordinate environmental research in the lower river and delta regions, but there is no accompanying commitment for research funds.[50] Although environmental groups on both sides of the border have called for a new treaty in which both nations agree to address these issues and not merely to study them, more science is at least a start in the right direction. If we learned nothing else from our experiences with the upper river, Grand Canyon, and lower river restoration programs, it is that movement is difficult and at times counterproductive unless we know where we are headed.

The absence of information about what will be needed to restore the ecosystems of the lower delta and Sea of Cortéz to some agreed upon goal leaves water users throughout the basin in a difficult position of uncertainty. It is one thing to commit to the idea of "one percent for the delta" with the assurance that one percent is all that will be required. But as new information becomes available about the impacts of reduced flows and restoration needs, one percent this year may lead to more next year and more the year after that. Once the holders of vested water rights in the United States agree to give up some amount of water beyond what is demanded by treaty, they implicitly acknowledge that the treaty no longer defines their maximum obligations, and they may believe that this leaves them vulnerable to much larger obligations.

Water is needed to restore the delta, and if we are serious about restoring this internationally significant resource, water *should* be allocated and released for that purpose. But completing the science necessary to define the extent of those needs may be essential to any agreement—by treaty amendment or otherwise—to make water available. To quantify appropriate restoration flows, we first need to define more clearly the boundaries within which restoration is sought, and the goals for restoration within those boundaries.

Some issues of boundaries surrounding Colorado River restoration should not, given our modern understanding of hydrology and ecology, be all that difficult to resolve. We can no longer continue to make decisions about the river that stop abruptly at the artificial international border. The Harmon Doctrine, in its many forms and by any other names, must be put to rest for good. More than thirty years ago the United States and Canada agreed to restore the Great Lakes according to ecosystem rather than political boundaries.[51] The politics of water stand in the way of a similar accord with respect to the Colorado River.

Other issues of boundary and scale are more difficult to resolve. In *The Edge of the Sea*, Rachel Carson explained the ephemeral qualities of natural boundaries: "For no two successive days is the shore line precisely the same. Not only do the tides advance and retreat in their own eternal rhythms, but the level of the sea itself is never at rest. It rises or falls as the glaciers melt or grow, as the floor of the deep ocean basins shifts under its increasing load of sediments, or as the earth's crust along the continental margins warps up or down in adjustment to strain and tension. Today a little more land may belong to the sea, tomorrow a little less. Always the edge of the sea remains *an elusive and indefinable boundary*."[52]

And so was the Colorado River before the dams. The river expanded and contracted over time. It meandered eastward to the Santa Clara depression and westward to the Salton Sink. But it was never everywhere at once. Once we began to engineer the river to train it in particular directions, some parts of the ecosystem dried up, while others were revived. To some degree, what is "natural" and what is "artificial" has become as elusive and indefinable as the boundaries of the river itself. We have yet to decide, in the context of the entire lower Colorado River and delta region, the boundary of this ecosystem we seek to restore. Until we do, our decisions will continue to be based more on politics than on ecology.

Another, perhaps more significant boundary, however, stands in the way of resolving this transboundary issue of water use and a whole series of related restoration problems up and down the river. To date, we have artificially separated discrete river restoration decisions from more fundamental choices about how we use the river for economic purposes. By expanding our focus to choices about water and power use and other issues, we can also broaden the nature of restoration efforts and expand the range of restoration tools. In the final chapter, we return to the controversy about the fate of the Glen Canyon Dam as a way to explore those issues.

CHAPTER NINE

The Lovely and the Usable: Toward a More Holistic Approach to Restoration

Very early in my research for this book I had lunch with Dr. Richard Ingebretsen, president of the Glen Canyon Institute (GCI). The GCI is a leading voice for decommissioning Glen Canyon Dam, an idea many consider radical if not downright ridiculous (chapter 1). But no one would mistake Rich Ingebretsen for a "radical environmentalist." Clean-cut, soft spoken, Mormon, a practicing internist and professor at the University of Utah Medical School. His ideas may cut, but certainly not his words or his tone.

Ingebretsen visited Glen Canyon when he was eight, while the dam was being built. He later hiked to Rainbow Bridge as a Boy Scout while the reservoir was filling, and was told that Bridge Canyon would soon be under water. He went back four or five years later, looked down, and his heart ached. At the construction site he asked "why?," but the official could not give him a clear answer. "There is no answer. No one knows why," said Ingebretsen, just as many people do not even know the dam is there. When he visited a Salt Lake City high school, several students thought Lake Powell was a natural lake. I asked, "Don't you mean 'Reservoir Powell?,'" and he smiled at his rare slip. "You're right," he said. I switched topics to the dam itself. "It's probably one of the most destructive objects man has ever built. If we get the canyon back, we'll never fill it again," he said with complete confidence and excitement. That was the only time in our conversation that his voice rose above a soft whisper.

We debated the merits of releasing warmer water from Glen Canyon Dam. Ingebretsen says they need to study all of the impacts and alternatives, including decommissioning. "What if [warmer water] would do more harm than good to native fish?" he asked. How would it help without restoring sediment flows? He expressed the view that temperature controls are being considered more to help sport fish than the endangered natives. I asked whether the institute's strategy is to prevent Band-Aid solutions, even if they might do some good, to force serious consideration of decommissioning? He got a twinkle in his eye, but stopped short of agreeing.

In December 2000, the GCI released its Citizens' Environmental Assessment (CEA) on the Decommissioning of Glen Canyon Dam, its preliminary case for a more serious analysis of the issue.[1] The CEA is just nineteen pages long, but is based on a series of studies prepared by a group of experts on various issues, including water, sediment, salinity, ecology, and socioeconomic impacts. The institute's strategy is to provide enough serious analysis of the concept so that the Interior Department will eventually evaluate the idea in a full-fledged environmental impact statement (EIS), which the department declined to do in the 1995 EIS and rejects as outside its purview. At the behest of Utah's Senator Orrin Hatch, Congress has included riders in the Department of Interior appropriations bill every year prohibiting the department from considering decommissioning. But evaluating the decommissioning proposal more seriously, whether or not it is ultimately adopted, could provide a much broader perspective on Colorado River restoration, and environmental restoration in general.

Brief as it is, the CEA makes some compelling arguments in favor of decommissioning the dam. As explained earlier, the main *raison d'etre* for the dam is political. The dam is the upper basin's main storage facility to meet an artificially defined delivery obligation at an artificially delineated dividing point along the river. According to models run by the environmental group Environmental Defense, storage in Lake Powell increases the reliability of water deliveries to the lower basin only marginally, and projected shortages in dry years would be small even if the dam is removed. The GCI believes that those shortages could be offset through improved water conservation. According to the institute's studies, the reservoir itself lost more than 30 million acre-feet (maf) through seepage and evaporation between 1963 and 1999. Even a portion of that evaporation loss could be used to restore large segments of the lower river and delta ecosystems. The reservoir is also silting up. It has already lost nearly 1 maf of its original storage capacity (between three and four percent), and will fill in completely in two to eight centuries, depending on sediment flow rates. The dam and reservoir also cause severe environmental impacts: increased salinity due to concentration of salts, drowned habitat for hundreds of species of plants and animals, and major changes to the river's downstream ecosystems.[2]

The CEA is somewhat more cautious when it comes to some of the other problems decommissioning would create. Glen Canyon Dam produces about 5,000 gigawatt-hours (5 million megawatt-hours) of electricity a year, much of it during peak power periods and at a low cost relative to coal-fired or gas-fired plants, and without accompanying air pollution and production of greenhouse gases. The institute argues that the dam still produces only three percent of the total power used in the Four Corners states (Utah, Colorado, New Mexico, and Arizona), and that the nearby coal-fired Navajo Power Plant generates almost twice as much power as the dam. The dam's generators would continue to operate during a ten- or fifteen-year decommissioning period, allowing sufficient time to find replacement sources of power, from renewable sources such as solar and wind, and by reducing power demand through energy efficiency improvements. The CEA also acknowledges that pollutants in sediment released during decommissioning could contaminate the Grand Canyon ecosystem, but proposes no solution other than monitoring.

A lot of the information relied on in the CEA, of course, is open to challenge. The water balance models were run before the most severe prolonged drought in the basin over the ensuing five years. Models are uncertain predictions of the future based on assumptions, simplifications, and other factors that can change or simply be wrong.[3] The fact that sediment might fill the reservoir within two to eight centuries is hardly a compelling argument; even 200 years is a long time to consider such a long-range problem. Perhaps the dam's power intakes will have to be modified in 150–200 years, and the dam itself may silt up completely several centuries later. That does not mean we should forgo the dam's benefits now, potentially hundreds of years before these problems occur. Coal and natural gas may be abundant, but it is ironic for an environmental group to suggest that we pump more pollutants into the air and more greenhouse gases into the atmosphere when global warming may constitute the most serious problem the planet has faced since the dawn of the industrial revolution. Water and energy conservation must be embraced in a profligate world that is fast confronting an era of limits. But continued projected growth in the Southwest could offset those savings, in which case they would not suffice to eliminate the need for the dam.

Neither Rich Ingebretsen nor anyone else at the GCI, however, argues that the issues they raise and the facts they present are irrefutable.

All they suggest is that those issues merit far more serious consideration. Are the suggested benefits of decommissioning convincing enough to offset the settled expectations of two nations, seven U.S. states, and millions of people who rely on the dam and the reservoir for water storage, power generation, flow regulation, and outdoor recreation? Those issues require us to think on wider scales of time and space.

The debate over the fate of Glen Canyon Dam allows us to evaluate the challenges facing the river as a whole in a larger context, with eyes to broader possible solutions. Perhaps we can avoid some of the key tradeoffs among river uses and values identified in earlier chapters by rethinking the basic ways in which we use and manage the river, and by developing a new, long-term vision for the river. We begin by pursuing the idea suggested in earlier chapters that more progress might be possible by restoring fundamental physical, hydrological, and biological processes than by just re-creating individual patches of restored habitats. We then ask whether that approach is compatible with the many legal and economic expectations created by the Law of the River, and whether some of those fundamental choices should be reconsidered in ways that might allow us to reconcile the many competing uses of the river.

Restoring Ecosystem Processes

In 1963, as Lake Powell began to fill, backpacking legend Colin Fletcher became the first person to hike the length of the Grand Canyon. Fletcher wrote of the deeper understanding he acquired by traversing the canyon from top to bottom and from end to end: "[A]s I sat there on the rock platform above the sparkling river, the pageant I saw spread out before me shone with a reality as rich as any I have ever caught in the beam of logic." Three decades later, Fletcher traveled the length of the Colorado River to the Sea of Cortéz. Beginning on foot in the headwaters of the Green River in Wyoming, and continuing by raft, he portaged only when stopped by a dam or other artificial impediment.[4]

It takes a similarly broad perspective to understand Colorado River restoration. Stepping back from the details of the specific restoration efforts up and down the river, one thing is abundantly clear. For the most part, they are not working. Populations of endangered species continue to decline, despite years of study and planning and millions of dollars spent. In the upper river, fish are being stocked in waters that lack the backwater habitats necessary for rearing and survival. Below Glen Canyon, efforts to restore beaches and sandbars appear futile because most of the necessary raw material remains trapped behind the dam. In the lower river, officials plan to plant relatively small patches of habitat for endangered birds while the flows necessary to support widespread, natural revegetation continue to decline as more water is diverted upriver. Restoration of the lower river and its delta in Mexico is left to chance, with replenishing flows occurring only in those rare years when reservoirs are full and "flood releases" are needed.

One common theme emerges. Existing restoration efforts seek to recover discrete resources to prescribed levels rather than restoring hydrological, physical, and biological *processes* by which ecosystems maintain health naturally. The upper river recovery program, for example, will be declared a success if the humpback chub population reaches defined levels for a predetermined amount of time. Incidental take permit conditions in the lower river will be met when about 8,000 acres of new habitat are "created." But upper river fish may not be able to reproduce and survive to adulthood until the river is reconnected to a significant portion of its floodplain. The aquatic ecosystem in the Grand Canyon will not recover its health until the flow of water, sediment, and other materials simulates natural seasonal patterns more closely. The lower river will remain a fundamentally artificial system if so much of it remains dammed and imprisoned within armored banks and levees. The delta cannot recover anything close to its natural lushness so long as Mexico receives just 1.5 maf of water a year.

Much of the Colorado River has been modified to the point where significant variation from what might be considered "natural" is inevitable. There is probably no way, for example, to eliminate exotic species from the river, and to restore purely endemic populations. Even if possible, many people would object to the idea of eliminating introduced trout. Still, restoration ecologists generally agree that our goal should be to restore natural processes (such as some flooding along the lower river), and not simply to freeze conditions at a predetermined "desirable" state. The consensus position of the Society for Ecological Restoration International (SERI) is that an ecosystem is restored "when it contains sufficient biotic and abiotic resources to continue its development without further assistance or subsidy. It will sustain itself structurally and functionally. It will demonstrate resilience to normal ranges of environmental stress and disturbance." SERI also notes that ecological processes and functions are "the dynamic attributes of ecosystems, including interactions among organisms and interactions between organisms and their environment," which are "the basis for self-maintenance in an ecosystem."[5]

Focusing on natural processes rather than a series of snapshot conditions and indicators would help to address some of the perplexing issues discussed in earlier chapters. There would be less need to worry about which specific conditions or baselines should constitute appropriate restoration targets. Once natural processes are in place, nature can take her own meandering course, as she did before major artificial changes interfered. That, however, might require us to abandon two of the dominant precepts that have either constrained or driven ecological restoration strategy in the Colorado River region and elsewhere. First, restoring natural processes in significant portions of a broad, shifting riparian zone will require a partial retreat from the water's edge, and acceptance that private property rights do not apply to public, or "unownable," resources within the land-water transition zone. That does not mean abandonment of private property rights. It just means that those rights should be limited to property that really is private. Second, allowing nature to take her own course suggests revision of restoration strategies that focus on individual species, rather than conditions that support natural assemblages of species. That does not mean abandoning the biodiversity protection focus of the Endangered Species Act (ESA), or the Clean Water Act goal of restoring and maintaining the chemical, physical, and biological integrity of the nation's waters. Restoring natural processes and habitats on a wider scale, however, is more likely to protect biodiversity and ecosystem integrity overall than a more fragmented, species-specific approach.

Restoring natural processes also would enable Colorado River ecosystems to become more self-sustaining. Current restoration strategies, even if they succeed in meeting specific goals, will require significant ongoing management. Restoring and maintaining beaches and eddy complexes within the Grand Canyon requires continuous monitoring of sediment deposits into the river below the dam, and carefully timed flood releases when conditions are ripe to move that material from the river channel to the sandbars. That can occur only when there is enough water in the reservoir, and when other river uses (such as water deliveries and power generation) will not be unduly affected. Replanted habitats along the lower river will have to be watched closely for tamarisk invasion and high soil salinity, both of which would require remedial action (tamarisk removal and soil flushing with irrigation water).

Given the highly modified state of some portions of the Colorado River, it is unrealistic to believe that completely self-sustaining ecosystems are possible in the foreseeable future. Continued monitoring, research, and ecosystem management will be needed, using the tools of adaptive management discussed in chapter 6. SERI distinguishes between ecological restoration, which promotes ecosystem recovery, and ecological management, which includes actions to maintain ongoing ecosystem health thereafter.[6] The closer we get to restoring natural chemical, physical, and biological processes and conditions, however, the less ongoing management will be needed. More importantly, restoring natural processes allows ecosystems to *evolve* over time, as they did before major artificial disturbance. This includes both evolutionary change and the ability to adapt to a range of natural disturbances. Rather than focusing on static target conditions, this restoration strategy is grounded in the principle that ecosystems change constantly in the natural world, but within particular ranges and rates of change.

In the Colorado River, existing efforts flirt with the goal of restoring natural ecosystem processes, but only where they will not interfere significantly with the Law of the River. In some cases, program managers assume that the Law of the River is sacrosanct, and that ecological restoration can occur only where consistent with those laws and institutions. At best, the programs purport unrealistically to be all things to all resources, that is, to comply fully with environmental restoration goals while still accommodating all existing water, power, and other objectives. In reality, restoration consistently takes second place to development.

Decommissioning Glen Canyon Dam, or perhaps one of the other major dams in the main stem of the river, would do far more than existing restoration efforts to bring back some of the river's natural processes. Connections would be restored between populations of endangered fish within the upper Colorado River metapopulations. Sediment would flow through the Grand Canyon, and high spring flows would reshape that material into beaches and backwater habitats. Warmer but more variable water temperatures would trigger spawning of humpback chub and other native species, and would avoid the temperature shock that impedes survival of young fish. Lake Mead would be full a greater percentage of the time, which would increase opportunities to release water for ecological restoration downstream to the delta. Many parts of the river's related ecosystems, in short, would benefit from a single although dramatic decision.

No silver bullet, of course, will restore endangered species and other lost attributes of the river's ecosystem. If warmer flows into Marble and Grand canyons also improve conditions for nonnative fish, we might have to expand and improve steps to control populations of those species. Efforts to return species now extirpated from the Grand Canyon and other reaches of the river may require the continued use of hatcheries, with careful attention to providing as much genetic diversity as possible. Riparian zones now blanketed with monocultures of tamarisk will have to be cleared to allow cottonwood, willow, and other native seeds carried by rejuvenated river flows to germinate and thrive. The salient question, however, is not whether decommissioning will succeed in restoring the river's environment absent other actions, but whether existing strategies will succeed without restoring the river's natural flows and other physical and biological processes to something closer to their native conditions.

During my research, several government officials privately (anonymously) expressed the view that Glen Canyon Dam probably should never have been built, and that decommissioning would do much more to achieve restoration than existing efforts. They rejected the possibility as unrealistic, however, on political and pragmatic grounds. The basin states and water and power interests have much too tight a grip on decisions about the fate of the river and its system of plumbing. Life as we know it in the American Southwest cannot be maintained without all of the dams built by an earlier generation of engineers and all of the interlocking documents drafted by a previous army of lawyers. Decommissioning something as large and as important as Glen Canyon Dam is nothing more than romantic folly, the subject of novels rather than reality.

Given the many potential benefits that decommissioning could bring, however, those assumptions must be revisited before reaching such definitive conclusions about the fate of the dam and the river. The logical place to start is with the Colorado River Compact and the rest of the Law of the River. The requirement that the upper basin states deliver minimum, guaranteed flows remains the dominant argument in favor of Glen Canyon Dam. Looming over the Colorado River basin, however, is the question of whether the compact is even resilient enough to address the many changes that have occurred since it was signed. If the compact must be revisited, perhaps it could be revised to facilitate rather than impede restoration.

Revisiting the Law of the River as a Restoration Strategy

Two natural springs symbolize the problems caused by holding the current generation to a water allocation deal struck in 1922. A display in the lobby of the Southern Nevada Water Authority provides a brief history of water use in the region. Spanish explorers arrived in 1776 at what was later named Las Vegas Springs, but ancient campsites show use by Pueblo Indians more than 3,000 years ago. Settlers pioneered the Spanish Trail from New Mexico to California in 1829 and 1830, and named the area *Las Vegas*, "the meadows." For more than a century, the springs supported travelers and small settlements. The first permanent settlement was established in 1867; the first water rights to the spring were filed in 1872; and the original Las Vegas town site was established in 1905. But water was available only for town residents, so smaller water companies drilled hundreds of private wells. This pumping war apparently did not provide adequate warning of the shortages to come. When Nevada ratified the Colorado River Compact knowing that it would receive a paltry 300,000 acre-feet of river water, little did anyone know what a thirsty metropolis would grow up around this modest spring. By 1947, the water basin was overdrawn. One of the nation's fastest-growing cities remains bound by a deal made when it was a small desert byway.

Not far to the southeast of Las Vegas is Pipe Spring National Monument, in the "Arizona Strip" region north of the Grand Canyon. The Dominguez-Escalante journals recorded this site as "the hill that seeped water." Mormon settlers used these springs beginning in the 1860s, and the Powell survey located its headquarters here. The springs used to flow at ten to thirty gallons per minute, but now they are dry (although nearby springs continue to flow). Although the reason for this change is not clear, the National Park Service speculates that the cause was an earthquake with its epicenter in Southern California. We could view this as a metaphor for the longstanding water dispute between California and other basin states. An earthquake in Southern California causes Arizona to lose water.

Recall that one main point of the compact was to prevent California from claiming the majority of the river under the prior appropriation doctrine before other basin states had time to develop. As a price for California's agreement to limit its ultimate use, other states had to make important concessions. The upper basin states agreed to their automatic delivery obligation, leaving them at much higher risk of shortage in times of drought. Although Arizona was given 2.8 maf of Colorado River water in the Boulder Canyon Project Act and the Supreme Court's decree in *Arizona v. California*, ultimately it agreed to subordinate its lower basin water rights to California to secure the Central Arizona Project, the massive system of pumps and canals that would bring Arizona's river share to the southern farms and cities where it was needed.[7]

Meanwhile, as explained in the last chapter, for many years California used as much as 1 maf more than its apportioned share of Colorado River water. This overuse was possible because Arizona and other states were not using all of their compact apportionments. Arizona cannot use all of its Colorado River water, Nevada faces shortages, and California continues to benefit from the surplus. Although California is now taking steps to eliminate this excess use, Arizona and Nevada recently devised a somewhat complicated "water banking" plan to alleviate some of the constraints they faced under the Law of the River.[8] Arizona will withdraw its full apportionment of water every year. But because it cannot put all of that water to immediate use, it will "bank" that water by storing it in aquifers. Arizona can withdraw some of this liquid currency in future years, when its demand for water increases or during times of drought. That way, Arizona benefits more fully from its lawful apportionment, rather than ceding some of its entitlement to its thirsty neighbor to the west. Nevada agreed to help pay for the water bank in return for rights to use some of the stored water as its needs increase. In those years, Nevada will withdraw some of Arizona's annual apportionment directly from Lake Mead (upstream of Lake Havasu, where Arizona ordinarily diverts water to the Central Arizona Project). Arizona will reduce its downstream withdrawals accordingly, so the water bank's account will balance.

This scenario illustrates just one example of ways in which incentives generated by the Law of the River adversely affect efforts to restore Colorado River ecosystems. Remember from the last chapter that part of California's plan to reduce its use within its compact apportionment means that more water will be withdrawn farther north, leaving less available to sustain aquatic ecosystems below. The same will be true when Nevada diverts some of Arizona's water from Lake Mead rather than Lake Havasu. And the whole point of the water bank is for Arizona to take more water from the river than it can actually use in a given year so that California cannot use it farther downstream. Changes are being designed not to meet existing water needs, but to maneuver for maximum future advantage within the artificial constraints of the Law of the River.

The more significant and equally artificial constraint on river restoration, however, derives from the upper basin's rolling ten-year obligation to deliver 75 maf of water at Lees Ferry, the dominant reason for the dam in Glen Canyon. Not water per se, not electric power, not recreation, but Article III(d) of the Colorado River Compact. Hardly any water is used directly from Lake Powell, except for the relatively small amounts used by the town of Page, Arizona (which was created out of whole cloth when the dam was built) and the Navajo Generating Station. Washington County, in southwestern Utah, proposes to build a pipeline from Lake Powell to Sand Hollow Reservoir to divert 100,000 acre-feet a year to meet burgeoning water use in that region. Those three, relatively small direct uses, however, hardly justify a reservoir with a capacity of nearly 30 maf! The real reason for the dam is to allow the upper basin to store enough water in wet years to meet ongoing delivery obligations without curtailing uses upriver.

The legal requirements that explain the location and existence of Glen Canyon Dam are the products of purely artificial arrangements. Lees Ferry is a logical place to divide the basin given the water allocation compromise struck in 1922, close to the artificial geopolitical borders between the states.[9] The 75-maf, ten-year delivery requirement was based on what we now know, from tree ring studies and other data, were faulty assumptions about the amount of water likely to be available in the basin over the long-term. The basic water allocation scheme in the compact also did not account fully for other significant needs, including the water ultimately guaranteed to Mexico by treaty and federal reserved water rights for tribes and federal lands, all of which come out of apportionments to individual states if they cannot be met with surplus water. Nor did compact drafters envision water to protect and sustain the unique aquatic ecosystems of the Colorado River, in the United States and in Mexico. Colorado River states and tribes have been betting that they will be able to develop their Colorado River water without running afoul of ESA or other environmental requirements. But no one has quantified the amount of water needed to meet environmental needs, or analyzed comprehensively when and where it will be needed and what constraints that will impose on existing and future human water uses.

Now, another wild card complicates this game of Colorado River poker: global warming. Current models suggest that climate change may bring even less runoff in the basin than suggested by tree ring analysis, although that prediction is fraught with uncertainty. Assuming a slightly different set of changes, some parts of the basin may have more runoff, and other parts less. Already, experts have been pondering how administration of the compact will change if less water continues to be available than anticipated when it was signed, and as basin uses continue to climb. In analysis conducted for drought management negotiations for the basin, the need for which was underscored by the prolonged, record drought that began in 2000, the Bureau of Reclamation predicted that lakes Powell and Mead may never again be full. If global warming reduces water supplies even more, it is reasonable to ask whether both reservoirs are needed or even desirable. Spreading much less water over the much larger surface area of two arid area reservoirs will result in more total losses to evaporation than if a single reservoir were kept full.[10]

All of this suggests a need to revisit the water allocation arrangement reached in 1922 in light of such significant new information and changed circumstances. An interstate agreement of this magnitude should not be revisited lightly. The compact has stood the test of time for eight decades, through periods of drought, flood, and dramatic socioeconomic changes in the Southwest. Significant investments and other decisions rest on the relative certainty provided by this landmark agreement and its related library of laws, regulations, court decisions, contracts, and other legal documents. But too much has changed to simply ignore the reality that the compact as currently structured needs to be revisited. The compact is already being stretched through arrangements such as the Arizona Water Bank. From the perspective of environmental restoration and even basic compliance with the ESA, it is simply not working.

This realization has significant implications for river restoration. If the debate about decommissioning Glen Canyon Dam remains bound within the existing paradigm known as the Law of the River, it is likely that the dam will stay. Indeed, the likelihood of worse water shortages in the future suggests that more upper basin storage will be needed to ensure that the basic compact delivery obligation can be met. But that would no longer be true if the upper basin's rigid delivery obligation is changed, or even the point along the river at which it must be satisfied. What if, as the GCI and others have suggested, the delivery point moves down to Hoover Dam?

The fact that Glen Canyon Dam was needed in 1963 to ensure compliance with an artificially defined water delivery schedule at an equally artificial place along the river cannot, standing alone, suffice to justify a structure that causes so much harm to the aquatic ecosystem in the Grand Canyon, and that inhibits efforts to restore habitats as far downstream as the Sea of Cortéz. If the dam stays, it must be justified on grounds independent of the compact, and if the compact does not reflect current realities, it should be reconsidered. The more difficult question is whether the dam is needed for purposes of water use and management. Do we need this much storage capacity to sustain the legitimate uses to which Colorado River water is put? Or, are there different or better ways to meet the end uses that water makes possible? Possible answers to those questions suggest the first example of ways in which the concept of environmental restoration might be broadened to include entirely different kinds of strategies.

Rethinking Water Use and Management as a Restoration Strategy

The Powell and Stanton expeditions and other early Colorado River runners rowed their boats the way you would on a flat lake. As described by river running historian David Lavender, "They sat with their backs to the bows, the normal position for oarsmen, and rowed blindly backward at full, thinking they could keep better control of the boats if they went faster than the current." That worked well for boaters who did not have to fear surging water, steep drops, and massive hidden boulders. They used this method because that was how it always had been done. Thirty years later, George Flavell decided to row facing forward so he could see and avoid rocks and other perils. Flavell made it from Green River, Wyoming, to Yuma, Arizona. Nathaniel Galloway and a generation of later river runners adopted this method, which is now standard practice in white water boating.[11]

The rigidity of western water law promotes a tendency to do things the way they have been done for a century and a half, and often to look backward rather than forward. Water rights under the prior appropriation doctrine are doled out as specific amounts of water, taken from identified sources at prescribed places and times, and designated for particular beneficial uses. Many observers have critiqued the lack of flexibility in this system, arguing that it promotes waste rather than efficiency. Under the "use it or lose it" principle, water which is not used might be reallocated and lost forever. That can inhibit innovative solutions that focus on end uses. No end user "needs" a particular amount of Colorado River water to be diverted at a particular place and time. People need water to flow from the tap when they turn the valve or to fill their irrigation canals when they open the gate. Likewise, no one needs x acre-feet of water to be stored in a particular reservoir. Water managers need sufficient system storage capacity to address the volatile nature of the climate. Focusing on real rather than perceived needs might help to devise creative solutions so that human uses can be met while keeping enough water in the river to restore and protect the health and integrity of the aquatic ecosystem. Looking forward may help us to see and avoid obstacles to more efficient water use and management.[12]

Two basic challenges dominate Colorado River water use: the total amount of water over time, and when and where it is available. Less water is available than is needed for all existing allocations, without even considering environmental restoration and protection. And because runoff volumes are seasonal and variable from year to year, storage is necessary to allow water to be used when and where it is needed. Ironically, however, those problems are exacerbated in several ways by the manner in which the river is currently used and managed. Transbasin diversions allow more water to be used at points of high demand, but have greater environmental impacts than in-basin uses because no return flows are available for use farther downstream, or to maintain higher flows in the river. Diverting water from Parker Dam to cities along the coast takes water from the system permanently, as do transbasin diversions from the headwaters in Rocky Mountain National Park and elsewhere downriver. Storing water in reservoirs facilitates year-round use, and keeps water in the bank for dry years. But in addition to the dramatic impacts of dams on the aquatic ecosystem, it also causes more water loss due to evaporation.

For several decades, the rigidity in western water law has slowly been changing. Marketing of water rights now allows water to be transferred from lower value agricultural uses to higher value urban areas, such as the transfers from Imperial Irrigation District to San Diego and Los Angeles discussed in chapter 8. In other parts of the West, increased flexibility has been used for environmental restoration.[13] Creative water management alternatives could, over time, free up some of the water now claimed under the Law of the River, obviate the need for some dams in the watershed, and better serve both human and environmental goals. We need to expand our concept of "restoration" to include those strategies. Let's look at a few examples.

The Bureau of Reclamation's large desalination plant at Yuma sits idle, while the fastest-growing uses of Colorado River water are adjacent to the salty Pacific Ocean, an essentially unlimited supply of water if it can be treated at a reasonable cost. Meanwhile, the federal and state governments propose to spend hundreds of millions of dollars to replant relatively small patches of native vegetation along the lower Colorado River. The U.S. Fish and Wildlife Service, as explained in chapter 7, deemed that modest amount of restoration sufficient under the Lower Colorado River Multi-Species Conservation Program to allow more river water to be diverted to the West Coast. Maybe this approach is backwards. Perhaps we should spend restoration dollars to build desalination plants to supply Southern California's growing cities with water from the ocean. For every gallon of water provided by a restoration program–funded desalination facility, California would forgo one gallon of its Colorado River apportionment. That water could be dedicated permanently to environmental purposes because it will be replaced with new water financed by the restoration program. Freeing up water to restore natural flows to portions of the lower river corridor, in both the United States and Mexico, might accomplish far more restoration than an equal amount spent on patchwork revegetation projects. Most important, it would keep more water in the river all the way to the delta.

Desalination is no panacea given current economic and other limits. Construction and energy costs are high, and safe disposal methods are needed for the highly saline waste brines. Yet workable, large-scale desalination methodology has been available since the late 1960s, and is now being used to supply potable water in more than 100 nations around the world. Desalination provides virtually all of Kuwait's freshwater. Costs are declining at a rate of about four percent a year, and experts predict that a technological revolution inevitably will result in even more dramatic cost reductions. Using Colorado River restoration programs to fund desalination plants and associated research and development might push that technological innovation forward more quickly. Some communities along the California coast are already building desalination plants, without subsidies, because water from competing sources is so expensive. Using restoration program dollars to increase the use of this technology will simply shift existing subsidies in a more environmentally beneficial way. Rather than subsidizing the transbasin diversion of Colorado River water and paying for the resulting environmental restoration costs, we could avoid more of the harm altogether by subsidizing the use of replacement water from desalination facilities.[14]

Another area in which we might row forward rather than backward is in groundwater storage, or "aquifer storage and recovery" (ASR). Like desalination, ASR alone cannot provide enough storage capacity in the short term to obviate the need for existing on-stream dams along the Colorado River. As an example of its *potential*, however, the water banking agreement between Arizona and Nevada first signed in 2001 and amended in 2004 guarantees that Arizona will store up to 1.25 maf of water for Nevada's benefit over the next decade. That arrangement is designed to ensure that Arizona can take full advantage of its compact apportionment while increasing the reliability of water supplies in southern Nevada. Absent the specific constraints of the compact, however, that same storage capacity could be used to replace an equivalent amount of reservoir capacity.

No one has comprehensively assessed the available aquifer storage capacity in the Colorado River basin to determine the overall potential for aquifer storage to replace some on-stream storage reservoirs over time. ASR has been used in Florida since 1983, and fifty ASR facilities (with a total of 333 wells and a capacity of 1.6 billion gallons a day, or 1.8 maf per year) have been built or are planned over the next twenty years as part of the Comprehensive Everglades Restoration Plan. That will make more freshwater available for the Everglades ecosystem. The State of Washington is pursuing ASR to augment municipal and other water supplies. Las Vegas itself has one of the largest active ASR projects in the country, with a capacity of 100 million gallons a day. As with desalination, however, ASR may have even more promise in the future as technologies improve.[15]

Replacing some reservoirs with ASR projects would eliminate barriers to the movement of water, sediment, nutrients, and aquatic life without losing the ability to store water for human use in times of drought and during dry seasons of the year. Water now lost to evaporation, minus seepage losses from storage aquifers, could be allocated to environmental restoration. The Bureau of Reclamation estimates that Lake Powell, for example, lost more than 0.5 maf of water between 1996 and 2000. (Those losses decreased during the ensuing drought, because of the smaller surface area in the reservoir.) The feasibility of using ASR to avoid those losses depends on technical factors such as the storage capacity and properties of particular aquifers, the rate at which water can be recovered, proximity to places of use or distribution, and quality of existing groundwater. Reservoir locations, however, also depended heavily on geology, hydrology, and other factors. From the early 1900s through the 1950s, the Bureau of Reclamation surveyed all major potential dam sites in the Colorado River basin, to coordinate "comprehensive water resources development" in the basin. A broader vision of restoration could include an equally comprehensive effort to identify potential locations for ASR, and to estimate how much reservoir capacity might feasibly be replaced at those sites.[16]

Another way to avoid the impact of dams on river systems is to locate reservoirs off-stream, especially where their main purpose is water storage rather than power generation. One example is Utah's Sand Hollow reservoir, which also involves aquifer recharge and recovery. Off-stream storage reservoirs do not solve the issue of evaporation, but if located in places that do not support other important habitats, they avoid many environmental problems of dams and reservoirs that block the flow of the river. And they continue to provide reservoir recreational opportunities without having to sacrifice recreation on free-flowing rivers. A long-term strategy to replace the storage capacity of Lake Powell could investigate potential locations for similar facilities where new storage capacity is needed to allow the upper basin states to develop their compact apportionments. Additional capacity in those facilities could partially offset storage space in existing in-stream reservoirs.

Reforms in how we store and distribute existing water supplies, of course, avoid the real "sacred cow" in western water use, which is the sacred cow. Agriculture is by far the largest user of Colorado River water. In 2000, irrigated agriculture accounted for about 7.3 maf out of 8.8 maf of Colorado River water consumptively used within the basin. (About 2.5

maf were lost to evaporation, and about 5.5 maf were exported out of the basin.) The predominant crops grown in the basin are heavy water users such as alfalfa, a hay crop used to feed cows and other livestock. Alfalfa uses fifty percent more water than the next thirstiest major crop in the region (grapefruit), and almost nine times as much as lettuce. In Arizona, alfalfa is the second largest irrigated crop by acreage, and the second largest consumer of water overall. But it has the lowest economic value per acre-foot of water used. Cotton is the largest water user both in total use and in acreage irrigated, and also has much lower economic value than food crops such as fruits and vegetables. If water from Bureau of Reclamation projects was not so heavily subsidized, this economically inefficient use of water would not be possible.[17]

Also because reclamation project water is so cheap, much of the irrigation in the region uses the least efficient methods. In Arizona, farmers irrigate about ninety percent of the acreage with flood or gravity systems, compared to nine percent with somewhat more efficient sprinklers and one percent using considerably more efficient drip systems that deliver water directly to the roots of individual plants. This suggests that improved water use efficiency should be considered more seriously as part of the Colorado River restoration effort, for which there is considerable precedent in the salinity control program. One study estimated that 1.2 maf of water could be saved in Arizona alone (nearly half of its entire Colorado River apportionment) through improvements in irrigation systems, shifts from high to low water-use crops, and fallowing or retirement of some lands currently planted in water-guzzling cotton and alfalfa. Likewise, urban water use could be reduced considerably through more aggressive conservation programs. One study estimated that a third of California's current urban water use, more than 2.3 maf per year, could be saved with existing technology, at lower costs than projected for new water supplies. Although the Bureau of Reclamation and others have devoted more attention to water conservation in recent years, in part due to new federal statutory programs and requirements, the potential to use improved efficiency to free up more water for the river and the delta has barely been tapped.[18]

Desalination; aquifer storage and recovery; off-stream storage reservoirs; and improved water efficiency. All potential ways to make more water available for the river, to reduce the need for on-stream reservoirs, and to restore some of the system's natural flows and connections. The concept of "restoration" should be viewed broadly enough to accommodate those kinds of longer view strategies. The precise contribution of each method cannot be known until it is studied in more detail, and some are likely to improve considerably over the next several decades. None of those proposals alone is likely to suffice to support decommissioning of Glen Canyon Dam. Together, however, they could improve the odds that the river and its habitats may recover ecological health. Planning and implementation of those strategies will likely take decades, just like it took a half century to plan and to build the river's current system of artificial plumbing. In the context of restoration programs that began thirty years ago and will continue for many decades to come, we should not reject them merely because they will take time and effort.

If enough water can be conserved, used more efficiently, replaced with other sources, and stored in aquifers or off-stream reservoirs, perhaps Glen Canyon could be decommissioned over time, with all of the accompanying benefits to downstream ecosystems. That would not, however, sit well with the millions of people each year who visit Lake Powell for outdoor recreation, and the many people who earn their living from that travel. Whether those secondary economic impacts of Glen Canyon Dam should stand in the way of proposed decommissioning is as much a function of individual values as of cold logic. We turn next to the impacts of decommissioning on recreation and other secondary benefits provided by the dam.

Rethinking Recreation as a Restoration Strategy

The vast region that hosts the Glen Canyon Dam is a land of stark contrasts. I toured this area with my family in part as a vacation and in part to research the region's spectacular recreational values. Descending from the Echo Cliffs, Navajo Mountain and the Navajo Generating Station come into view. The mountain is one of the most sacred natural places in Navajo tradition. The coal-fired electric power plant was built as part of the compromise that led to the construction of the Central Arizona Project, the massive system of pipes and canals that allows Arizona to convey its compact apportionment to farms and cities hundreds of miles from the river and thousands of feet higher in elevation. The power plant was built in part to supply electricity needed to pump Colorado River water to where it is needed. Edward Abbey attributed the following facetious (although not entirely accurate) thought to one of his famous Monkey Wrench Gang characters: "No wonder (thought Bonnie) they had to build a whole new power plant to supply energy to the power plant which was the same power plant the power plant supplied—the wizardry of reclamation engineers!"[19] But the plant also supplies power to others in the region, and provides badly needed jobs to the Navajo Nation, which faces massive unemployment problems. The tourism stimulated by Lake Powell adds more regional jobs.

As we continue toward Page, Lake Powell comes into view, a bright blue lake set off sharply against the desert sandstone. Or should I call it "Reservoir Powell," as urged by Rich Ingebretsen? Seeking a linguistic compromise between "lake" and "reservoir," all I can think of are "lesservoir" and "reservalake," which are hardly nonpartisan in sentiment. I confess that it is not easy for me to suppress my personal preference for natural rather than artificial places. This reflects my individual values, but other people clearly have different preferences, which are equally legitimate. As we approached the dam from the south, we first crossed an open desert plateau with a few scattered residences. Entering Page we found a very green golf course, a Wal-Mart, a Days Inn, strip malls, and chain stores. At least there is some humor. We passed the Dam Plaza, the Dam Bar & Grille, and the Dam Outlet.

We decided to buy tickets for the afternoon boat trip to Rainbow Bridge. Zane Gray captured the mystery and remoteness of the bridge in *The Rainbow Trail*, written when land access to the site was limited and arduous: "Ages before life had evolved upon the earth it had spread its grand arch from wall to wall, black and mystic at night, transparent and rosy in the sunrise, at sunset a flaming curve limned against the heavens. When the race of man had passed it would, perhaps, stand there still. It was not for many eyes to see. Only by toil, sweat, endurance, blood, could any man ever look at Nonnesoshe. So it would always be alone, grand, silent, beautiful, unintelligible."

Before the dam, reaching Rainbow Bridge demanded a fourteen-mile hike from the river. Among the early battles in Glen Canyon Dam history was a failed effort to build a smaller second dam or to limit reservoir levels to protect this sacred site from being flooded. Now, when full, the reservoir backs right up to the bridge, allowing easy access by tens of thousands of tourists a year (such as my family and me).[20]

Conflicting personal values about the recreational worth of Lake Powell became even more evident to me on this boat trip. On the observation deck at the top of the tour boat, we were joined by a group that seemed to ignore the scenery and talked instead about their hotel room with a hot tub and a huge-screen TV to watch tonight's (whatever) game. That attitude was not necessarily "wrong"; they were just there to have fun. Pretty sandstone formations rose above the unusually choppy water. Very nice, but we saw just the tips of what used to be soaring canyon walls. My objective mediator wife, Michele, consistently told me that this book should simply provide an honest, factual basis to help everyone reach his or her own conclusion about what are essentially value judgments. We had toyed with the idea of returning to rent a houseboat for a whole week just to see what it would be like (very uncharacteristic for our solitude-seeking clan). Now, she said she couldn't do it. She'd spend the whole week crying.

This afternoon boat ride was a microcosm of the massive tourism industry at Lake Powell. Two million people visit Glen Canyon National Recreation Area every year, giving it the third largest number of overnight stays in the national park system. They come to play in and on the water, and on the million acres of surrounding redrock wilderness. They fish for catfish, trout, and bass. They Jet Ski and water ski, climb, hike and bike, and spend lots of money, contributing an estimated $400 million a year to the local and regional economies. They use kayaks for solitude in small side canyons, sleek power boats for thrills, and for those seeking more luxury, massive houseboats for weeklong excursions. Leather furniture, wet bars, hardwood floors or plush carpets, full modern kitchens, stainless steel barbeque grills, satellite television, and Bose surround-sound entertainment systems. This is not for the low-budget traveler. In 2006, during peak season, a 46-foot "voyager class" rented for $4,000 a week. The 75-foot "odyssey class" would set you back $11,000. Not including fuel.[21]

As evidence of the magnitude of use and the conflicts that generates, in 2003 the park service completed an EIS to evaluate how to manage personal watercraft (Jet Ski) use in Glen Canyon National Recreation Area. The EIS was prompted in part by a lawsuit brought by the Bluewater Network, a group seeking to restrict or eliminate motorized recreation in national parks to reduce noise and other environmental impacts. Hundreds of power boats with leaking oil and gas cause significant pollution, as does the massive volume of human sewage and garbage. At times, the park service has closed beaches and other parts of the reservoir to swimming due to high bacteria levels caused by inappropriate dumping of human wastes from on-board toilets, despite the presence of "floating restrooms" and pump-out stations throughout the reservoir. Noise from powerboats and personal watercraft also can harm endangered species and other wildlife.[22]

The impressive visitation statistics suggest that large numbers of Americans view Lake Powell as a significant recreational amenity. But the dam was not built primarily for recreation, and Congress did not even list it as a major project purpose. Congress authorized recreational facilities as an incidental benefit to Glen Canyon Dam and other parts of the Colorado River Storage Project. Draining Lake Powell would not drain dollars from the wallets of American vacationers. Visitors would likely switch to Lake Mead or other locations for similar outdoor recreation. The economy of Page, Arizona, would suffer significantly, but Page did not even exist until it became the construction site for the dam. Harsh as that may sound, it is difficult to justify a facility that causes so much environmental harm solely on the basis of localized economic benefits that would likely be transferred elsewhere rather than lost entirely. I may no longer be welcome in Page after publishing this view, but as noted in chapter 1, we didn't avoid the shift from horses to automobiles because jobs would be lost in the livery stable industry.[23]

On the other hand, those who think that decommissioning would restore Glen Canyon to the solitary place it was before the dam rely on a romantic, unrealistic, and arguably selfish claim. Singer-songwriter Katie Lee expressed great annoyance at the "swarms of people" who descended on Glen Canyon as the 1950s progressed. Although devotion to special places has been the heart and soul of the American wilderness movement for decades, it only underscores charges of environmental elitism to write about "my river . . . mine, all mine!" and "I have dipped into a private treasure."[24] All of us who seek solace in wilderness periodically share those understandable feelings of irritation when others intrude on our private hideaways. But such places are just as "unownable" as the riparian zones along the lower Colorado claimed by farmers and homebuilders.

Both sides of the debate are probably wrong about the fate of Glen Canyon in a postdam era. The dam's supporters assert that Lake Powell provides recreation for millions of Americans rather than the elite few who boated Glen Canyon before the dam. Decommissioning advocates hark back to the solitude that early river runners found in "the place no one knew." It is deceptive for both sides, however, to compare the number of people who used the region for river recreation in the 1950s with the numbers who flocked to Lake Powell thirty to fifty years later, given the massive boom in outdoor recreation in the United States during that period. The National Park Service has struggled to regulate white water recreation in the Grand Canyon because there is so much more demand for river recreation than space to do so. Many river runners who wait decades for a permit to navigate Grand Canyon would use Glen Canyon to satisfy some of that unmet demand. Others who prefer not to endure the terrifying rapids of Cataract Canyon or Grand Canyon would seek the tamer but equally rewarding experience once found in floating Glen Canyon. At least some of the area's tourism dollars now provided by houseboats would be replaced by private and commercial river trips.

Rather than viewing river restoration as inconsistent with tourism and recreation, we might focus on a different vision of the river, compatible with both goals.

On the boat ride back from Rainbow Bridge I went below to chat with "Captain Bob," a former Merchant Marine who moved here in 1980 and has been piloting the lake ever since. I asked what he thought of draining the reservoir. After a long pause he said, "It's kinda silly. Will never happen in my lifetime. But I don't care; it doesn't matter to me. I'd just move to the coast and get a job there." Then we passed a large, very fast speed boat. Captain Bob raved that it goes 100 miles per hour standing still. "Listen to that power," he said. I couldn't tell whether he was being facetious. But power is another piece of this complex puzzle. We look next at the importance of hydropower generation, and how those needs might be met in ways that are compatible with river restoration.

Rethinking Power Generation as a Restoration Strategy

Tim Ulrich, the Bureau of Reclamation's area manager at Hoover Dam, was gracious enough to spend several hours giving me a tour of the dam and power facilities. I was in Las Vegas to run a marathon, but diverted south to learn more about this keystone facility in the Colorado River's water management system. Much of the massive machinery that generates electricity here has been in place since it was installed in the 1930s. The equipment reminds me of the machines you see in the Old Smithsonian Institution building in Washington, D.C.—relics of the 19th century. Especially compared to an era in which the laptop on which I now write will be obsolete in a couple of years, this stuff was built to last. The turbines steadily churn out more than 4 billion kilowatt-hours of power a year, fueled only by the water running through the dam—simple mechanics and hydraulics. Water flows in a circular route through the generators and out into the river. It is difficult for anything serious to go wrong, given the relatively simple technology and the huge mass of what is turning, making it a very dependable source of power.

The equipment has been modernized, however. From the 1980s through 1991, the bureau rewrapped the coils with thicker wire, which increased generating capacity. Ulrich indicated that they were in the process of refurbishing each generator, working on one a year. Replacing parts, cleaning, filling in corrosion holes and refinishing. At some point they will switch to more computerized equipment that will not need as much care and feeding. Obviously this will be more efficient and accurate, just as my marathon time was recorded by a chip tied to my shoelaces rather than an old-fashioned stopwatch. Undoubtedly the right move, but sad in a way.

Tim then took me to the control room, which by contrast has already jumped into the 21st century. This is one of the Bureau of Reclamation's three Colorado River command posts, which coordinate all water and power facilities on the river. (The other two are in Grand Junction, Colorado, and Yuma, Arizona.) Assistant area manager Gary Bryant showed me the Hoover Dam Load Control, a computerized, multicolor graphic that displays the status and operating characteristics of each generating unit at the facility. Only about half were running, but it was early February. Every two seconds the screen changed, showing the need for Hoover Dam power based on the instantaneous load on the entire electrical grid's transmission lines. Hoover plays a vital role in meeting peak power demand in the Southwest, especially during the summer when people run air conditioners during the hotter daytime hours.

Although hydropower is often touted as particularly important because generators at dams can be ramped up and down quickly to meet peak power demand, Glen Canyon Dam is now used more to meet baseload demand. In part, this is due to the new environmental restrictions that limit daily flow fluctuations, which makes it more difficult to accelerate power generation quickly. If Glen Canyon must be run largely as a base-load power plant to protect the Grand Canyon ecosystem, it is inappropriate to continue to rely on the peak power argument to justify the dam. On the other hand, although it has only eight generators compared to Hoover's seventeen, Glen Canyon generates more than 3,200 gigawatt-hours of power a year, out of about 4,340 generated by all of the facilities in the upper basin's Colorado River Storage Project.[25]

Like recreation, electrical power generation was never the principal reason for the Glen Canyon Dam. As noted in chapter 5, Congress directed the Bureau of Reclamation to operate the dams "for the generation of hydroelectric power, *as an incident of the [primary] purposes*" of those facilities.[26] Hydropower is unquestionably a useful incidental benefit of Glen Canyon and other large dams, and power revenues comprise the principal source of revenue to repay the federal treasury for the costs of the dam and to fund other programs (such as the salinity control program). But it was not why the dam was built. If decommissioning is the only viable way to restore Colorado River ecosystems, and if its primary water management purpose can be met in less damaging ways, the fact that the dam is a prolific cash register should not determine the result. That would, in effect, constitute a sort of institutionalized environmental bribe in which a highly damaging facility is retained in the face of more environmentally sustainable alternatives merely because it generates revenue. If Congress is willing to spend $8 billion to restore the Everglades ecosystem, it can include loan forgiveness on Glen Canyon Dam as part of a program to restore the Grand Canyon and the lower Colorado River.

Still, even the most passionate advocates for dam decommissioning should pause before arguing that power from Glen Canyon Dam could be replaced with fossil fuels. Coal, gas, or oil would emit significant quantities of both conventional air pollutants and greenhouse gases, and exacerbate America's addiction to fossil fuels as its major sources of electrical energy.[27]

Rich Ingebretsen nevertheless believes that hydropower is not a compelling reason to keep the dam in the long run, because "it is ludicrous to think we will continue to rely on this form of power a century from now." He is probably right. The United States in general and the Southwest in particular has not aggressively tapped available sources of renewable energy that could replace the power from Glen Canyon Dam and other existing sources. The United States added 2,500 megawatts of new wind energy capacity in 2005, about twice the generating capacity of Glen Canyon Dam. By 2020, the American Wind Energy Association predicts that wind could provide about six percent of the nation's power, roughly the equivalent of all hydroelectric power today. Solar power has an equally promising future. The sunny Southwest has the highest potential of any region in the country for development of solar electric power, and the U.S. Department of Energy predicts that photovoltaic power soon will be competitive in price with traditional sources of electricity. DOE predicts: "The enormous solar power potential of the Southwest—comparable in scale to the hydropower resource of the Northwest—will be realized. A desert area 10 miles by 15 miles could provide 20,000 megawatts of power, and the electricity needs of the entire United States could theoretically be met by a photovoltaic array within an area 100 miles on a side."[28]

Although decommissioning facilities such as Glen Canyon Dam will result in lost revenues, the switch to renewable energies, which the United States should do anyway to reduce its world-leading contribution to global warming and its dependence on foreign sources of energy, can compensate for those economic losses many times over. According to the Union of Concerned Scientists, an attainable goal of twenty percent reliance on renewable energy sources would save $49 billion in electric and gas bills, create 335,000 jobs, and generate $73 billion in new capital investment.[29] It is a serious mistake to assume, as part of a multidecade restoration program, that energy technology will remain static. A multidecade restoration focus can realistically assume significant continued improvement in renewable energy technology, with additional likely cost reductions.

A complete analysis of future energy sources is well beyond the scope of this book. Every source of energy has problems that must be addressed, such as the siting of wind farms and further technological development of photovoltaic technologies. As is true for water use, however, to address the real ecological needs of the Colorado River, we need to broaden our concept of restoration to include a far more serious search for replacement sources of power.

Likewise, a restoration program for the Colorado River that seriously considers decommissioning as a keystone strategy must include far more serious research and innovation in dam removal methods. Most dams that have been removed in the United States, for both environmental and dam safety reasons, have been small. Proposals to remove larger dams, such as several that block salmon migration on the Snake River, are considerably more controversial. Decommissioning a facility the size of Glen Canyon Dam will present significant technical challenges. For example, no one has ever dealt with the degree of sediment accumulation that lies beneath Lake Powell, especially given the likely degree of sediment contamination. Yet considerable progress has been made in dam removal science in recent years, and more is likely as experience is gained.[30]

The point is neither to trivialize these significant technical challenges, nor to presume to address them in depth here. But neither should those challenges eliminate fair consideration of decommissioning as part of a broader, long-term restoration strategy. Decommissioning might occur over a protracted period of time, with an adaptive management approach that allows corrections as more is learned about dam removal on the Colorado River and elsewhere. It may turn out, as some scientists have argued, that more ecological benefits could be obtained by decommissioning other facilities, such as Flaming Gorge, Fontenelle, or Laguna dams, because that would result in a much longer stretch of free-flowing river and have less significant effects on storage capacity, power generation, and other competing resources.[31]

Decommissioning Glen Canyon or another large Colorado River dam might have to await additional dam removal experience elsewhere. The simple fact that it has not been done on this scale, however, is not a compelling reason to exclude it from consideration if it is otherwise advisable or necessary. After all, nothing remotely like Glen Canyon

Dam even existed less than a century ago. Presuming that it is beyond our technological ability to remove those facilities seems unduly pessimistic given the remarkable technological advances we have made in so many other areas. And if hydrologists are correct that water shortages in the foreseeable future will continue to produce low reservoir levels, a partial test of decommissioning may be inevitable. It seems wiser to evaluate strategies for that scenario well in advance than to wait for it to occur and then to react in an emergency mode.

Challenges such as water management, energy alternatives, and dam removal will not be easy to solve, but are within the grasp of engineers and scientists armed with computers and other modern tools. Not all of the issues involved in deciding whether decommissioning is a good idea, however, are quite so amenable to objective thinking. As Henry Beston wrote around the time Hoover Dam was being built, "Poetry is as necessary to comprehension as science."[32] A little bit of poetry is also necessary to comprehend the passions that inspire those who seek to restore Glen Canyon and other parts of the Colorado River. This book has tackled a wide range of weighty and tangible issues and conflicts. Suckers and trout. Eagles and flycatchers. Tamarisk and cottonwoods. Water for millions of people and thousands of acres. Power for lights in Las Vegas and air conditioners in California. Treaties and regulations and decrees by the U.S. Supreme Court. All of that has ignored one key issue: beauty.

Restoration and Aesthetic Values

When Glen Canyon Dam was under construction, the Interior Department commissioned a series of studies of the natural, archaeological, and historical resources that would be inundated once Lake Powell began to fill. A wonderful picture book called *Ghosts of Glen Canyon*, compiled and written by the late C. Gregory Crampton, depicts some of the natural and historical treasures now buried beneath the reservoir. The Wasp House, the remnant of a small, mud and stone Anasazi Indian ruin at the foot of a massive canyon wall with a tapestry of mineral streaks. Robert Brewster Stanton's deserted mining dredge. Steps carved at Hole in the Rock by Mormon pioneers who precariously lined wagons down steep, slick rock canyon walls. The "Crossing of the Fathers" where Dominguez and Escalante forded the river on their way back to Sante Fe. Natural wonders such as Music Temple, a natural cliff-bound amphitheatre with magical acoustics, and Gregory Natural Bridge, the longest natural bridge in the world.

By spring 2005, the five-year drought in the Colorado River basin had lowered reservoir levels enough to resurrect some of those ghosts. *National Geographic* magazine published an article depicting many of the magnificent vistas and historical treasures that were uncovered by the receding waters. Twilight Canyon, Reflection Canyon, Fort Moqui, Tapestry Wall, and Cathedral in the Desert. I was fortunate enough to boat in to Cathedral in early May 2005. It was just beginning to submerge as spring runoff raised reservoir levels following the first relatively wet winter in six years. A friend and I were returning from a trail race in Colorado, and had just one day to divert to Bullfrog Marina, from which you can reach Cathedral in a day trip by small motorboat. After cruising down the reservoir, we turned right into the bay formed by the junction of Lake Powell and the Escalante River, and left into a smaller side canyon. Because of my amateur boating skills the motor kept cutting out as I tried to maneuver through the very shallow water and make the tight turn into an even smaller side canyon. Multicolored sandstone walls twist to the sky, where a small, irregularly shaped opening lets in streams of light that create shifting, shimmering reflections on water and the rock, creating a mixed feeling of serenity and wonder. Photographer Elliot Porter wrote: "Of all the phenomena of the side canyons, it is the light, even in the farthest depths of the narrowest canyon, that evokes the ultimate in awe."[33]

Great artists have expressed frustration about their inability to capture, in either pictures or words, the magic of Glen Canyon before the dam. Russell Martin explained, "How in the world could anyone reproduce in two dimensions the visual symphonies, the plays of light on stone, the splendid skies, the magic movement of water?" Elliott Porter decided that it was impossible to photograph this place in broad vistas, and that the only way to capture some of its spirit was to focus on "the visual details of delight," features such as branches, pools, rock walls, lichens, and mosses. Still, Katie Lee critiqued those images as "beautiful photos that utterly fail to capture the spirit of Glen Canyon." Georgia O'Keeffe accompanied Porter to Glen Canyon in summer 1961, and returned several times. Apparently she shared Porter's view, painting only microcosmic abstractions of water and rock, and finding only a few of them good enough to exhibit.[34]

Even Wallace Stegner, whose words illuminate as well as anyone's the beauty of the American West, wrote that "word pictures are about as inadequate as painted ones." But Stegner perhaps captured Glen Canyon better than anyone else: "As beautiful as any of the canyons, it is almost absolutely serene, an interlude for a pastoral flute." More than any other issue, of course, beauty varies with perspective. In his fore-word to the Bureau of Reclamation's glossy *Lake Powell, Jewel of the Colorado*, Interior Secretary Stewart Udall referred to the filling reservoir as "almost unbelievable beauty," in its "setting of incomparable grandeur." Stegner agreed, at least to some extent, that Lake Powell remained beautiful, and in some ways, perhaps even better. It was accessible to more people, and places that could be glimpsed only briefly or after a strenuous climb could now be seen by boat. Still, he lamented: "And yet, as vast and beautiful as it is, open now to anyone with a boat or the money to rent one, available soon (one supposes) to the quickie tour by floatplane and hydrofoil, democratically accessible and with its most secret beauties captured on color transparencies at infallible exposures, it strikes me, even in my exhilaration, with the consciousness of loss. In gaining *the lovely and the usable*, we have given up the incomparable."[35]

What role should aesthetics and emotion play in the debate over Glen Canyon and the restoration of the Colorado River? As with so much else in this web of conflicting interests, that becomes a matter of values. To those who wish to restore the magic and the mystery of this special place, it is everything. Pragmatists reply that restoring Glen Canyon simply for the sake of beauty is romantic folly. In a world that needs water and power and development, the sheer logic of economics, engineering, and science is more important. Arguably the greatest scientist of the 20th century, Albert Einstein, might disagree: "The most beautiful and most profound emotion we can experience is the sensation of the mystical. It is the source of all true science. He to whom this emotion is a stranger, who can no longer wonder and stand rapt in awe, is as good as dead."[36]

Stegner and others realized, though, that we lost much more than sheer beauty when Glen Canyon submerged beneath the reservoir: "Silt pockets out of reach of flood were gardens of fern and redbud; every talus and rockslide gave footing to cottonwood and willow and single-leafed ash; ponded places were solid with watercress; maidenhair hung from seepage cracks in the cliffs." Katie Lee sought in vain to list all of the wildlife she saw in Glen Canyon:

> quail, water ouzel, kingfisher, canyon wren, phainopepla, cliff swallow, western tanager, cardinal, snowy egret, blue heron, Canadian honker, duck, great horned owl, screech owl, redtail and other hawks, peregrine falcon, uncountable LBBs (little brown birds), and ravens-ravens-ravens! There have been deer, beavers, ringtail cats, pack rats, field mice, coyotes, bobcats,

lizards, chuckwallas, frogs, toads, raccoons, skunks, rabbits, chipmunks, snakes (rattle, water, grass, and gopher), catfish, chubs, minnows, pollywogs, little green turtles, a desert tortoise, those primordial-looking shrimp, bees, wasps, ants, moths, bats, and black widows.[37]

When the Bureau of Reclamation dammed Glen Canyon, we lost a living canyon much different from the cliff-bound channels of Cataract Canyon above and Marble and Grand canyons below. I realized this as we motored back from Cathedral in the Desert toward Bullfrog Marina, and stared at the water that now climbs the canyon walls, burying the sandbars and the deltas of side canyons that once provided habitat for so many species. This is part of what was different about this relatively calm, rapid-free stretch of the river. And that is important if we are to adopt a broader vision of Colorado River restoration. Existing restoration efforts assume that former habitats now submerged beneath reservoirs are lost forever. Some ecologists believe it significant that the more level flows through Grand Canyon allowed the formation of new riparian and marsh habitats that are so rare in the region.[38] Those habitats are now rare, however, only because so much was destroyed where it occurred naturally, such as Glen Canyon. In that sense, it is artificial to argue that we must choose between the health of in-stream aquatic habitats and the new riparian bounty within Grand Canyon (chapter 6). Restore the much larger areas of marshes and riparian zones that once lined Glen Canyon, and the artificially induced replacement habitats in Grand Canyon will no longer be so significant. It might take some time and some additional help, and they would not be exactly the same as before the dam, but over time those habitats would return.

If the reservoir is drained and the dam decommissioned, it might be possible to restore the natural habitats within Glen Canyon, *and* the natural ecosystem within the Grand Canyon, *and* provide the additional flows necessary to help restore portions of the lower river and the delta. That would be true, of course, only if the resulting increased flows are actually dedicated to restoration rather than being siphoned off as surplus for California, storage in the Arizona Water Bank, or groundwater recharge in Mexico. Decommissioning would constitute a dramatic step with significant economic and other consequences. That can be dismissed as infeasible and impracticable, unless we look beyond the usual scope of environmental restoration tasks and adopt a more inclusive concept of what changes in economic policies and resource uses are appropriate as a part of environmental restoration programs, and perhaps a much broader vision of what the river itself can be.

I do not intend to argue necessarily that decommissioning Glen Canyon Dam is the "right" thing to do. Perhaps decommissioning would be more feasible or produce even greater environmental benefits at Flaming Gorge or Fontenelle dam upriver, or at Laguna Dam below. Maybe another restoration approach altogether would provide a better balance between human needs and environmental health. The main point is that existing analytical approaches to Colorado River restoration have been confined just as much as the water held behind the dams. We approach restoration decisions in the wrong order by allowing legal and institutional decisions made long ago to constrain choices about our goals for the river for future generations, and about the best means to achieve those goals. The 1922 Colorado River Compact, laws passed in the 1950s and 1960s, and other aspects of the Law of the River define a rigid set of boundaries within which current restoration efforts must function. As a result, new decisions are limited by legal and other constraints that may no longer be valid. Likewise, because existing river uses and policies are taken as a given, physical restoration approaches are restricted to discrete modifications to existing facilities and rehabilitation of isolated patches of habitat, rather than potentially more sweeping changes in the fundamental ways in which we use and relate to the river and its surroundings, such as changes in water use and distribution, power generation and consumption, and land development at the water's edge.

Instead, we should begin our analysis with fundamental choices about the uses and values for which the Colorado River should be restored, managed, and protected, what we might call a new vision for the river. If the Colorado River Compact and other legal and institutional arrangements are no longer consistent with that new vision, perhaps they need to be changed before more effective restoration approaches (such as dam decommissioning) can be evaluated more seriously. If that newer vision affects the manner in which we use water and power and other resources, or if the restoration methods needed to achieve them are not possible given existing technologies and river management practices, those choices should be revisited as well. A much broader range of strategies for river use and restoration is then possible. Just as Colorado River restoration efforts should not be constrained by artificial political boundaries, they should also not be limited by artificial legal and policy decisions made by past generations. Then, perhaps we can obtain both the lovely *and* the usable.

Into New Dimensions

Thinking back on the phenomenal complexity of efforts to restore Colorado River ecosystems, Clarence Dutton's impression from the North Rim of the Grand Canyon seems all the more appropriate: "Dimensions mean nothing to the senses, and all that we are conscious of in this respect is *a troubled sense of immensity.*"[1]

The river and its ecosystems are impaired by hundreds of discrete insults. Artificial modifications have changed not only individual ecosystem components, but fundamental physical, hydrological, and ecological processes and conditions. Some of those changes are likely to be irreversible, especially along the most severely modified reaches of the river. Others will require considerably more research and experimentation before effective restoration methods are identified.

Using current approaches, we are making progress in restoring some of the ecological integrity of the river and its associated habitats. In the upper river, we are removing some barriers to the movement of fish and other resources, reconnecting parts of the river to its natural floodplain and backwater habitats, and restocking native fish populations with fish reared in hatcheries and artificial ponds and backwaters. Below Glen Canyon Dam, we are trying to recover native fish populations with modified flows, timed to make use of new sediment that enters the river from tributaries following storms. We are replanting some lands along the lower river with native species designed to provide habitat for southwestern willow flycatchers and other species.

Unfortunately, existing efforts are falling short of what is needed to restore some of the key ecological and aesthetic attributes that once characterized the majestic Colorado River. Time may be running short for some components of Colorado River ecosystems, especially the four species of endangered big river fish. Current programs are limited in funding, other resources, and sometimes political support. Ecological restoration programs for the Colorado are plagued by scientific uncertainty, difficult choices between competing river uses and values, and entrenched legal and political institutions. Almost every existing or proposed restoration strategy has impacts on other river uses and values, which trigger opposition from various user groups, delays in implementing effective restoration methods, and sometimes complete policy gridlock.

One response to an undertaking that is overwhelming in its complexity is to narrow your focus, break things down into their component parts, and work on them in smaller pieces. "In the face of immensity it may be that the mind, to preserve its sanity, seeks relief in something small and comprehensible."[2] That works for some kinds of endeavors, especially those for which action in one area has little or no impact on others, or where those impacts are predictable enough to know how to proceed on one front without causing irreparable damage elsewhere. It is also an effective strategy where solutions are relatively well known, but difficult to implement. None of those conditions are true for Colorado River ecosystem restoration. Yet we have still proceeded with restoration with an unduly narrow focus and by addressing discrete aspects of the problem independently.

Current environmental laws and regulations place an almost insurmountable burden on scientists and managers to prove that new restoration strategies will succeed, and that no adverse impacts will occur to other resources. Where potentially effective solutions to one problem might adversely affect other uses and values, the manner and extent to which we have insisted on collaborative process and full consensus among interest groups can do more to preserve the status quo than to promote effective, innovative solutions. Existing programs remain fragmented along different reaches of the river, with little effort to consider impacts among programs. Those programs are also unduly narrow in both geographic and conceptual scope. They consider restoration within the main stem of the river but not in critical adjacent habitats, in part because private property rights are protected absolutely at the expense of ecological integrity, even within land-water transition zones where public needs should transcend private development interests. They address ecological needs within the United States, but not across the U.S.-Mexico border despite severe, demonstrated effects of U.S. decisions within Mexico, and even though there may be greater potential to restore ecological values south of the international border than immediately to the north.

Most tellingly, restoration solutions to date have been limited strictly to those consistent with the Law of the River and the entrenched legal and economic expectations that have developed based on a deal struck in 1922 by representatives of only some of the interests that now use or value the river and its resources. We have not yet had the wisdom to recognize that ecological, economic, and political conditions have changed dramatically since 1922, and that greater scientific understanding of the basin's current and potential future hydrology and climate suggests that the Colorado River Compact and other aspects of the Law of the River should be reconsidered.

Ironic though it may seem, a better approach to the immense complexity of Colorado River restoration may be to broaden rather than narrow our focus, and to consider a wider range of issues and potential solutions than are reflected in existing efforts and institutions. Existing environmental laws such as the National Environmental Policy Act and the Endangered Species Act should be interpreted or amended to encourage rather than impede sound experimentation with new environmental restoration strategies. Perhaps a new federal Environmental Restoration Act is needed to affirmatively promote and support restoration efforts on the Colorado River and elsewhere. Adaptive management is a potentially effective approach to scientific uncertainty in restoration programs, but not when endless process does more to impede than to facilitate the experimentation that is at the heart of the adaptive management idea. Collaborative stakeholder programs can be used to make fundamental decisions about restoration goals and choices, but not to micromanage program implementation in ways that render those goals and choices impossible to attain. Where key choices among competing uses and values cannot be made through collaborative process, they must be made at higher political levels. Although such choices can be frightening, the absence of decisions (gridlock) can be even more destructive.

More broadly focused restoration programs can also consider strategies that have been ignored, deemed not viable, or not even considered in the past. Major dam decommissioning is one of potentially many restoration strategies that might become viable if we rethink the basic economic uses to which we put the Colorado River, such as water, power generation, and recreation, and alternative ways of fulfilling those needs. The entire concept of environmental restoration should be broad enough to encompass reshaping and revitalizing our economy in ways that are more compatible with environmental health and integrity. In the long run, perhaps both the economy and the environment will be better off if we do so.

Geologist Clarence Dutton lamented the difficulties of comprehending the physical immensity he witnessed from the rim of the Grand Canyon. More recently, from below the canyon rim, naturalist Ann Zwinger reminded us of the human capacity to transcend the unimaginable by adopting a new focus: "1.2 billion missing years right at eye level, a change between rock types so unmistakable and spectacular, so conceptually overwhelming, that once it engages your attention, you watch it weave through the canyon, a reminder of the human ability to conjecture about what has gone before, the briefness of life on earth compared to the eons of this rock, and the need to stretch the little gray cells *into new dimensions.*"[3] We never may have known the natural wonders of the remote reaches of the Colorado River had intrepid explorers not braved unknown canyons and rapids to find them. It will take a different kind of courage to explore new dimensions of law, science, and policy so that we can find and achieve a new vision for the Colorado River, and for other disturbed ecosystems around the world.

Endnotes

Preface

1. Philip L. Fradkin, *A River No More: The Colorado River and the West* (Berkeley: University of California Press, 1995), xxvi.
2. John McPhee, *Encounters with the Archdruid* (New York: Farrar, Straus, and Giroux, 1971), 161.
3. Clarence E. Dutton, *Tertiary History of the Grand Cañon District* (University of Arizona Press, 2001), 149–150, first published in 1882 by the U.S. Geological Survey (emphasis added).

Chapter 1

1. Lynn White, "The Historical Roots of Our Ecologic Crisis," *Science* 155 (1967): 1203.
2. Robert B. Stanton, in *Down the Colorado*, ed. Dwight L. Smith (Norman: University of Oklahoma Press, 1965), 19.
3. Russell Martin, *A Story That Stands Like a Dam: Glen Canyon and the Struggle for the Soul of the West* (New York: Henry Holt, 1989); Philip L. Fradkin, *A River No More: The Colorado River and the West* (Berkeley: University of California Press, 1995); Colin Fletcher, *The Man Who Walked Through Time* (New York: Vintage Books, 1989), 144 ("flicker of a nod"); John C. Schmidt, "The Colorado River," in *Large Rivers*, ed. A. Gupta (London: John Wiley, in press).
4. Schmidt, "The Colorado River"; Dale Pontius, *Colorado River Basin Study* (Denver: Western Water Policy Review Advisory Commission, 1997); Michael Collier et al., "Dams and Rivers: A Primer on the Downstream Effects of Dams," U.S. Geological Survey Circular 1126 (1996); Michael Collier, "Mags Hot, Master On: Flying for the Vision," in *Water, Earth and Sky: The Colorado River Basin*, ed. Michael Collier et al. (Salt Lake City: University of Utah Press, 1999); Gordon A. Mueller and Paul C. Marsh, *Lost, a Desert River and Its Native Fishes: A Historical Perspective of the Lower Colorado River*, Information and Technology Report USGS/BRD/ITR-2002-0010 (2002).
5. Donald Worster, *Rivers of Empire: Water, Aridity and the Growth of the American West* (New York: Oxford University Press, 1985), 34; Fradkin, *A River No More*, 21.
6. Pontius, *Colorado River Basin Study*.
7. U.S. Department of Energy, Energy Information Administration, "Basic Electricity Statistics," http://www.eia.doe.gov/neic/quickfacts/quickelectric.html (accessed April 19, 2006).

8. 56 *Federal Register* 54,957 (Oct. 23, 1991) (razorback sucker); 45 *Federal Register* 27,710 (Apr. 23, 1980) (bonytail chub); 32 *Federal Register* 4001 (Mar. 11, 1967) (humpback chub and Colorado squawfish [pikeminnow]).

9. Robert H. Webb et al., *The Controlled Flood in Grand Canyon* (Washington, D.C.: American Geophysical Union, 1999); Steven Carothers and Richard A. Valdez, "The Aquatic Ecosystem of the Colorado River in Grand Canyon," *Grand Canyon Data Integration Project Synthesis Report* (Prepared for the Bureau of Reclamation by SWCA, Inc., Flagstaff, AZ, July 1, 1998); Steven Carothers and C.O. Minckley, *A Survey of the Fishes, Aquatic Invertebrates and Aquatic Plants of the Colorado River and Selected Tributaries from Lees Ferry to Separation Rapids* (Prepared for the Bureau of Reclamation by the Museum of Northern Arizona, Flagstaff, AZ, 1981); Steven Carothers and Bryan T. Brown, *The Colorado River Through Grand Canyon* (Tucson: University of Arizona Press, 1991).

10. Worster, *Rivers of Empire*, 19–21; Charles F. Wilkinson, "The Headwaters of the Public Trust: Some Thoughts on the Source and Scope of the Traditional Doctrine," *Environmental Law* 19 (1989): 431–435.

11. Robert A. Abell et al., *Freshwater Ecoregions of North America: A Conservation Assessment* (Washington, D.C.: Island Press, 2000), 17–20, 62–70; Robert W. Adler, "The Two Lost Books in the Water Quality Trilogy: The Elusive Objectives of Physical and Biological Integrity," *Environmental Law* 33 (2003): 71; National Research Council, *Riparian Areas: Functions and Strategies for Management* (Washington, D.C.: National Academy Press, 2002), 8–13.

12. Joseph L. Sax, "The New Age of Environmental Restoration," *Washburn Law Journal* 41 (2001): 1; Storm Cunningham, *The Restoration Economy: The Greatest New Growth Frontier* (San Francisco: Berrett-Koehler, 2002).

13. Dade W. Moeller, *Environmental Health* (Cambridge: Harvard University Press, 2005); Robert W. Adler et al., *The Clean Water Act 20 Years Later* (Washington, D.C.: Island Press, 1993); Stewart Udall, *The Quiet Crisis* (New York: Holt, Reinhart, 1963); John Volkman, *A River in Common: The Columbia River, the Salmon Ecosystem, and Water Policy* (Denver: 1997), 38; Michael C. Blumm, *Sacrificing the Salmon: A Legal and Political History of the Decline of the Columbia Basin Salmon* (Den Bosch, Netherlands: Bookworld Publications, 2002); Cunningham, *The Restoration Economy*, 68.

14. Ronald H. Coase, "The Problem of Social Cost," *Journal of Law and Economics* 3 (1960): 1; William F. Baxter, *People or Penguins: The Case for Optimal Pollution* (New York: Columbia University Press, 1974); Arthur P. Hurtur et al., "Benefit-Cost Analysis and the Common Sense of Environmental Policy," in *Cost-Benefit Analysis and Environmental Regulations: Politics, Ethics, and Methods*, ed. Daniel Swartzman et al. (Washington, D.C.: The Conservation Foundation, 1982); Frank Ackerman and Lisa Heinzerling, *Priceless: On Knowing the Price of Everything and the Value of Nothing* (New York: W.W. Norton and Company, 2004); David M. Driesen, *The Economical Dynamics of Environmental Law* (Cambridge: MIT Press, 2003).

15. National Environmental Policy Act, *U.S. Code* 42, §§ 4321 et seq.; Endangered Species Act, *U.S. Code* 16, §§ 1532 et seq.; Comprehensive Environmental Response, Compensation, and Liability Act (Superfund), *U.S. Code* 42, §§ 9601 et seq.; Resource Conservation and Recovery Act, *U.S. Code* 42, §§ 6901 et seq.; Joel S. Hirschorn, "Pollution Prevention Comes of Age," *Georgia Law Review* 29 (1995): 325.

16. Eric Higgs, *Nature by Design* (Cambridge: MIT Press, 2003); Donald A. Falk et al., *Foundations of Restoration Ecology* (Washington, D.C.: Island Press, 2006); Society for Ecological Restoration International, Science and Policy Working Group, *The SER International Primer on Ecological Restoration* (Tucson: Society for Ecological Restoration International, 2004); Cunningham, *The Restoration Economy* (also using the holistic medicine analogy).

17. John C. Schmidt et al., "Science and Values in River Restoration in the Grand Canyon," *BioScience* 48, no. 9 (1998): 735.

18. Eric Katz, "Another Look at Restoration: Technology and Artificial Nature," in *Restoring Nature: Perspectives from the Social Sciences and Humanities*, ed. Paul H. Gobster and R. Bruce Hill (Covelo, CA: Island Press, 2000), 37; Frank Chessa, "Endangered Species and the Right to Die," *Environmental Ethics* 27 (2005): 23–41; William R. Jordan III, "Restoration, Community, and Wilderness," in *Restoring Nature: Perspectives from the Social Sciences and Humanities*, ed. Paul H. Gobster and R. Bruce Hill (Covelo, CA: Island Press, 2000), 23; Andrew Light, "Ecological Restoration and the Culture of Nature: A Pragmatic Perspective," in *Restoring Nature: Perspectives from the Social Sciences and Humanities*, ed. Paul H. Gobster and R. Bruce Hill (Covelo, CA: Island Press, 2000), 49; Cunningham, *The Restoration Economy*.

19. National Research Council, *Restoration of Aquatic Ecosystems* (Washington, D.C.: National Academy Press, 1992), 17; Carothers and Valdez, "Aquatic Ecosystem of the Colorado River"; Edward Dolnick, *Down the Great Unknown: John Wesley Powell's 1868 Journey of Discovery and Tragedy Through the Grand Canyon* (New York: Harper Collins, 2001), 265; Society for Ecological Restoration International, *Primer on Ecological Restoration*, 3.

20. Society for Ecological Restoration International, *Primer on Ecological Restoration*, 2, 9; Dave Egan and Evelyn A. Howell, "Introduction," in *The Historical Ecology Handbook: A Restorationist's Guide to Reference Ecosystems*, ed. Dave Egan and Evelyn A. Howell (Washington, D.C.: Island Press, 2001), 1; Tabatha J. Wallington et al., "Implications of Current Ecological Thinking for Biodiversity Conservation: A Review of the Salient Issues," *Ecology and Science* 10, no. 1 (2005): 15.

21. Higgs, *Nature by Design*, 10, 38; Tabatha J. Wallington and Susan A. Moore, "Ecology, Values and Objectivity: Advancing the Debate," *BioScience* 55, no. 10 (2005): 873–878; Schmidt et al., "Science and Values in River Restoration."

22. Mary Austin, *The Land of Little Rain* (New York: Houghton, Mifflin and Company, 1903), 35–37 (emphasis added).

23. U.S. Department of the Interior, Bureau of Reclamation, *Area Searches at the Cibola Nature Trail Restoration Site and the Pratt Restoration Site, Breeding Season 2004*.

24. Society for Ecological Restoration International, *Primer on Ecological Restoration*, 2; Katharine N. Suding and Katherine L. Gross, "The Dynamic Nature of Ecological Systems: Multiple States and Restoration Trajectories," in Falk et al., *Foundations of Restoration Ecology*, 190.

25. Lawrence J. Macdonnell, *From Reclamation to Sustainability: Water, Agriculture, and the Environment of the American West* (Niwot, CO: University Press of Colorado, 1999); The H. John Heinz III Center for Science, Economics and the Environment, *Dam Removal, Science and Decision Making* (Washington, D.C.: 2002); American Rivers, Friends of the Earth, and Trout Unlimited, http://www.americanrivers.org/site/PageServer?pagename=AMR_content_8cf8 (accessed December 26, 2006).

26. John McPhee, *Encounters with the Archdruid* (New York: Farrar, Straus, and Giroux, 1971); Roderick Nash, *Wilderness and the American Mind* (New Haven: Yale University Press, 1967); Byron E. Pearson, *Still the Wild River Runs: Congress, the Sierra Club, and the Fight to Save Grand Canyon* (Tucson: University of Arizona Press, 2002); David Brower, "Let the River Run Through It," *Sierra Club Magazine*, March 1997.

27. Edward Abbey, *Desert Solitaire* (Tucson: University of Arizona Press, 1988), 165 ("unsung hero" quote); Edward Abbey, *The Monkey Wrench Gang* (Philadelphia: Lippincott, 1975), xxii (Thoreau quote); Jim Lichatowich, *Salmon Without Rivers: A History of the Pacific Salmon Crisis* (Washington, D.C.: Island Press, 1999).

28. Glen Canyon Institute, *Citizens' Environmental Assessment (CEA) on the Decommissioning of Glen Canyon Dam, Report on Initial Studies* (December 2000); David L. Wegner, "Looking Toward the Future: The Time Has Come to Restore Glen Canyon," *Arizona Law Review* 42 (2000): 239; Scott Miller, "Undamming Glen Canyon: Lunacy, Rationality, or Prophecy?" *Stanford Environmental Law Journal* 19 (2000): 121.

29. House Committee on Resources, Subcommittee on National Parks and Public Lands and Subcommittee on Water and Power, *Sierra Club's Proposal to Drain Lake Powell or Reduce Its Water Storage Capability*, 105th Cong., 1st sess., 1997 (all quotes); Martin, *A Story That Stands Like a Dam*, 170–171; Katie Lee, *All My Rivers Are Gone: A Journey of Discovery Through Glen Canyon* (Boulder: Johnson Books, 1998), 29, 64.

30. House Committee on Resources, *Sierra Club's Proposal to Drain Lake Powell*; Aldo Leopold, *A Sand County Almanac* (London: Oxford University Press, 1968), 158; Nash, *Wilderness and the American Mind*, 262–263.

31. Brower, "Let the River Run Through It"; House Committee on Resources, *Sierra Club's Proposal to Drain Lake Powell*; Lee, *All My Rivers Are Gone*, 249.

32. John McPhee, "Farewell to the Nineteenth Century, The Breaching of Edwards Dam," *The New Yorker*, April 1999.

33. Doug Bereuter, "Eighth Report on the Hong Kong Transition," August 1, 2000, http://www.house.gov/international_relations/ap/hk8.htm (accessed April 11, 2001); "Modern Hong Kong History," *Inside China Today*, wysiwyg://52/http://www.insidechina.com/hktrans/hkhist.php (accessed April 13, 2001); "Hong Kong Special Administrative Region," *Inside China Today*, wysiwyg://52/http:// www.insidechina.com/ hktrans/hkreg.php (accessed April 13, 2001); "Everything You Wanted to Know About the Handover (But Were Afraid to Ask)," *Time Asia*, http://www.time.com/time/hongkong/ special/handover.html (accessed April 12, 2001); Edward A. Gargan, "Hong Kong's Legal System Braces for Chinese Rule," *The New York Times Hong Kong*, http://www.nytimes.com/specials/hongkong/archive/050797hong-kong-law.html (accessed April 19, 2001); Graeme R. Halford and Kwung-Yee Cheung, "Hong Kong After the Transition of Sovereignty," http://www.insol.org/newinsol-world/nov97/ NLnov97pthk.html (accessed April 12, 2001); Donald Tong, "Hong Kong: 7 Months After the Handover," presented at the David See-Chai Lam Centre for International Communication Pacific Region Forum on Business and Management Communication on February 24, 1998, http://www.cic.sfu.ca/forum/DonaldTongMar301998.html (accessed April 11, 2001).

34. Colorado River Compact, *Utah Code Annotated*, § 73-12a-2; Boulder Canyon Project Act of 1928, 45 Stat. 1064, *U.S. Code* 43, § 617l(a); Ray L. Wilbur and Northcutt Ely, *The Hoover Dam Documents*, H.R. Doc. No, 717, 80th Cong., 2d sess. 22 (Washington, D.C., 1948).

35. *Arizona v. California*, 373 U.S. 546 (1963); *Navajo Nation v. United States Department of the Interior, et al.*, Civil No. 03-CV-507 (District of Arizona, filed March 14, 2003).

36. Upper Colorado River Compact, Act of April 6, 1949, c. 48, 63 Stat. 31; Boulder Canyon Project Act, *U.S. Code* 43, § 617 et seq.; *Arizona v. California*, 373 U.S. 546 (1963); "Treaty for the Utilization of Waters of the Colorado, Tijuana and Rio Grande Rivers," Feb. 13, 1944, U.S.-Mex., art. 10, 59 Stat. 1219.

37. Norris Hundley, *Water and the West: The Colorado River Compact and the Politics of Water in the American West* (Berkeley: University of California Press, 1975); Charles J. Meyers, "The Colorado River," *Stanford Law Review* 19 (1966): 1; Charles J. Meyer, "The Colorado River: The Treaty with Mexico," *Stanford Law Review* 19 (1967): 367; Felix L. Sparks, "Synopsis of Major Documents and Events Relating to the Colorado River," *University of Denver Water Law Review* 3 (2000): 339–356 (with Introduction by James S. Lochhead, originally published in 1976); Joseph Sax et al., *Legal Control of Water Resources: Cases and Materials* (St. Paul: West Pub Co., 3rd ed., 2000).

38. *Wyoming v. Colorado*, 259 U.S. 419 (1922).

39. U.S. Department of the Interior, *Colorado River Reservoirs: Coordinated Long-Range Operation*, 35 *Federal Register* 8951 (June 10, 1970).

40. *U.S. Code* 43, § 620(f), 1552(a)(3).

41. Carothers and Brown, *The Colorado River Through Grand Canyon*.

Chapter 2

1. Wallace Stegner, *Mormon Country* (New York: Hawthorn Books, 1970), 293, 297.
2. R. Dana Ono et al., *Vanishing Fishes of North America* (Washington, D.C.: Stonewall Press, 1983), 87–91; R.R. Miller, "Origin and Affinities of the Freshwater Fish Fauna of Western North America," *Zoogeography*, Publication 51 (1958); Richard S. Wydoski and John Hamill, "Evolution of a Cooperative Recovery Program for Endangered Fishes in the Upper Colorado River Basin," in *Battle Against Extinction: Native Fish Management in the American West*, ed. W.L. Minckley and James E. Deacon (Tucson: University of Arizona Press, 1991); Michael Collier et al., "Dams and Rivers: A Primer on the Downstream Effects of Dams," U.S. Geological Survey Circular 1126 (1996); Richard Valdez, "Of Suckers, Chubs, and 100-Pound Minnows," in *Water, Earth and Sky: The Colorado River Basin*, ed. Michael Collier et al. (Salt Lake City: University of Utah Press, 1999), 71 (quote).
3. Robert T. Muth et al., *Flow and Temperature Recommendations for Endangered Fishes in the Green River Downstream of Flaming Gorge Dam* (Upper Colorado River Endangered Fish Recovery Program, September 2000); U.S. Fish and Wildlife Service, *Colorado Pikeminnow* (Ptychocheilus lucius) *Recovery Goals: Amendment and Supplement to the Colorado Squawfish Recovery Plan* (Denver: U.S. Fish and Wildlife Service, Mountain-Prairie Region [6], 2002); Harold M. Tyus, "Ecology and Management of Colorado Squawfish," in *Battle Against Extinction*, ed. W.L. Minckley and James E. Deacon (Tucson: University of Arizona Press, 1991), 379; Fred Quartarone, *Historical Accounts of Upper Colorado River Basin Endangered Fish* (Information and Education Committee of the Recovery Program for Endangered Fish of the Upper Colorado River Basin, Final Report, Revised Edition, September 1995) (all quotes); Wydoski and Hamill, "Evolution of a Cooperative Recovery Program for Endangered Fish," in *Battle Against Extinction*, ed. Minckley and Deacon, 123.
4. Robert B. Stanton, *Down the Colorado*, ed. Dwight L. Smith (Norman: University of Oklahoma Press, 1965), 108–109; Steven Carothers and Bryan T. Brown, *The Colorado River Through Grand Canyon* (Tucson: University of Arizona Press, 1991), 97 (Kolb quote); Valdez, "Of Suckers, Chubs, and 100-Pound Minnows," in *Water, Earth and Sky*, Collier et al., 71, 83.
5. Tyus, "Ecology and Management of Colorado Squawfish," in *Battle Against Extinction*, ed. Minckley and Deacon, 384, 399 ("Loss of the Colorado squawfish"); Quartarone, *Historical Accounts*, 5; U.S. Fish and Wildlife Service, *Pikeminnow Recovery Goals*, A-12.
6. Robert E. Ricklefs, *Ecology* (Newton, MA: Chiron Press, 1973), 529–569.
7. Fish and Wildlife Service, *Pikeminnow Recovery Goals*, A-4, A-8–A-10; Minckley and Deacon, *Battle Against Extinction*, ed. Minckley and Deacon, 3; R.R. Miller, "Man and the Changing Fish Fauna of the American Southwest," *Michigan Academy of Science, Arts, and Letters* 46 (1961): 365.
8. Tyus, "Ecology and Management of Colorado Squawfish," in *Battle Against Extinction*, ed. Minckley and Deacon, 395–396; Jim Lichatowich, *Salmon Without Rivers: A History of the Pacific Salmon Crisis* (Washington, D.C.: Island Press, 1999), 76.
9. Tyus, "Ecology and Management of Colorado Squawfish," in *Battle Against Extinction*, ed. Minckley and Deacon, 396; John Volkman, *A River in Common: The Columbia River, the Salmon Ecosystem, and Water Policy* (Denver: Western Water Policy Review Advisory Commission, 1997).
10. Robert H. Webb et al., *Climatic Fluctuations, Drought, and Flow in the Colorado River Basin*, U.S. Geol. Surv. Fact Sheet 2004-3062.
11. Tyus, "Ecology and Management of Colorado Squawfish," in *Battle Against Extinction*, ed. Minckley and Deacon, 383; Fish and Wildlife Service, *Pikeminnow Recovery Goals*, A-7.
12. Fish and Wildlife Service, *Pikeminnow Recovery Goals*, A-3, A-6.

13. Tyus, "Ecology and Management of Colorado Squawfish," in *Battle Against Extinction*, ed. Minckley and Deacon, 379; W.L. Minckley et al., "Management Toward Recovery of the Razorback Sucker," in *Battle Against Extinction*, ed. W.L. Minckley and James E. Deacon (Tucson: University of Arizona Press, 1991), 303.

14. Tyus, "Ecology and Management of Colorado Squawfish," in *Battle Against Extinction*, ed. Minckley and Deacon, 384–395.

15. Anthony A. Echelle, "Conservation Genetics and Genetic Diversity in Freshwater Fishes of Western North America," in *Battle Against Extinction*, ed. W.L. Minckley and James E. Deacon (Tucson: University of Arizona Press, 1991), 151.

16. R.M. Hirsch et al., "The Influence of Man on Hydrologic Systems," in *Surface Water Hydrology*, ed. M.G. Wolman and H.C. Riggs, series *The Geology of North America*, vol. O-1 (Boulder, CO: Geologic Society of North America, 1990), 329–359; Dale Pontius, *Colorado River Basin Study* (Denver: Western Water Policy Review Advisory Commission, 1997).

17. Marc Reisner, *Cadillac Desert: The American West and Its Disappearing Water* (New York: Viking, 1986).

18. Michael Collier, "Mags Hot, Master On: Flying for the Vision" in *Water, Earth and Sky: The Colorado River Basin*, ed. Collier et al. (Salt Lake City: University of Utah Press, 1999), 19.

19. Pontius, *Colorado River Basin Study*; Collier, "Mags Hot, Master On," *Water, Earth and Sky: The Colorado River Basin*, ed. Collier et al., 29 ("significantly impede").

20. Collier, "Mags Hot, Master On," *Water, Earth and Sky: The Colorado River Basin*, ed. Collier et al., 17, 19–20.

21. Philip L. Fradkin, *A River No More: The Colorado River and the West* (Berkeley: University of California Press, 1995), xxii; John C. Schmidt, "The Colorado River," in *Large Rivers*, ed. A. Gupta (London: John Wiley, in press), 20; Jennifer Pitt et al., "Two Nations, One River: Managing Ecosystem Conservation in the Colorado River Delta," *Natural Resources Journal* 40 (2000): 819; Robert Meredith, *A Primer on Climatic Variability and Change in the Southwest* (Tucson, AZ: University of Arizona, 2000), 3; Environmental Defense Fund, *A Delta Once More: Restoring Riparian and Wetland Habitat in the Colorado River Delta* (Washington, D.C.: Environmental Defense Fund, 1999), 4.

22. Timothy Hoffnagle et al., "Fish Abundance, Distribution, and Habitat Use," in *The Controlled Flood in Grand Canyon*, Robert H. Webb et al. (Washington, D.C.: American Geophysical Union, 1999), 273; Minckley et al., "Management Toward Recovery of the Razorback Sucker," in *Battle Against Extinction*, ed. Minckley and Deacon, 327; Muth et al., *Flow and Temperature Recommendations*; Environmental Defense Fund, *A Delta Once More*; R.D. Ohmart et al., *The Ecology of the Lower Colorado River from Davis Dam to the Mexico-United States International Boundary: A Community Profile*, U.S. Fish and Wildlife Service Report 85 (7.10) (1988); Edward Glenn et al., "Ecology and Conservation Biology of the Colorado River Delta, Mexico," *Journal of Arid Environments* 49 (2001): 5–15; Gordon A. Mueller and Paul C. Marsh, *Lost, a Desert River and Its Native Fishes: A Historical Perspective of the Lower Colorado River*, Information and Technology Report USGS/BRD/ITR-2002-0010 (2002).

23. Thomas V. Cech, *Principles of Water Resources: History, Development, Management, and Policy* (New York: Wiley, 2003), 75; Carothers and Brown, *The Colorado River Through Grand Canyon*, 22.

24. E.D. Andrews, "Wet River, Dry River," in *Water, Earth and Sky*, ed. Michael Collier et al. (Salt Lake City: University of Utah Press, 1999), 61–63.

25. Carothers and Brown, *The Colorado River Through Grand Canyon*, 179; Steven Carothers and C.O. Minckley, *A Survey of the Fishes, Aquatic Invertebrates and Aquatic Plants of the Colorado River and Selected Tributaries from Lees Ferry to Separation Rapids* (Prepared for the Bureau of Reclamation by the Museum of Northern Arizona, Flagstaff, AZ, 1981), 2 ("markedly affected"); David Lavender, *River Runners of the Grand Canyon* (Grand Canyon, AZ: Grand Canyon Natural History Association, 1986), 132.

26. Wydoski and Hamill, "Cooperative Recovery Program," in *Battle Against Extinction*, ed. Minckley and Deacon, 129; Minckley et al., "Management Toward Recovery of the Razorback Sucker," in *Battle Against Extinction*, ed. Minckley and Deacon, 327; Tyus, "Ecology and Management of Colorado Squawfish," in *Battle Against Extinction*, ed. Minckley and Deacon, 394; Carothers and Brown, *The Colorado River Through Grand Canyon*, 67.

27. Alice Boulton, Director, Silt Historical Society, personal communication, March 29–30, 2004; John Frank Dawson, *Place Names in Colorado—Why 700 Places Were So Named, 150 of Spanish or Indian Origin* (Denver: J.F. Dawson Pub. Co., 1954).

28. Joseph C. Ives, *Report Upon the Colorado River of the West* (New York: Da Capo Press, 1969), 26, 29.

29. Edmund D. Andrews, "Sediment Transport in the Colorado River Basin," *Colorado River Ecology and Dam Management*, National Academy of Sciences, Committee to Review the Glen Canyon Environmental Studies, Proceedings of a Symposium (Washington, D.C.: National Academy Press, 1991), 63; Carothers and Minckley, *A Survey of Fishes*, 26; National Academy of Sciences, National Research Council, Committee on Water, *Water and Choice in the Colorado Basin: An Example of Alternatives in Water Management* (Washington, D.C.: National Academy of Sciences, 1968), 8; John C. Schmidt et al., *System-Wide Characteristics in the Distribution of Fine Sediment in the Colorado River Between Glen Canyon Dam and Bright Angel Creek, Arizona: Final Report to Grand Canyon Monitoring and Research Center* (Logan, UT: Utah State University, October 2004).

30. Godfrey Sykes, *The Colorado River Delta* (American Geographical Society, 1937), 80; Environmental Defense Fund, *A Delta Once More*, 2; U.S. Department of the Interior, Bureau of Reclamation, *The Colorado River: A Natural Menace Becomes a Natural Resource*, Project Planning Report No. 34-8-1, The Colorado River (1945), 35; Fradkin, *A River No More*, 182 ("too thick to drink").

31. Russell Martin, *A Story That Stands Like a Dam: Glen Canyon and the Struggle for the Soul of the West* (New York: Henry Holt, 1989), 33; Andrews, "Sediment Transport," 60–61; Sykes, *Report on the Lower Colorado River*, 7, 134.

32. Zane Grey, *The Rainbow Trail* (New York: Grosset and Dunlap, 1915); John Wesley Powell, *The Exploration of the Colorado River and Its Canyons* (New York: Dover Publications, 1961), 17, 20, 227.

33. Ives, *Report Upon the Colorado River*, 40; John Steinbeck, *The Log from the Sea of Cortez* (New York: Viking Press, 1951), 89; Ann Zwinger, *Run, River, Run: A Naturalist's Journey Down One of the Great Rivers of the West* (New York: Harper and Row, 1975), 86.

34. Mark Twain, *Roughing It* (New York: Penguin, 1981), 322.

35. Ives, *Report Upon the Colorado River*, 39, 66, 78; Aldo Leopold, *A Sand County Almanac* (London: Oxford University Press, 1968), 151–154; R.D. Ohmart et al., *The Ecology of the Lower Colorado River* .

36. Environmental Defense Fund, *A Delta Once More*, 1–3.

37. U.S. Department of the Interior, National Park Service, *A Survey of the Recreational Resources of the Colorado River Basin* (Washington, D.C.: GPO, 1950), 58; Zwinger, *Run, River, Run*, 111; Powell, *The Exploration of the Colorado*, 231–233; Lee, *All My Rivers Are Gone*, 112.

38. Ives, *Report Upon the Colorado River*, 52, 66, 73; U.S. Department of the Interior, *Survey of Recreational Resources*, 90–92; Jared Farmer, *Glen Canyon Dammed: Inventing Lake Powell and the Canyon Country* (Tucson: University of Arizona Press, 1999), 214–215; Gregory C. Crampton, *Ghosts of Glen Canyon: History Beneath Lake Powell* (St. George, UT: Publisher's Place, 1986); Powell, *Exploration of the Colorado River*, 59 ("little patch"); Lisa H. Kearsley et al., "Changes in the Number and Size of Campsites as Determined by Inventories and Measurements," in *The Controlled Flood in Grand Canyon*, ed. Robert H. Webb et al. (Washington, D.C.: American Geophysical Union, 1999), 147.

39. Fradkin, *A River No More*, 182, 243; Bureau of Reclamation, *The Colorado River: A Natural Menace Becomes a Natural Resource*, 191.

40. President's Water Resources Policy Commission, *Ten Rivers in America's Future*, Vol. 2, (Washington, D.C.: 1950), 2:400; Bureau of Reclamation, *The Colorado River: A Natural Menace Becomes a Natural Resource*, 191; Pontius, *Colorado River Basin Study*, 173.

41. William L. Graf, *The Colorado River, Instability, and Basin Management* (Washington, D.C.: Association of American Geographers, 1985), 55; National Research Council, Committee on Water, *Water and Choice in the Colorado River: An Example of Alternatives in Water Management*, 2:433; Schmidt et al., *System-Wide Characteristics in the Distribution of Fine Sediment*.

42. Stanton, *Down the Colorado*, 90–91; Clarence E. Dutton, *Tertiary History of the Grand Cañon District* (University of Arizona Press, 2001), 236–237, first published in 1882 by the U.S. Geological Survey; Robert H. Webb, *A Century of Change: Rephotography of the 1889–1890 Stanton Expedition* (Tucson: University of Arizona Press, 1996), 126–127, 138 ("knocking over the trees").

43. Webb, *A Century of Change*, 132–133, 144–146.

44. Robin L. Vannote et al., "The River Continuum Concept," *Canadian Journal of Fish and Aquatic Sciences* 37 (1980), citing G.W. Burton and E.P. Odom, "The Distribution of Stream Fish in the Vicinity of Mountain Lake, Virginia," *Ecology* 26 (1945): 182–194.

45. Vannote, "The River Continuum Concept," *Canadian Journal of Fish and Aquatic Sciences* 37 (1980): 130–137; Jack A. Stanford and James V. Ward, "Dammed Rivers of the World: Symposium Rationale," in *The Ecology of Regulated Streams: First International Symposium on Regulated Streams*, ed. James V. Ward and Jack A. Stanford (New York: Plenum Press, 1979).

46. Vannote, "The River Continuum Concept," 32; James T. Brock et al., "Periphyton Metabolism: A Chamber Approach," in *The Controlled Flood in Grand Canyon*, ed. Robert H. Webb et al. (Washington, D.C.: American Geophysical Union, 1999), 217–233.

47. Vannote, "The River Continuum Concept," 133; Kenneth Cummings, "The Natural Stream Ecosystem," in *The Ecology of Regulated Streams*, ed. Stanford and Ward, 8–9.

48. Stanford and Ward, "Dammed Rivers of the World," in *The Ecology of Regulated Streams*, ed. Stanford and Ward, 2–3.

49. Jack A. Stanford and James V. Ward, "The Serial Discontinuity Concept of Lotic Ecosystems," in *Dynamics of Lotic Ecosystems*, ed. Thomas D. Fontaine III and Steven M. Bartell (Ann Arbor, MI: Science Publishers, 1983): 30; Vannote, "The River Continuum Concept," 134.

50. Carothers and Brown, *The Colorado River Through Grand Canyon*, 10.

51. Steven Carothers and Richard A. Valdez, "The Aquatic Ecosystem of the Colorado River in Grand Canyon," *Grand Canyon Data Integration Project Synthesis Report* (Prepared for the Bureau of Reclamation by SWCA, Inc., Flagstaff, AZ, July 1, 1998), 23, 35–36; Carothers and Minckley, *A Survey of Fishes*, 104–105.

52. Richard A. Valdez et al., "Biological Implications of the 1996 Controlled Flood," in *The Controlled Flood in Grand Canyon* (Washington, D.C.: American Geophysical Union, 1999), 343, 345; Carothers and Valdez, "The Aquatic Ecosystem of the Colorado River," 38; Carothers and Brown, *The Colorado River Through Grand Canyon*, 63–65, 68, 75 ("veritable underwater forests"); Brock et al., "Periphyton Metabolism," in *The Controlled Flood*, ed. Robert H. Webb et al., 219.

53. Carothers and Brown, *The Colorado River Through Grand Canyon*, 11; Carothers and Valdez, "The Aquatic Ecosystem of the Colorado River," 53.

54. Roderic A. Parnell et al., "Mineralization of Riparian Vegetation Buried by the 1996 Controlled Flood," in *The Controlled Flood in Grand Canyon*, ed. Robert H. Webb et al. (Washington, D.C.: American Geophysical Union, 1999), 227; Carothers and Brown, *The Colorado River Through Grand Canyon*, 69–71; Stanford and Ward, "Dammed Rivers of the World," ed. Stanford and Ward, 85, 90.

55. Ohmart et al., *The Ecology of the Lower Colorado River*, 51.

56. Sykes, *The Colorado Delta*, 48, 87, 105–106.

57. U.S. Department of Energy, Office of Environmental Management, *Remediation of the Moab Uranium Mill Tailings, Grand and San Juan Counties, Utah: Final Environmental Impact Statement* (DOE/EIS0-0355, 1995); Graf, *The Colorado River, Instability, and Basin Management*, 63–65.

58. Graf, *The Colorado River, Instability, and Basin Management*, 63–64; Jack A. Stanford and James V. Ward, "Limnology of Lake Powell and the Chemistry of the Colorado River," in *Colorado River Ecology and Dam Management: Proceedings of a Symposium* (1991), 90; Theresa S. Presser et al., "Bioaccumulation of Selenium from Natural Geologic Sources in Western States and Its Potential Consequences," *Environmental Management* 18 (1994): 423–436; Theresa S. Presser, "The Kesterson Effect," *Environmental Management* 18 (1994): 437–454; Dianne R. Nielson, Executive Director, Utah Department of Environmental Quality, Mercury in Utah, Presentation to the Natural Resources, Agriculture, and Environment Interim Committee, May 17, 2006, http://www.deq.utah.gov/Issues/Mercury/docs/LEG_HG_051706.pdf (accessed December 27, 2006).

59. Ohmart et al., *The Ecology of the Lower Colorado River*, 47, 51, 54–55; Environmental Defense Fund, *A Delta Once More*, 16, 30.

60. U.S. Department of the Interior, Quality of Water, *Colorado River Basin, Progress Report No. 18* (1997), 5, 8.

61. U.S. Environmental Protection Agency, *The Mineral Quality Problem in the Colorado River* (1971); Loretta C. Lohman et al., U.S. Department of the Interior, Bureau of Reclamation, *Estimating Economic Impacts of Salinity of the Colorado River* (1988).

62. Edward Abbey, *Desert Solitaire* (Tucson: University of Arizona Press, 1988), 171.

63. A.C. Behnke, "A Perspective on North America's Vanishing Streams," *Journal of the North American Benthological Society* 91 (1991): 7–88; R.D. Judy et al., "1982 National Fisheries Survey, Volume I Technical Report: Initial Findings" (U.S. Fish and Wildlife Service, U.S. Department of the Interior, 1984).

64. Mats Dynesius and Christer Nilsson, "Fragmentation and Flow Regulation of River Systems in the Northern Third of the World," *Science* 266 (November 1994): 753–762; Patrick McCully, *Silenced Rivers: The Ecology and Politics of Large Dams* (Atlantic Highlands, NJ: Zed Books, 1996); Stanford and Ward, "Dammed Rivers of the World," in *The Ecology of Regulated Streams*, ed. Stanford and Ward, 4.

65. Fish and Wildlife Service, *Pikeminnow Recovery Goals*, A-1.

66. Fish and Wildlife Service, *Pikeminnow Recovery Goals*, 10; Robert Rush Miller et al., "Ichthyological Exploration of the American West: The Hubbs-Miller Era, 1915–1950," in *Battle Against Extinction: Native Fish Management in the American West*, ed. W.L. Minckley and James E. Deacon (Tucson: University of Arizona Press, 1991), 36; Richard Valdez, "Of Suckers, Chubs, and 100-Pound Minnows," in *Water, Earth and Sky: The Colorado River Basin*, ed. Michael Collier et al. (Salt Lake City: University of Utah Press, 1999), 82–84 (quote).

Chapter 3

1. U.S. Fish and Wildlife Service, *Biological and Conference Opinion on Lower Colorado River Operations and Maintenance—Lake Mead to Southern International Boundary* (April 30, 1997), 116.

2. U.S. Fish and Wildlife Service, *Colorado Pikeminnow* (Ptychocheilus lucius) *Recovery Goals: Amendment and Supplement to the Colorado Squawfish Recovery Plan* (Denver: U.S. Fish and Wildlife Service, Mountain-Prairie Region [6], 2002); U.S. Fish and Wildlife

Service, *Razorback Sucker* (Xyrauchen texanus) *Recovery Goals: Amendment and Supplement to the Razorback Sucker Recovery Plan* (Denver: U.S. Fish and Wildlife Service, Mountain-Prairie Region (6), 2002); W.L. Minckley et al., "Management Toward Recovery of the Razorback Sucker," in *Battle Against Extinction: Native Fish Management in the American West*, ed. W.L. Minckley and James E. Deacon (Tucson: University of Arizona Press, 1991), 303; Robert T. Muth et al., *Flow and Temperature Recommendations for Endangered Fishes in the Green River Downstream of Flaming Gorge Dam*; 56 *Federal Register* 54,957 (October 23, 1991); 59 *Federal Register* 13,374 (March 21, 1994).

3. Minckley et al., "Management Toward Recovery of the Razorback Sucker," in *Battle Against Extinction*, ed. Minckley and Deacon, 303; W.L. Minckley et al., "A Conservation Plan for Native Fishes of the Lower Colorado River," *BioScience* 53, no. 3 (2003): 219–234.

4. Minckley et al., "A Conservation Plan for Native Fishes."

5. U.S. Fish and Wildlife Service, *Policy and Guidelines for Planning and Coordinating Recovery of Endangered and Threatened Species* (Washington, D.C.: U.S. Department of the Interior, 1990); Fish and Wildlife Service, *Razorback Sucker Recovery Goals*, 4; *U.S. Code* 16, § 1532(3); Michael J. Scott et al., "Recovery of Imperiled Species Under the Endangered Species Act: The Need for a New Approach," *Frontiers in Ecology and the Environment* 3, no. 7 (2005): 383–389.

6. R.D. Ohmart et al., *The Ecology of the Lower Colorado River from Davis Dam to the Mexico-United States International Boundary: A Community Profile*, U.S. Fish and Wildlife Service Report 85 (7.10) (1988); U.S. Fish and Wildlife Service, *Biological and Conference Opinion*; Lawrence Garrett et al., *A Review of the Second Administrative Draft Conservation Plan for the Lower Colorado River Multi-Species Conservation Program: Final Report* (Olathe, CO: M3 Research, January 2003); Gordon A. Mueller and Paul C. Marsh, *Lost, a Desert River and Its Native Fishes: A Historical Perspective of the Lower Colorado River*; Information and Technology Report USGS/BRD/ITR-2002-0010 (2002); Godfrey Sykes, *The Lower Colorado Delta* (American Geographical Society, 1937), 66, 78.

7. J. Grinnell, "An Account of the Mammals and Birds of the Lower Colorado Valley, with Special Reference to the Distributional Problems Present," *University of California Publications in Zoology* 12 (1914): 51–294; Ohmart et al., *The Ecology of the Lower Colorado River*, 9–12; Environmental Defense Fund, *A Delta Once More: Restoring Riparian and Wetland Habitat in the Colorado River Delta* (Washington, D.C.: Environmental Defense Fund, 1999), 2.

8. 60 *Federal Register* 10,694 (February 27, 1995).

9. Ohmart et al., *The Ecology of the Lower Colorado River*, 16.

10. U.S. Fish and Wildlife Service, *Biological and Conference Opinion*, 104; Garrett et al., *Review of the Second Administrative Draft Conservation Plan*; Lower Colorado River Multi-Species Conservation Program, *Lower Colorado River Multi-Species Conservation Program, Volume II: Habitat Conservation Plan: Final* (Sacramento, CA: J&S 00450.00., 2004), 3–4.

11. Lower Colorado River Multi-Species Conservation Program, *Habitat Conservation Plan*, 93; Mueller and Marsh, *Lost, a Desert River*.

12. U.S. Fish and Wildlife Service, *Biological and Conference Opinion*, 59, 64.

13. 65 *Federal Register* 43,031 (July 12, 2000); Lower Colorado River Multi-Species Conservation Program, *Habitat Conservation Plan*; Ives, *Report on the Lower Colorado River*, 48.

14. Edward Glenn et al., "Ecology and Conservation Biology of the Colorado River Delta, Mexico," *Journal of Arid Environments* 49 (2001); Raymond M. Turner et al., *Changing Mile Revisited: An Ecological Study of Vegetation Change with Time in the Lower Mile of an Arid and Semiarid Region* (Tucson: University of Arizona Press, 2003), 5.

15. Southern Utah Wilderness Alliance, http://www.suwa.org.

16. President's Water Resources Policy Commission, *Ten Rivers in America's Future* (1950), 2:421 (soils undisturbed); James O. Pattie, *Personal Narrative of James O. Pattie*, ed. Richard Batman (Missoula, MT: Mountain Press Publishing Co., 1988).

17. U.S. Department of the Interior, National Park Service, *A Survey of the Recreational Resources of the Colorado River Basin* (Washington, D.C.: GPO, 1950), 58; Ann Zwinger, *Run, River, Run: A Naturalist's Journey Down One of the Great Rivers of the West* (New York: Harper and Row, 1975), 111.
18. U.S. Department of the Interior, *Survey of Recreational Resources*, 58.
19. Wallace Stegner, *Beyond the 100th Meridian* (Lincoln: University of Nebraska Press, 1982); U.S. Department of the Interior, *Survey of Recreational Resources*, 59.
20. U.S. Department of the Interior, *Survey of Recreational Resources*, 59.
21. Turner et al., *Changing Mile Revisited*, 26–28.
22. Ohmart et al., *The Ecology of the Lower Colorado River*, 8, 21, 26.
23. R.R. Miller, "Man and the Changing Fish Fauna of the American Southwest," *Michigan Academy of Science, Arts, and Letters* 46 (1961); C.L. Armour et al., "The Effects of Livestock Grazing on Riparian and Stream Ecosystems," *AFS Position Statement*, January–February 1991; Conservation Agreement and Strategy for Colorado River Cutthroat Trout (*Oncorhynchus clarki pleuriticus*) in the States of Colorado, Utah, and Wyoming, http://mountain-prairie.fws.gov/species/fish/crct/crctfinl.pdf (accessed December 29, 2006); *Virgin River Fishes Recovery Plan* (Denver: U.S. Fish and Wildlife Service Region [6], 1995); Leo Lentsch et al., *Virgin Spinedace, Conservation Agreement and Strategy*, Utah Department of Natural Resources, Publication No. 95-13I.
24. Ann Zwinger, *Down Canyon* (Tucson: University of Arizona Press, 1995), 9; Jon Krakauer, *Under the Banner of Heaven: A Story of Violent Faith* (New York: Doubleday, 2003), 209–225; Will Bagley, *Blood of the Prophets: Brigham Young and the Massacre at Mountain Meadows* (Norman: University of Oklahoma Press, 2002).
25. U.S. Department of the Interior, *Survey of Recreational Resources*, 60; Turner, *Changing Mile Revisited*, 34; President's Water Resources Policy Commission, *Ten Rivers in America's Future*, 421.
26. Debra Donahue, *The Western Range Revisited: Removing Livestock From Public Lands to Conserve Native Biodiversity* (Norman: University of Oklahoma Press, 1999), 18, 31; Armour et al., "The Effects of Livestock Grazing"; Turner et al., *Changing Mile Revisited*, 29–30.
27. Taylor Grazing Act, *U.S. Code* 43, §§ 315 et seq.; Donahue, *The Western Range Revisited*, 34–45, 287.
28. U.S. Department of the Interior, *Survey of Recreational Resources*, 63.
29. John C. Schmidt, personal communication.
30. William L. Graf, *The Colorado River, Instability, and Basin Management* (Washington, D.C.: Association of American Geographers, 1985), 29.
31. Ibid., 30–31.
32. U.S. Department of the Interior, *Survey of Recreational Resources*, 64–65; President's Water Resources Policy Commission, *Ten Rivers in America's Future*, 422 ("washed away farmhouses").
33. Miller, "Man and the Changing Fish Fauna," 368–369.
34. H.S. Colton, "Some Notes on the Original Condition of the Little Colorado River: A Side Light on the Problem of Erosion," *Museum Notes of the Museum of Northern Arizona*, 17 (1937).
35. Ibid.
36. Ibid., 20.
37. President's Water Resources Policy Commission, *Ten Rivers in America's Future*; Dale Pontius, *Colorado River Basin Study* (Denver: Western Water Policy Review Advisory Commission, 1997), 367–368.
38. President's Water Resources Policy Commission, *Ten Rivers in America's Future*, 420–421; Turner et al., *Changing Mile Revisited*, 29–34 ("Grazing unquestionably weakened"); Graf, *The Colorado River, Instability, and Basin Management*, 36.

39. Graf, *The Colorado River, Instability, and Basin Management*, 32.
40. Graf, *The Colorado River, Instability, and Basin Management*, 36–37; President's Water Resources Policy Commission, *Ten Rivers in America's Future*, 420; Rosanne D'Arrigo and Gordon Jacoby, "A 1000 Year Record of Winter Precipitation from Northwestern New Mexico, U.S.A.: A Reconstruction from Tree Rings and Its Relation to El Niño and the Southern Oscillation," *The Holocene* 1, 2 (1991): 95–101.
41. Robert J. Naiman, "Ecosystem Alteration of Boreal Forest Streams by Beaver," *Ecology* 67 (1986): 1254.
42. President's Water Resources Policy Commission, *Ten Rivers in America's Future*, 416; Rich Olsen and Wayne A. Hubert, *Beaver: Water Resources and Riparian Habitat Manager* (Laramie, WY: University of Wyoming, 1994), 15; Joseph C. Ives, *Report Upon the Colorado River of the West* (New York: Da Capo Press, 1969), 196.
43. David J. Weber, *The Taos Trappers: The Fur Traders in the Far Southwest* (Norman: University of Oklahoma Press, 1971); Pattie, *Personal Narrative of James O. Pattie*, 130; Zwinger, *Run, River, Run*, 62–63; Olsen and Hubert, *Beaver: Water Resources and Riparian Habitat Manager*.
44. Turner et al., *Changing Mile Revisited*, 257–258.
45. Suzanne Fouty, personal communication.
46. L.L. Apple et al., "The Use of Beavers for Riparian/Aquatic Habitat Restoration of Cold Desert, Gully Cut Stream Systems in Southwestern Wyoming," *Investigations of Beavers* 4 (1985): 123–132; Suzanne Fouty, "Forest Service Employees for Environmental Ethics," *Forest Magazine* (August 9, 2004).
47. National Research Council, *Restoration of Aquatic Ecosystems* (Washington, D.C.: National Academy Press, 1992), 524; Robert W. Adler, "Addressing Barriers to Watershed Protection," *Environmental Law* 25 (1995): 973, 982, n.32, n.33.
48. Aldo Leopold, *A Sand County Almanac* (London: Oxford University Press, 1968), 116 (emphasis added).
49. Fray Angelico, trans., *The Dominquez-Escalante Journal: Their Expedition Through Colorado, Utah, Arizona, and New Mexico in 1776*, ed. J. Warner (Provo, UT: BYU Press, 1976), 44.

Chapter 4

1. Clarence E. Dutton, *Tertiary History of the Grand Cañon District* (University of Arizona Press, 2001), 150, first published in 1882 by the U.S. Geological Survey.
2. Dutton, *Tertiary History of the Grand Cañon District*, 13; Steven Carothers and C.O. Minckley, *A Survey of the Fishes, Aquatic Invertebrates and Aquatic Plants of the Colorado River and Selected Tributaries from Lee's Ferry to Separation Rapids* (Prepared for the Bureau of Reclamation by the Museum of Northern Arizona, Flagstaff, AZ, 1981), 1; Theodore S. Melis et al., *Adaptive Management of the Colorado River Ecosystem Below Glen Canyon Dam, Arizona: Using Science and Modeling to Resolve Uncertainty in River Management*, Adaptive Management of Water Resources, American Water Resources Assocation Summer Specialty Conference (in press).
3. Stephen J. Pyne, *How the Canyon Became Grand* (New York: Viking, 1998), 140.
4. Robert H. Webb, *A Century of Change: Rephotography of the 1889–1890 Stanton Expedition* (Tucson: University of Arizona Press, 1996), 71.
5. Ibid., 72.
6. Ibid., 71, n. 21.
7. Ibid., 72, 76–77.
8. Michael J. O'Brien, "Archaeology, Paleoecosystems, and Ecological Restoration," in *The Historical Ecology Handbook: A Restorationist's Guide to Reference Ecosystems*, ed. Dave Egan and Evelyn A. Howell (Washington, D.C.: Island Press, 2005), 1; Webb, *A Cen-*

tury of Change, 72, 80–81; Paul S. Martin, "Thinking Like a Canyon: Wild Ideas and Wild Burros," in *A Century of Change*, ed. Webb, 82.

9. Martin, "Thinking Like a Canyon," in *A Century of Change*, ed. Webb, 82–83.

10. Steven W. Carothers, "Feral Burros: Old Arguments and New Twists," in *A Century of Change*, ed. Webb, 84–85.

11. Wild Free-Roaming Horses and Burros Act, *U.S. Code* 16, §§ 1331-40; *Kleppe v. New Mexico*, 426 U.S. 529, 531, 535 (1976).

12. Wild Free-Roaming Horses and Burros Act, *U.S. Code* 16, § 1331; Kenneth P. Pitt, "The Wild Free-Roaming Horses and Burros Act: A Western Melodrama," *Environmental Law* 15 (1985): 503.

13. Paul H. Gobster and R. Bruce Hill, ed., *Restoring Nature: Perspectives from the Social Sciences and Humanities* (Covelo, CA: Island Press, 2000); Eric Higgs, *Nature By Design* (Cambridge: MIT Press, 2003); Society for Ecological Restoration International, Science and Policy Working Group, *The SER International Primer on Ecological Restoration* (Tucson: Society for Ecological Restoration International, 2004); Martin, "Thinking Like a Canyon," quoting A.S. Leopold et al., "Wildlife Management in the National Parks," *Transactions of the North American Wildlife and Natural Resources Conference* 24 (1963): 29–44, in *A Century of Change*, ed. Webb, 82 ("condition that prevailed"); Webb, *A Century of Change*, 81; Joanna Behrens and John Brooks, "Wind in Their Wings: The Condor Recovery Program," *Endangered Species Bulletin* 25, no. 3 (May/June 2000): 8–9.

14. Willa Cather, *Death Comes for the Archbishop* (New York: Vintage Books, 1990), 201–202 (emphasis added).

15. Ibid., 17, 263; P.B. Shafroth et al., "Control of Tamarisk in the Western United States: Implications for Water Salvage, Wildlife Use, and Riparian Restoration," *Environmental Management* 35, no. 3 (2005): 231–246; Benjamin L. Everitt, "Chronology of the Spread of Tamarisk in the Central Rio Grande," *Wetlands* 18 (1998): 4; T.W. Robinson, "Saltcedar in the Western States," *Studies of Evapotranspiration* 19.16:491-A, A3 (1965); David R. Harris, "Recent Plant Invasions in the Arid and Semi-Arid Southwest of the United States," *Plant Invasions* (1966): 418–420 ("earliest definitive record"); William L. Graf, "Tamarisk and River-Channel Management," *Environmental Management* 6 (1982): 283–296.

16. Blanche H. Gelfant, introduction to *O Pioneers*, Willa Cather (Penguin Books, reissued 1994); Cather, *Death Comes for the Archbishop*, 7.

17. Fray Angelico, trans., *The Dominquez-Escalante Journal: Their Expedition Through Colorado, Utah, Arizona, and New Mexico in 1776*, ed. J. Warner (Provo, UT: BYU Press, 1976), 80; Earl M. Christensen, "The Rate of Naturalization of Tamarisk in Utah," *The American Midland Naturalist* 68 (1962); Robinson, "Saltcedar in the Western States," A3.

18. Ellen Meloy, "The Silk That Hurls Us Down Its Spine," in *Water, Earth, and Sky: The Colorado River Basin*, ed. Michael Collier et al. (Salt Lake City: University of Utah Press, 1999), 105; Edward Abbey, *Desert Solitaire* (Tucson: University of Arizona Press, 1988), 158, 164; Elliot Porter and David Brower, ed., *The Place No One Knew: Glen Canyon on the Colorado* (San Francisco: Sierra Club, 1963), 13; Donald Worster, *Rivers of Empire: Water, Aridity and the Growth of the American West* (New York: Oxford University Press, 1985), 324.

19. Shafroth et al., "Control of Tamarisk"; Robinson, "Saltcedar in the Western States," A3–A6; Graf, "Tamarisk and River-Channel Management," 22–23; Harris, "Recent Plant Invasions," 418–420; Philip L. Fradkin, *A River No More: The Colorado River and the West* (Berkeley: University of California Press, 1995), 183.

20. Christensen, "The Rate of Naturalization of Tamarisk in Utah," 53–55; Robinson, "Saltcedar in the Western States," A7; Webb, *A Century of Change*, 100–101, 112–113.

21. Webb, *A Century of Change*, 100 (quotes); Lower Colorado River Multi-Species Conservation Program, *Lower Colorado River Multi-Species Conservation Program, Volume II: Habitat Conservation Plan: Final* (Sacramento, CA: J&S 00450.00., 2004).

22. Robinson, "Saltcedar in the Western States," A6–A10; Harris, "Recent Plant Invasions," 418.

23. Robinson, "Saltcedar in the Western States," A6; Jerome S. Horton and John E. Flood, "Taxonomic Notes on *Tamarix pentandra* in Arizona," *The Southwestern Naturalist* 7 (June 1962): 23–28; Theodore Kerpez and Norman S. Smith, *Saltcedar Control for Wildlife Habitat Improvement in the Southwestern United States*, U.S. Fish and Wildlife Service Resource Publication 169 (1987), 2; Harris, "Recent Plant Invasions," 418 (quote).

24. Robinson, "Saltcedar in the Western States," A10; Horton and Flood, "Notes on *Tamarix pentandra*," 125–126; Kerpez and Smith, "Saltcedar Control," 2–3.

25. Kerpez and Smith, "Saltcedar Control," 3; Graf, "Tamarisk and River-Channel Management," 286–288; Harris, "Recent Plant Invasions," 420.

26. Jeffrey P. Cohn, "Tiff over Tamarisk: Can a Nuisance Be Nice, Too?" *BioScience* 55, no. 8 (2005): 648–654; Shafroth et al., "Control of Tamarisk."

27. Kerpez and Smith, "Saltcedar Control," 4; Shafroth et al., "Control of Tamarisk," 236.

28. Cohn, "Tiff over Tamarisk"; Shafroth et al., "Control of Tamarisk"; Horton and Flood, "Notes on *Tamarix pentandra*," 124; U.S. Fish and Wildlife Service, *Southwestern Willow Flycatcher Recovery Plan* (Albuquerque: 2002); U.S. Fish and Wildlife Service, *Final Rule for Determining Endangered Status for the Southwestern Willow Flycatcher*, 60 *Federal Register* 10694 (February 27, 1995); Webb et al., "Downstream Effects of Glen Canyon Dam on the Colorado River in Grand Canyon: A Review," in *The Controlled Flood in Grand Canyon*, ed. Robert H. Webb et al. (Washington, D.C.: American Geophysical Union, 1999), 12; Kerpez and Smith, "Saltcedar Control," 5.

29. T.M. Allred and J.C. Schmidt, "Channel Narrowing by Vertical Accretion Along the Green River near Green River, Utah," *Geological Society of America Bulletin* 111, no. 12 (1999): 1757–1772; P.E. Grams and J.C. Schmidt, "Geomorphology of the Green River in the Eastern Uinta Mountains, Dinosaur National Monument, Colorado and Utah," in *Varieties of Fluvial Form*, ed. A.J. Miller and A. Gupta (John Wiley & Sons), 81–111; Kerpez and Smith, "Saltcedar Control," 3; Graf, "Tamarisk and River-Channel Management," 283; Richard A. Valdez et al., "Biological Implications of the 1996 Controlled Flood," in *The Controlled Flood in Grand Canyon*, Webb et al., 348; Benjamin L. Everitt and William L. Graf, "Fluvial Adjustments to the Spread of Tamarisk in the Colorado Plateau Region: Discussion and Reply," *Geological Society of America Bulletin* 90 (1979): 1183–1184.

30. Kerpez and Smith, "Saltcedar Control," 5–6; Shafroth et al., "Control of Tamarisk"; Graf, "Tamarisk and River-Channel Management," 283.

31. Joseph C. Ives, *Report Upon the Colorado River of the West* (New York: Da Capo Press, 1969), 58.

32. R.B. Payne, "Brood Parasitism in Birds: Strangers in the Nest," *BioScience* 48 (1998): 377–386; S.I. Rothstein, "A Model System for Co-evolution: Avian Brood Parasitism," *Annual Review of Ecology and Systematics* 21 (1990): 481–508.

33. Payne, "Brood Parasitism in Birds"; Rothstein, "A Model System for Co-evolution"; P.R. Ehrlich, D.S. Dobkin, and D. Wheye, *The Birder's Handbook* (New York: Simon & Schuster, 1988), 289.

34. National Geographic Society, *Field Guide to the Birds of North America* (Washington, D.C.: National Geographic Society, 1987).

35. Rothstein, "A Model System for Co-evolution."

36. Harris, "Recent Plant Invasions," 408, 411, 413, 417; Raymond M. Turner, *Changing Mile Revisited: An Ecological Study of Vegetation Change with Time in the Lower Mile of an Arid and Semiarid Region* (Tucson: University of Arizona Press, 2003).

37. R.R. Miller, "*Gila cypha*: A Remarkable New Species of Cyprinid Fish from the Colorado River in Grand Canyon, Arizona," *Journal of the Washington Academy of Sciences* 36 (1946): 409–415; Steven Carothers and Bryan T. Brown, *The Colorado River Through Grand Canyon* (Tucson: University of Arizona Press, 1991), 95; U.S. Fish and Wildlife Service, *Humpback Chub* (Gila cypha) *Recovery Goals: Amendment and Supplement to the*

Humpback Recovery Plan (Denver: U.S. Fish and Wildlife Service, Mountain-Prairie Region (6), 2002), A-2.

38. Fish and Wildlife Service, *Humpback Chub Recovery Goals*, A-1, A-5; U.S. Department of the Interior, Native Fish and Wildlife, Endangered Species, 32 *Federal Register* 4001 (March 11, 1967).

39. Fish and Wildlife Service, *Humpback Chub Recovery Goals*, A-1; Richard Valdez, "Of Suckers, Chubs, and 100-Pound Minnows," in *Water, Earth and Sky: The Colorado River Basin*, ed. Michael Collier et al. (Salt Lake City: University of Utah Press, 1999), 71; Robert T. Muth et al., *Flow and Temperature Recommendations for Endangered Fishes in the Green River Downstream of Flaming Gorge Dam* (Upper Colorado River Endangered Fish Recovery Program, September 2000).

40. Steven Carothers and Richard A. Valdez, "The Aquatic Ecosystem of the Colorado River in Grand Canyon," *Grand Canyon Data Integration Project Synthesis Report* (Prepared for the Bureau of Reclamation by SWCA, Inc., Flagstaff, AZ, July 1, 1998), 48.

41. W.L. Minckley and Michael E. Douglas, "Discovery and Extinction of Western Fishes: A Blink of the Eye in Geologic Time," in *Battle Against Extinction: Native Fish Management in the American West*, ed. W.L. Minckley and James E. Deacon (Tucson: University of Arizona Press, 1991), 11.

42. Jim Lichatowich, *Salmon Without Rivers: A History of the Pacific Salmon Crisis* (Washington, D.C.: Island Press, 1999), 124–126.

43. Ann Zwinger, *Down Canyon* (Tucson: University of Arizona Press, 1995), 14 (emphasis added).

44. U.S. Department of the Interior, Bureau of Reclamation, *The Colorado River: A Natural Menace Becomes a Natural Resource*, Project Planning Report No. 34-8-1, The Colorado River (1945), 17; President's Water Resources Policy Commission, *Ten Rivers in America's Future*, Vol. 2 (Washington, D.C: 1950), 2:378.

45. Holmes Rolston III, "Fishes in the Desert: Paradox and Responsibility," in *Battle Against Extinction: Native Fish Management in the American West*, ed. W.L. Minckley and James E. Deacon (Tucson: University of Arizona Press, 1991), 105; Valdez, "Of Suckers, Chubs, and 100-Pound Minnows," in *Water, Earth and Sky*, Collier et al., 85; Carothers and Valdez, "The Aquatic Ecosystem of the Colorado River," 48; Jack A. Stanford and James V. Ward, "Limnology of Lake Powell and the Chemistry of the Colorado River," in *Colorado River Ecology and Dam Management: Proceedings of a Symposium* (1991), 94; R.D. Ohmart et al., *The Ecology of the Lower Colorado River from Davis Dam to the Mexico-United States International Boundary: A Community Profile*, U.S. Fish and Wildlife Service Report 85 (7.10) (1988), 27.

46. Carothers and Minckley, *A Survey of Fishes*, 88–89, 172–173, 258.

47. Zwinger, *Down Canyon*, 39; R.R. Miller, "Man and the Changing Fish Fauna of the American Southwest," *Michigan Academy of Science, Arts, and Letters* 46 (1961): 390.

48. W.L. Minckley et al., "A Conservation Plan for Native Fishes of the Lower Colorado River," *BioScience* 53, no. 3 (2003): 220; Gordon A. Mueller and Paul C. Marsh, *Lost, a Desert River and Its Native Fishes: A Historical Perspective of the Lower Colorado River*, Information and Technology Report USGS/BRD/ITR-2002-0010 (2002), 2 (quote); Ohmart et al., *The Ecology of the Lower Colorado River*.

49. Carothers and Minckley, *A Survey of Fishes*, 2, 112–114; Webb et al., *The Controlled Flood in Grand Canyon*, 15, 274; Melis et al., *Adaptive Management of the Colorado River*.

50. W.L. Minckley, "Native Fishes of the Grand Canyon Region: An Obituary?" in *Colorado River Ecology and Dam Management*, National Academy of Sciences, Committee to Review the Glen Canyon Environmental Studies, Proceedings of a Symposium (Washington, D.C.: National Academy Press, 1991), 125–128; U.S. Fish and Wildlife Service, *Bonytail (Gila elegans) Recovery Goals: Amendment and Supplement to the Bonytail Chub Recovery Plan* (Denver: U.S. Fish and Wildlife Service, Mountain-Prairie Region [6], 2002), A-1; Carothers and Brown, *The Colorado River Through Grand Canyon*, 96.

51. Webb et al., *The Controlled Flood*, 5; Abbey, *Desert Solitaire*, 180; Colin Fletcher, *The Man Who Walked Through Time* (New York: Vintage Books, 1989), 115; Minckley et al., "A Conservation Plan for Native Fishes," 219; Robert Rush Miller et al., "Ichthyological Exploration of the American West: The Hubbs-Miller Era, 1915–1950," in *Battle Against Extinction: Native Fish Management in the American West*, ed. W.L. Minckley and James E. Deacon (Tucson: University of Arizona Press, 1991), 23.

52. Fish and Wildlife Service, *Humpback Chub Recovery Goals*, 10, A-10–11; Paul C. Marsh and Michael E. Douglas, "Predation by Introduced Fishes on Endangered Humpback Chub and Other Native Species in the Little Colorado River, Arizona," *Trans. American Fish Society* 126 (1997): 343–346.

53. W.L. Minckley et al., "Management Toward Recovery of the Razorback Sucker," in *Battle Against Extinction: Native Fish Management in the American West*, ed. W.L. Minckley and James E. Deacon (Tucson: University of Arizona Press, 1991), 331 (quote); Miller, "Man and the Changing Fish Fauna"; Mueller and Marsh, *Lost, a Desert River*, 41, 62.

54. W.L. Minckley et al., "Management Toward Recovery of the Razorback Sucker," in *Battle Against Extinction*, ed. Minckley and Deacon, 328; R.R. Miller, "Origin and Affinities of the Freshwater Fish Fauna of Western North America," *Zoogeography, Publication 51* (1958): 208; Miller et al., "Ichthyological Exploration of the American West," in *Battle Against Extinction*, ed. Minckley and Deacon, 19; Minckley et al., "A Conservation Plan for Native Fishes," 220; Mueller and Marsh, *Lost, a Desert River*, 50.

55. Mueller and Marsh, *Lost, a Desert River*; Richard S. Wydoski and John Hamill, "Evolution of a Cooperative Recovery Program for Endangered Fishes in the Upper Colorado River Basin," in *Battle Against Extinction: Native Fish Management in the American West*, ed. W.L. Minckley and James E. Deacon (Tucson: University of Arizona Press, 1991), 129; Carothers and Valdez, "The Aquatic Ecosystem of the Colorado River," xi.

56. Cather, *Death Comes for the Archbishop*, 221–222; Ann Zwinger, *Run, River, Run: A Naturalist's Journey Down One of the Great Rivers of the West* (New York: Harper and Row, 1975), 220; Abbey, *Desert Solitaire*, 114 (emphasis added).

57. David Lavender, *River Runners of the Grand Canyon* (Grand Canyon, AZ: Grand Canyon Natural History Association, 1986), 7–9; Robert B. Stanton, *Colorado River Controversies* (New York: Dodd, Mead, and Company, 1932); Russell Martin, *A Story That Stands Like a Dam: Glen Canyon and the Struggle for the Soul of the West* (New York: Henry Holt, 1989), 246.

58. Lichatowich, *Salmon Without Rivers*, 43, 129.

59. Stewart L. Udall, forward to *Battle Against Extinction: Native Fish Management in the American West*, ed. W.L. Minckley and James E. Deacon (Tucson: University of Arizona Press, 1991), ix; Edwin Philip Pister, "The Desert Fishes Council: Catalyst for Change," in *Battle Against Extinction: Native Fish Management in the American West*, ed. W.L. Minckley and James E. Deacon (Tucson: University of Arizona Press, 1991), 64 ("Native species"); Rolston, "Fishes in the Desert," in *Battle Against Extinction*, ed. Minckley and Deacon, 98; Miller et al., "Ichthyological Exploration of the American West," in *Battle Against Extinction*, ed. Minckley and Deacon, 40 ("What we must avoid").

Chapter 5

1. Edward Dolnick, *Down the Great Unknown: John Wesley Powell's 1868 Journey of Discovery and Tragedy Through the Grand Canyon* (New York: Harper Collins, 2001), 172.

2. John Wesley Powell, *The Exploration of the Colorado River and Its Canyons* (New York: Dover Publications, 1961), 247 (emphasis added); Wallace Stegner, *Beyond the 100th Meridian* (Lincoln: University of Nebraska Press, 1982).

3. Henry J. Pollack, *Uncertain Science–Uncertain World* (New York: Cambridge University Press, 2003).

4. B. Everman and C. Rutter, "The Fishes of the Colorado Basin," *Bulletin of the U.S. Fish Commissioner* 14 (1895), cited in "Managing River Resources: Lessons from Glen Canyon Dam," Jeffrey W. Jacobs and James L. Wescoat, Jr., *Environment* 44, no. 2 (2002): 8–19.

5. R.R. Miller, "Man and the Changing Fish Fauna of the American Southwest," *Michigan Academy of Science, Arts, and Letters* 46 (1961); Paul B. Holden, "Ghosts of the Green River: Impacts of Green River Poisoning on Management of Native Fishes," in *Battle Against Extinction: Native Fish Management in the American West*, ed. W.L. Minckley and James E. Deacon (Tucson: University of Arizona Press, 1991), 43; Todd Hartman, "Fish Story," a three-part series in the Denver *Rocky Mountain News*, December 3–5, 2000.

6. Robert Rush Miller et al., "Ichthyological Exploration of the American West: The Hubbs-Miller Era, 1915–1950," in *Battle Against Extinction: Native Fish Management in the American West*, ed. W.L. Minckley and James E. Deacon (Tucson: University of Arizona Press, 1991), 21.

7. *The Daily Rocket*, September 7, 1962, cited in Fred Quartarone, *Historical Accounts of Upper Colorado River Basin Endangered Fish* (Information and Education Committee of the Recovery Program for Endangered Fish of the Upper Colorado River Basin, Final Report, Revised Edition, September 1995).

8. Holden, "Ghosts of the Green River," in *Battle Against Extinction*, ed. Minckley and Deacon, 51.

9. Udall Directive (March 25, 1963), quoted in Holden, "Ghosts of the Green River," in *Battle Against Extinction*, ed. Minckley and Deacon, 51.

10. U.S. Department of the Interior, Native Fish and Wildlife, Endangered Species, 32 *Federal Register* 4001 (March 11, 1967); Holden, "Ghosts of the Green River," in *Battle Against Extinction*, ed. Minckley and Deacon, 54.

11. National Environmental Policy Act (NEPA), *U.S. Code* 42, §§ 4221 et seq.

12. NEPA, *U.S. Code* 42, § 4331(a); *Robertson v. Methow Valley Citizens Council*, 490 U.S. 332, 350 (1989); Lynton Caldwell, *The National Environmental Policy Act: An Agenda for the Future* (Bloomington: University of Indiana Press, 1998).

13. Endangered Species Act (ESA), *U.S. Code* 16, §§ 1532–44.

14. ESA, *U.S. Code* 16, §§ 1532(19), 1539(a).

15. *Tennessee Valley Authority v. Hill*, 437 U.S. 153 (1978); Richard J. Lazarus, "Restoring What's Environmental About Environmental Law in the Supreme Court," *UCLA Law Review* 47 (2003): 703.

16. Rick Krause, "Time for a New Look at NEPA," *The Environmental Forum* (May/June 2005): 38; Bradley C. Karkkainen, "Wither NEPA," *NYU Environmental Law Journal* 12 (2003): 333; Bradley C. Karkkainen, "Toward a Smarter NEPA: Monitoring and Managing Government's Environmental Performance," *Columbia Law Review* 102 (2002): 903; Executive Office of the President, Council on Environmental Quality, *The National Environmental Policy Act, A Study of Its Effectiveness After Twenty-five Years* (January 1997), http://ceq.eh.doe.gov/nepa/nepa25fn.pdf (accessed December 27, 2006).

17. A. Brower et al., "Consensus Versus Conservation in the Upper Colorado River Basin Recovery Implementation Program," *Conservation Biology* 15, no. 4 (2003): 1001–1007.

18. John Keys, Commissioner of Reclamation, Address to the Colorado River Water User's Association (December 2005); Robert H. Webb et al., *Climatic Fluctuations, Drought, and Flow in the Colorado River Basin*, U.S. Geol. Survey Fact Sheet 2004-3062; C.A. Woodhouse et al., "Updated Streamflow Reconstructions for the Upper Colorado River Basin," *Water Resources Research* 42, WO5415, doi:10:1029/2005SWR004455 (2006), http://www.agu.org/pubs/crossref/2006/2005WR004455.shtml (accessed December 27, 2006).

19. David Brandon, Chief Hydrologist in Charge, National Oceanic and Atmospheric Administration, Address to the Colorado River Water User Association, December 2004.

20. *Natural Resources Defense Council v. Houston*, 146 F.3d 1118 (9th Cir. 1998); *Pacific Coast Federation of Fisherman's Associations v. U.S. Bureau of Reclamation*, 138 F.Supp.2d 1228 (N.D. Cal. 2001).

21. Charles W. Stockton and Gordon C. Jacoby, "Long Term Surface Water Supply and Streamflow Trends in the Upper Colorado River Basin," *National Science Foundation, Lake Powell Research Project Bulletin No. 18* (1976).

22. *Fifty-Sixth Annual Report of the Upper Colorado River Commission*, 25 (Salt Lake City, UT: Upper Colorado River Commission, 2005).

23. Stockton and Jacoby, "Long Term Surface Water Supply and Streamflow Trends"; Robert Meredith, *A Primer on Climatic Variability and Change in the Southwest* (Tucson: University of Arizona, 2000), 3, 10; Woodhouse et al., "Updated Streamflow Reconstructions" (quote).

24. Philip L. Fradkin, *A River No More: The Colorado River and the West* (Berkeley: University of California Press, 1995); David H. Getches, "From Askhabad, to Wellton-Mohaw, to Los Angeles: The Drought in Water Policy," *University of Colorado Law Review* 64 (1993): 523; Dale Pontius, *Colorado River Basin Study* (Denver: Western Water Policy Review Advisory Commission, 1997), 6; Gregory Hobbs, "History of Colorado River Law, Development, and Use: A Primer and Look Forward," *Address at Natural Resources Law Center, University of Colorado School of Law Symposium, Hard Times on the Colorado River: Drought, Growth and the Future of the Compact* (June 8–10, 2005); Norris Hundley, *Water and the West: The Colorado River Compact and the Politics of Water in the American West* (Berkeley: University of California Press, 1975); E.C. LaRue, *Water Power and Flood Control of Colorado River Below Green River, Utah*, U.S. Geological Survey, Water-Supply Paper 556 (1925).

25. Kenneth Strzepek and Davis N. Yates, "Assessing the Effects of Climate Change on the Water Resources of the Western United States," in *Water and Climate in the Western United States*, ed. William M. Lewis, Jr. (2003), 161–170; Niklas S. Christensen et al., "The Effects of Climate Change on the Hydrology and Water Resources of the Colorado River Basin," *Climatic Change* 62 (2004): 337–363; Gregory Hobbs, "The Role of Climate Change in Shaping Western Water Institutions," *University of Denver Water Law Review* 7 (2003): 1.

26. Hundley, *Water and the West*; Evan R. Ward, *Border Oasis: Water and the Political Ecology of the Colorado River Delta, 1945–1975* (Tucson: University of Arizona Press, 2003); "Treaty for the Utilization of Waters of the Colorado, Tijuana and Rio Grande Rivers," Feb. 13, 1944, U.S.-Mex., Art. 10, 59 Stat. 1219.

27. *Winters v. United States*, 207 U.S. 564 (1908); Hundley, *Water and the West*; Colorado River Compact, Art. VII.

28. *Arizona v. California*, 376 U.S. 340 (1964); *Navajo Nation v. United States Department of the Interior, et al.*, Civil No. 03-CV-507 (District of Arizona, filed March 14, 2003); Pontius, *Colorado River Basin Study*.

29. 32 *Federal Register* 4001 (March 11, 1967); Richard S. Wydoski and John Hamill, "Evolution of a Cooperative Recovery Program for Endangered Fishes in the Upper Colorado River Basin," in *Battle Against Extinction: Native Fish Management in the American West*, ed. W.L. Minckley and James E. Deacon (Tucson: University of Arizona Press, 1991), 123.

30. U.S. Department of the Interior, Fish and Wildlife Service, *Recovery Implementation Program for Endangered Fish Species in the Upper Colorado River Basin*, revised March 8, 2000.

31. U.S. Department of the Interior, *Recovery Implementation Program*; James V. Hansen, "Endangered Economies: An Alternative Approach to Compliance with the Endangered Species Act Is Restoring Endangered Fish in the Colorado and San Juan Rivers," *Forum for Applied Research and Public Policy* (Spring 2001): 45–51; Federico Cheever, "The Road to Recovery: A New Way of Thinking About the Endangered Species Act,"

Ecology Law Quarterly 23 (1996): 1, 71–72 ("subordinates" and "creates a significant danger"); Brower et al., "Consensus Versus Conservation."

32. Robert Keiter, *Keeping Faith with Nature: Ecosystems, Democracy, and America's Public Lands* (New Haven: Yale University Press, 2003).

33. Federal Guidance for the Establishment, Use and Operation of Mitigation Banks, 60 *Federal Register* 58,065 (November 28, 1995); William W. Sapp, "The Supply-Side and Demand-Side of Wetlands Mitigation Banking," *Oregon Law Review* 74 (1995): 951; *U.S. Code* 33, §§ 1331, 1344; 40 C.F.R. § 230.10(a); National Research Council, *Restoration of Aquatic Ecosystems* (Washington, D.C.: National Academy Press, 1992); Michael Le Desma, "A Sound of Thunder: Problems and Prospects in Wetland Mitigation Banking," *Columbia Journal of Environmental Law* 19 (1994): 497; Robert W. Adler, "Economic Incentives for Wetlands and Water Quality Protection: A Public Perspective," *The Environmental Counselor* 110 (1997): 2.

34. U.S. Department of the Interior, *Recovery Implementation Program*, 1.

35. Robert T. Muth et al., *Flow and Temperature Recommendations for Endangered Fishes in the Green River Downstream of Flaming Gorge Dam* (Upper Colorado River Endangered Fish Recovery Program, September 2000); Fish and Wildlife Service, *Humpback Chub* (Gila cypha) *Recovery Goals Amendment and Supplement to the Humpback Recovery Plan* (Denver: U.S. Fish and Wildlife Service, Mountain-Prairie Region, 2002).

36. Cheever, "The Road to Recovery"; Brower et al., "Consensus Versus Conservation."

37. Powell, *The Exploration of the Colorado River*; Stegner, *Beyond the 100th Meridian*; Robert B. Stanton, *Colorado River Controversies* (New York: Dodd, Mead, and Company, 1932); Dolnick, *Down the Great Unknown*; Jon Krakauer, *Under the Banner of Heaven: A Story of Violent Faith* (New York: Doubleday, 2003).

38. ESA § 4(f); *U.S. Code* 16, § 1533(f).

39. U.S. Fish and Wildlife Service, *Notice of Availability of Recovery Goals for Four Endangered Fishes of the Colorado River Basin*, 67 *Federal Register* 55,270 (August 28, 2002); U.S. Fish and Wildlife Service, *Colorado Pikeminnow* (Ptychocheilus lucius) *Recovery Goals: Amendment and Supplement to the Colorado Squawfish Recovery Plan* (Denver: U.S. Fish and Wildlife Service, Mountain-Prairie Region [6], 2002).

40. U.S. Fish and Wildlife Service, *Pikeminnow Recovery Goals*; Daniel Rohlf, "Six Biological Reasons Why the Endangered Species Act Doesn't Work—And What to Do About It," *Conservation Biology* 5 (1991): 273; ESA § 2(b); *U.S. Code* 16, § 1531(b).

41. U.S. Fish and Wildlife Service, *Notice of Availability of Recovery Goals*.

42. U.S. Fish and Wildlife Service, Determination of Threatened Status for the Mojave Population of Desert Tortoise, 55 *Federal Register* 12,178 (April 2, 1990).

43. Christopher Smart, "Rare Tortoises May Have Been Hit Hard by Fires," *Salt Lake Tribune*, July 1, 2005.

44. Mark Shaffer, "Minimum Viable Populations: Coping with Uncertainty," in *Viable Populations for Conservation*, ed. Michael E. Soulé (New York: Cambridge University Press, 1987).

45. Illkka A. Hanski and Daniel Simberloff, "The Metapopulation Approach: Its History, Conceptual Domain, and Application to Conservation," in *Metapopulation Biology, Ecology, Genetics, and Evolution*, ed. Illkka A. Hanski and Michael E. Gilpen (Academic Press, 1997).

46. Shaffer, "Minimum Viable Populations."

47. Michael E. Soulé, "Introduction," in *Viable Populations for Conservation*, ed. Michael E. Soulé (New York: Cambridge University Press, 1987).

48. Eric T. Freyfogle and Julianne Lutz Newton, "Putting Science in Its Place," *Conservation Biology* 16, no. 4 (2002): 863–873; Shaffer, "Minimum Viable Populations."

49. Shaffer, "Minimum Viable Populations."

50. Ian Robert Franklin, "Evolutionary Change in Small Populations," in *Conservation Biology: An Evolutionary Ecological Perspective*, ed. Michael E. Soulé and Bruce A. Wilcox

(Sunderland, MA: Sinauer Associates, 1980); Russell Lande and George F. Barraclough, "Effective Population Size, Genetic Variation, and Their Use in Population Management," in *Viable Populations for Conservation,* ed. Michael E. Soulé (New York: Cambridge University Press, 1987).

51. Lande and Barraclough, "Effective Population Size"; Shaffer, "Minimum Viable Populations."

52. Rick Johnson, letter to Dr. Ralph O. Morgenweck, Regional Director, U.S. Fish and Wildlife Service (April 26, 2002); R. Frankham, "Effective Population Size/Adult Population Size Ratios in Wildlife: A Review," *Genetical Research* 66 (1995): 95–107.

53. W.L. Minckley et al., "A Conservation Plan for Native Fishes of the Lower Colorado River," *BioScience* 53, no. 3 (2003). Applying the 3/1 male to female ratio produces a total of 6,665 fish to generate an effective population size of 5,000. Using a breeding ratio of 0.1 increases the total number of fish needed to 66,650. Adding the mortality factor of 15 percent used by FWS increases the total to 76,648.

54. Hanski and Simberloff, "The Metapopulation Approach," in *Metapopulation Biology, Ecology, Genetics, and Evolution,* ed. Hanski and Gilpen; William G. Conway, "An Overview of Captive Propagation," in *Conservation Biology, An Evolutionary-Ecological Perspective,* ed. Soulé and Wilcox, 199–207.

55. Carl Walters, Statement at meeting of the Glen Canyon Dam Adaptive Management Workgroup, May 29, 2003.

56. Michael E. Soulé, "Conservation Biology and the 'Real World,'" in *Conservation Biology: The Science and Scarcity of Diversity,* Michael E. Soulé (Sunderland, MA: Sinauer Associates, 1986), 7 ("quest"); David Faigman, *Legal Alchemy: The Use and Misuse of Science in the Law* (New York: W.H. Freeman and Co., 1999); Rosemary J. Erickson and Rita J. Simon, *The Use of Social Science Data in Supreme Court Decisions* (Urbana: University of Illinois Press, 1998); Sheila Jasanoff, *Science at the Bar: Law, Science, and Technology in America* (Cambridge: Harvard University Press, 1995); Steven Goldberg, *Culture Clash: Law and Science in America* (New York: New York University Press, 1994).

57. John P. Dwyer, "The Pathology of Symbolic Legislation," *Ecology Law Quarterly* 17 (1990): 233; David Schoenbrod, "Goals Statutes or Rules Statutes: The Case of the Clean Air Act," *UCLA Law Review* 30 (1983): 740.

58. Michael Soulé, "Where Do We Go From Here," in *Viable Populations for Conservation,* ed. Michael E. Soulé (New York: Cambridge University Press, 1987), 176; Pollack, *Uncertain Science.*

59. *Grand Canyon Trust v. Norton,* 2006 WL 167560 (D. Ariz. 2006).

60. U.S. Department of the Interior, Bureau of Reclamation, *Operation of Flaming Gorge Dam Final Environmental Impact Statement* (September 2005), http://www.usbr.gov/uc/ envdocs/eis/ fgFEIS/index.html (accessed December 27, 2006).

61. Jack A. Stanford, *Instream Flows to Assist the Recovery of Endangered Fishes of the Upper Colorado River Basin* (Washington, D.C.: U.S. Department of the Interior Biological Survey, 1994).

62. Muth et al., *Flow and Temperature Recommendations.*

63. U.S. Department of the Interior, Bureau of Reclamation, *Record of Decision on Operation of Flaming Gorge Dam: Final Environmental Impact Statement* (February 16, 2006) http://www.usbr.gov/uc/envdocs/eis/fgFEIS/index.html.

64. Pollack, *Uncertain Science,* 3.

Chapter 6

1. Russell Martin, *A Story That Stands Like a Dam: Glen Canyon and the Struggle for the Soul of the West* (New York: Henry Holt, 1989); Charles F. Wilkinson, *Fire on the Plateau: Conflict and Endurance in the American Southwest* (Washington, D.C.: Island Press, 1999).

2. John C. Schmidt et al., "Origins of the 1996 Controlled Flood in Grand Canyon," in *The Controlled Flood in Grand Canyon*, ed. Robert H. Webb et al. (Washington, D.C.: American Geophysical Union, 1999), 34; Bruce Babbitt, forward to *The Colorado River Through Grand Canyon*, Steven Carothers and Bryan T. Brown (Tucson: University of Arizona Press, 1991), vii.

3. John C. Schmidt et al., "Science and Values in River Restoration in the Grand Canyon," *BioScience* 48, no. 9 (1998): 735.

4. Kai Lee, *Compass and Gyroscope: Integrating Science and Politics for the Environment* (Washington, D.C.: Island Press, 1993); Robert Keiter, *Keeping Faith with Nature: Ecosystems, Democracy, and America's Public Lands* (New Haven: Yale University Press, 2003); Carl J. Walters and C.S. Holling, "Large-Scale Management Experiments and Learning by Doing," *Ecology* 71 (1990): 2060–2068; Theodore S. Melis et al., *Adaptive Management of the Colorado River Ecosystem Below Glen Canyon Dam, Arizona: Using Science and Modeling to Resolve Uncertainty in River Management*, Adaptive Management of Water Resources, American Water Resources Association Summer Specialty Conference (in press).

5. Schmidt et al., "Science and Values in River Restoration"; Jeffrey W. Jacobs and James L. Wescoat, Jr., "Managing River Resources: Lessons from Glen Canyon Dam," *Environment* 44, no. 2 (2002); Holly Doremus and A. Dan Tarlock, "Science, Judgment, and Controversy in Natural Resource Regulation," *Public Land & Resources Law Review* 26 (2006): 1; Eric T. Freyfogle and Julianne Lutz Newton, "Putting Science in Its Place," *Conservation Biology* 16, no. 4 (2002).

6. Fray Angelico, trans., *The Dominquez-Escalante Journal: Their Expedition Through Colorado, Utah, Arizona, and New Mexico in 1776*, ed. J. Warner (Provo, UT: BYU Press, 1976), 71.

7. James C. Davis, *The Human Story* (Harper Collins, 2004).

8. Willa Cather, *Death Comes to the Archbishop* (New York: Vintage Books, 1990), 276.

9. William E. Smythe, *The Conquest of Arid America* (New York: Harper and Brothers, 1900); Katie Lee, *All My Rivers Are Gone: A Journey of Discovery Through Glen Canyon* (Boulder: Johnson Books, 1998); Roderick Nash, *Wilderness and the American Mind* (New Haven: Yale University Press, 1967).

10. Colorado River Storage Project Act of April 11, 1956, c. 203, 70 Stat. 105 (codified at *U.S. Code* 43, § 620); Nash, *Wilderness and the American Mind*.

11. Colorado River Storage Project Act, *U.S. Code* 43, §§ 620(f)-(g).

12. Colorado River Storage Project Act of 1968, Public Law 90-537, 82 Stat. 886 et seq. (September 30, 1968).

13. Byron E. Pearson, *Still the Wild River Runs: Congress, the Sierra Club, and the Fight to Save Grand Canyon* (Tucson: University of Arizona Press, 2002).

14. U.S. Department of the Interior, Colorado River Reservoirs, Coordinated Long-Range Operation, 35 *Federal Register* 8951–8952 (June 10, 1970); U.S. Department of the Interior, Bureau of Reclamation, *Annual Operating Plan for Colorado River Reservoirs* (2006).

15. Robert Dolan et al., "Man's Impact on the Colorado River in the Grand Canyon," *American Scientist* 62 (1974): 392–401.

16. National Research Council, *Downstream, Adaptive Management of Glen Canyon Dam and the Colorado River Ecosystem* (Washington, D.C.: National Academy Press, 1999); National Research Council, *River Resource Management in the Grand Canyon* (Washington, D.C.: National Academy Press, 1987).

17. *Southwest Center for Biological Diversity v. Bureau of Reclamation*, 143 F.3d 515 (9th Cir. 1998).

18. Grand Canyon Protection Act, Public Law 102-575, October 30, 1992, 106 Stat. 4600.

19. National Park Service Organic Act, *U.S. Code* 16, § 1 (emphasis added).

20. Grand Canyon Protection Act, § 1809.

21. Lisa H. Kearsley et al., "Changes in the Number and Size of Campsites as Determined by Inventories and Measurements," in *The Controlled Flood in Grand Canyon*, ed. Robert

H. Webb et al. (Washington, D.C.: American Geophysical Union, 1999), 147, 150; R. Roy Johnson, *Synthesis and Management Implications of the Colorado River Research Program*, Technical Report No. 17, Grand Canyon National Park Colorado River Research Series No. 47 (National Park Service, 1977), 1; E.D. Andrews et al., "Topographic Evolution of Sand Bars," in *The Controlled Flood in Grand Canyon*, ed. Robert H. Webb et al. (Washington, D.C.: American Geophysical Union, 1999), 117–118; Joseph E. Hazel, Jr., et al., "Topographic and Bathymetric Changes at Thirty-Three Long-Term Study Sites," in *The Controlled Flood in Grand Canyon*, ed. Robert H. Webb et al. (Washington, D.C.: American Geophysical Union, 1999), 161.

22. Richard W. Stoffle et al., "Piapaxa Uipi (Big Canyon River): Southern Paiute Ethnographic Resource Inventory and Assessment for Colorado River Corridor, Glen Canyon National Recreation Area, Utah and Arizona, and Grand Canyon National Park, Arizona, Final Report," for NPS and BOR in Conjunction with Glen Canyon Environmental Studies (June 1994).

23. John R. Thomas, "Navajo Nation Position Paper," *Glen Canyon Dam Environmental Impact Statement* (June 1993).

24. S.P. Gloss et al., ed. U.S. Department of the Interior, U.S. Geological Survey, *The State of the Colorado River Ecosystem in Grand Canyon: A Report of the Grand Canyon Monitoring and Research Center 1991–2004*, USGS Circular 1282 (2005); John C. Schmidt et al., *System-Wide Characteristics in the Distribution of Fine Sediment in the Colorado River Between Glen Canyon Dam and Bright Angel Creek, Arizona: Final Report to Grand Canyon Monitoring and Research Center* (October 2004).

25. Gloss et al., ed., *The State of the Colorado River Ecosystem*.

26. Kearsley et al., "Changes in the Number and Size of Campsites," in *The Controlled Flood*, ed. Webb et al., 158.

27. Dolan et al., "Man's Impact on the Colorado"; E.M. Laursen et al., "On Sediment Transport Through the Grand Canyon," *Proc. Third Federal Inter-Agency Sedimentation Conference*, March 22, 1976, Denver, Water Resources Council, 4-76–4-87; Gloss et al., ed., *The State of the Colorado River Ecosystem*, 18, 21.

28. Mark J. Brouder et al., "Changes in Number, Sediment Composition, and Benthic Invertebrates of Backwaters," in *The Controlled Flood in Grand Canyon*, ed. Robert H. Webb et al. (Washington, D.C.: American Geophysical Union, 1999), 241; Timothy L. Hoffnagle et al., "Fish Abundance, Distribution, and Habitat Use," in *The Controlled Flood in Grand Canyon*, ed. Robert H. Webb et al. (Washington, D.C.: American Geophysical Union, 1999), 273; John C. Schmidt et al., "Origins of the 1996 Controlled Flood in Grand Canyon," in *The Controlled Flood*, ed. Webb et al., 24; Kearsley et al., "Changes in the Number and Size of Campsites," in *The Controlled Flood*, ed. Webb et al., 158.

29. Babbitt, forward to *The Colorado River Through Grand Canyon*, Carothers and Brown, xiv; Schmidt et al., "Science and Values in River Restoration."

30. Gloss et al., ed., *The State of the Colorado River Ecosystem*, 27.

31. Ibid., 37; Melis et al., *Adaptive Management of the Colorado River*.

32. Hoffnagle et al., "Fish Abundance, Distribution, and Habitat Use," in *The Controlled Flood*, ed. Webb et al., 281.

33. Richard A. Valdez et al., "Biological Implications of the 1996 Controlled Flood," in *The Controlled Flood in Grand Canyon*, ed. Robert H. Webb et al. (Washington, D.C.: American Geophysical Union, 1999), 349.

34. Hoffnagle et al., "Fish Abundance, Distribution, and Habitat Use," in *The Controlled Flood*, ed. Webb et al., 274; Valdez et al., "Biological Implications of the 1996 Controlled Flood," in *The Controlled Flood*, ed. Webb et al., 348.

35. U.S. Department of the Interior, *Proposed Experimental Releases from Glen Canyon Dam and Removal of Nonnative Fish, Environmental Assessment* (2002).

36. U.S. Department of the Interior, *Finding of No Significant Impact, Proposed Modification to Removal of Non-Native Fish from the Colorado River in Grand Canyon* (August 12, 2003); Gloss et al., ed., *The State of the Colorado River Ecosystem*, 218; Melis et al., *Adaptive Management of the Colorado River*.

37. Valdez et al., "Biological Implications of the 1996 Controlled Flood," in *The Controlled Flood*, ed. Webb et al., 348.

38. U.S. Department of the Interior, Bureau of Reclamation, Upper Colorado Region, *Glen Canyon Dam, Modifications to Control Downstream Temperatures, Plan and Draft Environmental Assessment* (January 1999).

39. Adaptive Management Program Science Advisors, *Evaluating a Glen Canyon Dam Temperature Control Device to Enhance Native Fish Habitat in the Colorado River: A Risk Assessment* (July 2003).

40. U.S. Department of the Interior, *Modifications to Control Downstream Temperatures*, ii (quote); Adaptive Management Program Science Advisors, *Evaluating a Glen Canyon Dam Temperature Control Device*, 3.

41. Carl Walters, Statement at May 29–30, 2003, meeting of the Glen Canyon Dam Adaptive Management Workgroup, Phoenix, Arizona.

42. Steven Carothers and Bryan T. Brown, *The Colorado River Through Grand Canyon* (Tucson: University of Arizona Press, 1991), 4, 39.

43. Betty Leavengood, *Grand Canyon Women: Lives Shaped by Landscape* (Pruett Publishing Co., 1999); David Lavender, *River Runners of the Grand Canyon* (Grand Canyon, AZ: Grand Canyon Natural History Association, 1986); E.U. Clover and L. Jotter, "Floristic Studies in the Canyon of the Colorado and Its Tributaries," *American Midland Naturalist* 32 (1944): 618.

44. Robert H. Webb et al., "Downstream Effects of Glen Canyon Dam on the Colorado River in Grand Canyon: A Review," in *The Controlled Flood in Grand Canyon*, ed. Webb et al. (Washington, D.C.: American Geophysical Union, 1999), 12–13.

45. Ann Zwinger, *Down Canyon* (Tucson: University of Arizona Press, 1995); Dennis Kubly, personal communication.

46. Zwinger, *Down Canyon*.

47. Babbitt, forward to *The Colorado River Through Grand Canyon*, Carothers and Brown, xiii; Lavender, *River Runners of the Grand Canyon*, 132.

48. Zwinger, *Down Canyon*, 122, 142–144; Carothers and Brown, *The Colorado River Through Grand Canyon*, 32–33, 40–42, 45; Webb et al., *The Controlled Flood in Grand Canyon*.

49. Gloss et al., ed., *The State of the Colorado River Ecosystem*, 17–33.

50. Quoted in Stephen J. Pyne, *How the Canyon Became Grand* (New York: Viking, 1998), 159.

51. Grand Canyon Protection Act §§ 1803(b), 1804(c)(3).

52. Gloss et al., ed., *The State of the Colorado River Ecosystem*, 9; U.S. Department of the Interior, *Strategic Plan, Glen Canyon Dam Adaptive Management Program, Final Draft* (August 17, 2001), http://www.usbr.gov/uc/rm/amp/strategic_plan.html (accessed December 28, 2006).

53. National Research Council, *Downstream*; Jacobs and Wescoat, "Managing River Resources"; Schmidt et al., "Science and Values"; Carothers and Brown, *The Colorado River Through Grand Canyon*, 172.

54. Walters and Holling, "Large-Scale Management Experiments and Learning by Doing"; Gloss et al., ed., *The State of the Colorado River Ecosystem*, 6; National Research Council, *Downstream*, 53; Lee, *Compass and Gyroscope*, 6–11.

55. G. Richard Marzolf et al., "Flood Releases From Dams as Management Tools: Interactions Between Science and Management," in *The Controlled Flood in Grand Canyon*, ed. Robert H. Webb et al. (Washington, D.C.: American Geophysical Union, 1999), 363–364.

56. National Research Council, *Downstream;* Hobbs, "Setting Effective and Realistic Restoration Goals."

Chapter 7

1. Mark Resnick, *Kirinyaga: A Fable of Utopia* (New York: Ballantine Books, 1998).
2. U.S. Department of the Interior, Bureau of Reclamation, *The Colorado River: A Natural Menace Becomes a Natural Resource,* Project Planning Report No. 34-8-1, The Colorado River (1945), 1.
3. R.D. Ohmart et al., *The Ecology of the Lower Colorado River from Davis Dam to the Mexico-United States International Boundary: A Community Profile,* U.S. Fish and Wildlife Service Report 85 (7.10) (1988), 20–21; Gordon A. Mueller and Paul C. Marsh, *Lost, a Desert River and Its Native Fishes: A Historical Perspective of the Lower Colorado River,* Information and Technology Report USGS/BRD/ITR-2002-0010 (2002), 5, 8.
4. Bureau of Reclamation, *The Colorado River: A Natural Menace Becomes a Natural Resource,* 34; Philip L. Fradkin, *A River No More: The Colorado River and the West* (Berkeley: University of California Press, 1995), 187, 244–245, 270; Donald Worster, *Rivers of Empire: Water, Aridity and the Growth of the American West* (New York: Oxford University Press, 1985), 196–208; Godfrey Sykes, *Report on the Lower Colorado River* (American Geographical Society, 1937), 39, 80, 82, 108 et seq.; William L. Graf, *The Colorado River, Instability, and Basin Management* (Washington, D.C.: Association of American Geographers, 1985), 1, 18–19; Norris Hundley, *Water and the West: The Colorado River Compact and the Politics of Water in the American West* (Berkeley: University of California Press, 1975); E.C. LaRue, *Water Power and Flood Control of Colorado River Below Green River, Utah,* U.S. Geological Survey, Water-Supply Paper 556 (1925); Mueller and Marsh, *Lost, a Desert River,* 53, figure 63.
5. Robert H. Webb et al., *Climatic Fluctuations, Drought, and Flow in the Colorado River Basin,* U.S. Geol. Survey Fact Sheet 2004-3062, 3, Fig. 3; Mueller and Marsh, *Lost, a Desert River,* 8; Toni Linenberger, *Dams, Dynamos, and Development: The Bureau of Reclamation's Power Program and Electrification of the West* (Washington, D.C.: Bureau of Reclamation, 2002); Fradkin, *A River No More,* 243; Worster, *Rivers of Empire,* 196–208.
6. Mueller and Marsh, *Lost, a Desert River,* 20, 29 ("water delivery canal"); T.H. Mooser and W.D. Sears, "Sediment Control at Imperial Dam," *Proceedings of the Third Federal Inter-Agency Sedimentation Conference,* March 22–25, 1976, Denver, Water Resources Council; Eldon L. Johns, "Sediment Problems in the Mohave Valley—A Case History," Proceedings of the Third Federal Inter-Agency Sedimentation Conference, March 22–25, 1976, Denver, Water Resources Council; Ohmart et al., *The Ecology of the Lower Colorado River.*
7. Mueller and Marsh, *Lost, a Desert River,* 2, 9; R.R. Miller, "Man and the Changing Fish Fauna of the American Southwest," *Michigan Academy of Science, Arts, and Letters* 46 (1961): 374–375.
8. BIO-WEST Inc., *Colorado River Backwater Enhancement Species Report* to U.S. Bureau of Reclamation, Lower Colorado Region, Boulder City, Nevada (October 2005); Ohmart et al., *The Ecology of the Lower Colorado River;* Mueller and Marsh, *Lost, a Desert River,* 2.
9. Fradkin, *A River No More,* 236, 263–264.
10. Clean Water Act, *U.S. Code* 33, §§ 1251(a), 1362(19).
11. Mueller and Marsh, *Lost, a Desert River,* 2.
12. S. Rep. No. 92-414, 76 (1972); H.R. Rep. No. 92-911, 76–77 (1972) (emphasis added); J.R. Karr and D.R. Dudley, "Ecological Perspectives on Water Quality Goals," *Environmental Management* 5 (1981): 55.
13. Robert W. Adler and Michele Straube, "Lessons from Large Watershed Programs: A Comparison of the Colorado River Basin Salinity Control Program with the San Francisco Bay-Delta Program, Central and South Florida, and the Chesapeake Bay Program,"

National Academy of Public Administration, Learning from Innovations in Environmental Protection, Research Paper Number 10 (Washington, D.C.: National Academy of Public Administration, 2000); Robert W. Adler and Michele Straube, "Watersheds and the Integration of U.S. Water Law and Policy: Bridging the Great Divides," *William and Mary Environmental Law and Policy Review* 25 (2000): 1.

14. Fradkin, *A River No More*, 245–247.

15. Mueller and Marsh, *Lost, a Desert River*, 2 ("future is grim"); BIO-WEST Inc., *Backwaters Enhancement Report* ("other researchers").

16. Lower Colorado River Multi-Species Conservation Program, *Lower Colorado River Multi-Species Conservation Program, Volume II: Habitat Conservation Plan: Final* (Sacramento: J&S 00450.00., 2004), 1–2.

17. Ibid., 1–3 (emphasis added); Endangered Species Act (ESA), *U.S. Code* 16, § 1533(f); U.S. Fish and Wildlife Service, *Southwestern Willow Flycatcher Recovery Plan* (Albuquerque: 2002).

18. ESA, *U.S. Code* 16, §§ 1538(a), 1532(19); *Babbitt v. Sweet Home Chapter of Communities for a Greater Oregon*, 515 U.S. 687 (1995).

19. ESA § 10(a); *U.S. Code* 16, § 1539(a); U.S. Fish and Wildlife Service, Federal Fish and Wildlife Permit TE-086834-0 (April 4, 2005) ("ESA section 10 permit").

20. ESA Section 10 Permit ¶ N-3 and Table 7-1. Funding and Implementation Agreement ¶ 8.3.

21. Lower Colorado River Multi-Species Conservation Program, *Habitat Conservation Plan*, chapter 7.

22. Ibid.; U.S. Department of the Interior, Bureau of Reclamation, Lower Colorado River Multi-Species Conservation Program, *Cibola Valley Conservation Area: Phase I Planting, Management, and Monitoring Draft Plan* (January 2005), 9.

23. 63 *Federal Register* 8,859 (Feb. 23, 1998); *Spirit of the Sage Council v. Norton*, 411 F.3d 225 (D.C. Cir. 2005).

24. Lower Colorado River Multispecies Conservation Program Implementing Agreement, ¶ 13.2.2.

25. Brent M. Hadad, *Rivers of Gold: Designing Markets to Allocate Water in California* (Washington, D.C.: Island Press, 2000).

26. U.S. Fish and Wildlife Service, *Biological and Conference Opinion on the Lower Colorado River Multi-Species Conservation Program, Arizona, California, and Nevada* (March 4, 2005), 136.

27. Ibid., 11–13.

28. 50 C.F.R. § 402.02.

29. U.S. Bureau of Reclamation, Lower Colorado Region, Lower Colorado River Multi-Species Conservation Program, http://www.usbr.gov/lc/lcrmscp/index.html (accessed December 28, 2006).

30. 50 C.F.R. § 402.03.

31. *Southwest Center for Biological Diversity v. United States Bureau of Reclamation*, 143 F.3d 515 (9th Cir. 1998).

32. Lower Colorado River Multi-Species Conservation Program, *Habitat Conservation Plan*, 5-12–5-16.

33. Ibid., 3-4.

34. Ibid., 5-1.

35. National Research Council, *Restoration of Aquatic Ecosystems* (Washington, D.C.: National Academy Press, 1992); National Research Council, *Wetlands: Characteristics and Boundaries* (Washington, D.C.: National Academy Press, 1995).

36. Revegetation and Wildlife Management Center, Inc., *River Mile 31 Revegetation Project: 2000*, submitted to U.S. Bureau of Reclamation (December 2000).

37. Ohmart et al., *The Ecology of the Lower Colorado River*; Jeffrey P. Cohn, "Tiff over Tamarisk: Can a Nuisance Be Nice, Too?" *BioScience* 55, no. 8 (2005).

38. Mary Austin, *The Land of Little Rain* (New York: Houghton, Mifflin and Company, 1903), 26; U.S. Bureau of Reclamation, *Imperial National Wildlife Refuge Imperial Native Fish Habitat Reconstruction, Final Report* (July 11, 2005), 6.

39. Lower Colorado River Multi-Species Conservation Program, *Habitat Conservation Plan*, 5-3, 5-11; U.S. Bureau of Reclamation, *Area Searches at the Cibola Nature Trail Restoration Site and the Pratt Restoration Site, Breeding Season 2004* (September 2004); BIO-WEST Inc., *Backwaters Enhancement Report*, 35, 56–59 ("clear and thorough knowledge").

40. Revegetation and Wildlife Management Center, Inc., *River Mile 31 Revegetation Project*; Lower Colorado River Multi-Species Conservation Program, *Habitat Conservation Plan*, 5-6; U.S. Department of the Interior, Bureau of Reclamation, *Beal Lake Habitat Restoration* (April 2005), 9; Bureau of Reclamation, *Cibola Valley Conservation Area: Phase I*, 9; U.S. Department of the Interior, Bureau of Reclamation, Lower Colorado River Multi-Species Conservation Program, *Cibola Valley Conservation Area: Draft Report* (November 2005), 5; Bureau of Reclamation, *Beal Lake Habitat Restoration*, 1.

41. Lower Colorado River Multi-Species Conservation Program, *Habitat Conservation Plan*, 5-16, 5-17.

42. W.L. Minckley et al., "A Conservation Plan for Native Fishes of the Lower Colorado River," *BioScience* 53, no. 3 (2003).

43. U.S. Bureau of Reclamation, *Imperial National Wildlife Refuge*.

44. Minckley et al., "A Conservation Plan for Native Fishes."

45. U.S. Fish and Wildlife Service, *Biological and Conference Opinion* (March 4, 2005), 103.

46. U.S. Fish and Wildlife Service, *Recovery Implementation Program for Endangered Fish Species in the Upper Colorado River Basin* (March 8, 2000); U.S. Fish and Wildlife Service, Upper Colorado River Endangered Fish Recovery Program and San Juan River Basin Recovery Implementation Program, *Program Highlights 2004–2005*.

47. Barbara Kingsolver, *High Tide in Tucson: Essays from Now or Never* (New York: Harper Collins Publishers, 1999), 26, 33 (emphasis added); Ralph Waldo Emerson, in *The Norton Book of Nature Writing*, ed. Robert Finch and John Elder (New York: W.W. Norton & Co., 2002), 143; Henry David Thoreau, in *The Norton Book of Nature Writing*, ed. Finch and Elder, 187; William Kittredge, in *The Norton Book of Nature Writing*, ed. Finch and Elder, 713; Robert W. Adler, "The Law at the Water's Edge: Limits to 'Ownership' of Aquatic Ecosystems," in *Wet Growth: Should Water Law Control Land Use?*, ed. Craig Anthony Arnold (Washington, D.C.: Environmental Law Inst., 2005).

48. John Locke, *The Second Treatise of Government: An Essay Concerning the Original Extent and End of Civil Government* (New York: Macmillan, 1968, originally published 1698), § 34; Eric T. Freyfogle, "Ethics, Community, and Private Land," *Ecology* 23 (1996): 631, 633–634; Eric T. Freyfogle, "Ownership and Ecology," *Case Western Reserve Law Review* 43 (1993): 1269, 1284; Terry W. Frazier, "Protecting Ecological Integrity with the Balancing Function of Property Law," *Environmental Law* 28 (1998): 53, 56–57; Joseph Sax, "Property Rights and the Economy of Nature: Understanding *Lucas v. South Carolina Coastal Council*," *Stanford Law Review* 45 (1993): 1433, 1442.

49. Craig Anthony Arnold, "The Reconstitution of Property: Property as a Web of Interests," *Harvard Environmental Law Review* 26 (2002): 281, 286; Freyfogle, "Ownership and Ecology," 1269, 1292, 1295–1297; Freyfogle, "Ethics, Community, and Private Land," 640, n. 30; Daniel H. Cole and Peter Z. Grossman, "The Meaning of Property 'Right': Law v. Economics," *Land Economics* 78 (2002): 317, 320; Carol M. Rose, "Givenness and Gift: Property and the Quest for Environmental Ethics," *Environmental Law* 24 (1994): 1, 14–15; Charles F. Wilkinson, "The Headwaters of the Public Trust: Some Thoughts on the Source and Scope of the Traditional Doctrine," *Environmental Law* 19 (1989): 430, n. 27; Alison Rieser, "Ecological Preservation as a Public Property Right: An Emerging Doctrine in Search of a Theory," *Harvard Environmental Law Review* 15 (1991): 393; Joseph Sax, "The Limits of Private Rights in Public Waters," *Environmen-*

tal Law 19 (1989): 473; Gretchen C. Daily, ed., *Nature's Services: Societal Dependence on Natural Ecosystems* (Washington, D.C.: Island Press, 1997).

50. Freyfogle, "Ethics, Community, and Private Land," 649; Cole and Grossman, "The Meaning of Property 'Right,'" 317; Rose, "Given-ness and Gift," 4; *United States v. Riverside Bayview Homes*, 474 U.S. 121, 132 (1985) (quote).

51. Joseph Sax et al., *Legal Control of Water Resources: Cases and Materials* (St. Paul, MN: West Pub. Co., 3rd ed., 2000); Sax, "The Limits of Private Rights"; Harrison C. Dunning, "Revolution (and Counter-Revolution) in Western Water Law: Reclaiming the Public Character of Water Resources," *Fordham Environmental Law Journal* 8 (1997): 439.

52. Richard J. Lazarus, "Restoring What's Environmental About Environmental Law in the Supreme Court," *UCLA Law Review* 47 (2003): 633–636; Ralph W. Johnson, "Water Pollution and the Public Trust Doctrine," *Environmental Law* 19 (1989): 485, 490; Lloyd R. Cohen, "The Public Trust Doctrine: An Economic Perspective," *California Western Law Review* 29 (1992): 240, 249–252; Charles F. Wilkinson, "The Headwaters of the Public Trust: Some Thoughts on the Source and Scope of the Traditional Doctrine," *Environmental Law* 19 (1989): 429; *Martin v. Waddell's Lessee*, 41 U.S. (16 Pet.) 367 (1842); *Arnold v. Mundy*, 6 N.J.L. 1 (N.J. 1821).

53. Joseph Sax, "The Public Trust Doctrine in Natural Resources Law: Effective Judicial Intervention," *Michigan Law Review* 68 (1970): 471; Joseph Sax, "Liberating the Public Trust Doctrine from Its Historical Shackles," *U.C. Davis Law Review* 14 (1980): 185; *National Audubon Society v. Superior Court*, 658 P.2d 709 (Cal. 1983); *Gould v. Greylock Reservation Commission*, 215 N.E.2d 114 (1966); *Paepke v. Public Building Commission*, 263 N.E.2d 11 (Ill. 1970); *Van Ness v. Borough of Deal*, 393 A.2d 571 (1978); *Marks v. Whitney*, 491 P.2d 374 (Cal. 1971); *Wade v. Kramer*, 459 N.E.2d 1025 (Ill. 1984); *Save Ourselves, Inc. v. Louisiana Environmental Control Commission*, 452 So.2d 1154 (La. 1984); *Orion Corp. v. Washington*, 747 P.2d 1062 (Wash. 1987).

54. Rieser, "Ecological Preservation," 397–398, n. 25, citing W.W. Buckland, *Textbook of Roman Law from Augustus to Justinian*, 182–183 (1966); *Pierson v. Post*, 3 Cai. R. 175 (N.Y. Sup. Ct. 1805).

55. *Missouri v. Holland*, 252 U.S. 416 (1920); *Geer v. Connecticut*, 161 U.S. 519, 529 (1896); *Lacoste v. Department of Conservation of State of Louisiana*, 263 U.S. 545, 549 (1924); *Toomer v. Witsell*, 334 U.S. 385 (1948); *Douglas v. Seacoast Products, Inc.*, 431 U.S. 265 (1977); *Hughes v. Oklahoma*, 441 U.S. 322, 324–325 (1979).

56. Aldo Leopold, *A Sand County Almanac* (London: Oxford University Press, 1968); Freyfogle, "Ownership and Ecology," 1276, 1280–1281, citing Roderick Nash, *The Rights of Nature: A Field Guide to the Literature* (1989); Arnold, "The Reconstitution of Property," 305; Sax, "Property Rights and the Economy of Nature," 1448; Christopher Stone, "Should Trees Have Standing? Toward Legal Rights for Natural Objects," *Southern California Law Review* 45 (1972): 450; *Sierra Club v. Morton*, 405 U.S. 727 (1972); Daily, ed., *Nature's Services*.

57. Peter Raven and Joel Cracraft, "Seeing the World as It Really Is: Global Stability and Environmental Change," in *The Living Planet in Crisis*, ed. Joel Cracraft and Francesca T. Grifo (1999); Marjorie Reaka-Kudla et al., *Biodiversity II: Understanding and Protecting Our Biological Resources* (1997).

58. Thomas A. Campbell, "The Public Trust, What's It Worth?," *Natural Resources Journal* 34 (1994): 86 (quote); Freyfogle, "Ownership and Ecology," 1288; Sax, "Property Rights and the Economy of Nature," 1445; Sax, "The Limits of Private Rights," 475; *Clajon Prod. Corp. v. Petera*, 854 F. Supp. 843 (D. Wyo. 1994).

59. Sax, "Property Rights and the Economy of Nature," 1446–1450; Sax, "The Limits of Private Rights," 478; Arnold, "The Reconstitution of Property," 328–329, 347–348; James H. Archer and Terrence W. Stone, "The Interaction of the Public Trust and the

'Takings' Doctrine: Protecting Wetlands and Critical Coastal Areas," *Vermont Law Review* 20 (1995): 81, 95, n. 79, 108–109; Sax, "The Public Trust Doctrine," 257; *Stevens v. Cannon Beach*, 835 P.2d 940, 942 (Ore. App. 1992).

60. U.S. Department of the Interior, Bureau of Reclamation, *Operation of Flaming Gorge Dam Final Environmental Impact Statement* (September 2005), http://www.usbr.gov/uc/envdocs/eis/fgFEIS/index.html (accessed December 27, 2006).

Chapter 8

1. Joseph C. Ives, *Report Upon the Colorado River of the West* (New York: Da Capo Press, 1969), 42–43.

2. John Steinbeck, *The Log from the Sea of Cortez* (New York: Viking Press, 1951), 2–3; John Van Dyke, *The Desert*, quoted in *Colorado River Controversies*, Robert B. Stanton (New York: Dodd, Mead, and Company, 1932), xxxviii; Philip L. Fradkin, *A River No More: The Colorado River and the West* (Berkeley: University of California Press, 1995), 319; Ives, *Report Upon the Colorado*, 49.

3. Aldo Leopold, *A Sand County Almanac* (London: Oxford University Press, 1968), 151.

4. Ives, *Report Upon the Colorado*, 26; Environmental Defense Fund, *A Delta Once More: Restoring Riparian and Wetland Habitat in the Colorado River Delta* (Washington, D.C.: Environmental Defense Fund, 1999); Jennifer Pitt et al., "Two Nations, One River: Managing Ecosystem Conservation in the Colorado River Delta," *Natural Resources Journal* 40 (2000); Evan R. Ward, *Border Oasis: Water and the Political Ecology of the Colorado River Delta, 1945–1975* (Tucson: University of Arizona Press, 2003); James O. Pattie, *Personal Narrative of James O. Pattie*, ed. Richard Batman (Missoula, MT: Mountain Press Publishing Co., 1988); Richard Batman, *James Pattie's West: The Dream and the Reality* (Norman: University of Oklahoma Press, 1986).

5. Environmental Defense Fund, *A Delta Once More*, 4; Ives, *Report Upon the Colorado*, 28; Edward Glenn et al., "Ecology and Conservation Biology of the Colorado River Delta, Mexico," *Journal of Arid Environments* 49 (2001); Edward P. Glenn et al., "Effects of Water Management on the Wetlands of the Colorado River Delta, Mexico," *Conservation Biology* 10, no. 4 (1996): 1175–1186.

6. Bridget Kellogg, "The Dam Controversy: Does the Endangered Species Act Apply Internationally to Protect Foreign Species Harmed by Dams on the Colorado River?" *Journal of Transnational Law and Policy* 13 (2004): 447; Miguel A. Cisneros-Mata et al., "Life History and Conservation of *Totoaba macdonaldi*," *Conservation Biology* 9 (1995): 806.

7. Centro de Estudios de Almejas Muertas (C.E.A.M.) Web site, http://www.geo.arizona.edu/ceam/ (accessed December 28, 2006).

8. Irwin Sliber and Fred Sliber, ed., *Folksinger's Wordbook* (New York: Oak Publications, 1973).

9. Michal Kowaleski et al., "Dead Delta's Former Productivity: Two Trillion Shells at the Mouth of the Colorado River," *Geology* 28, no. 12 (December 2000): 1059–1062.

10. "Treaty for the Utilization of Waters of the Colorado, Tijuana and Rio Grande Rivers," Feb. 13, 1944, U.S.-Mex., Art. 10, 59 Stat. 1219; Charles J. Meyers, "The Colorado River: The Treaty with Mexico," *Stanford Law Review* 19 (1967): 367; Ward, *Border Oasis*, xxvi.

11. Russell Martin, *A Story That Stands Like a Dam: Glen Canyon and the Struggle for the Soul of the West* (New York: Henry Holt, 1989), 30 (Roosevelt quote).

12. William Eno DeBuys, *Salt Dreams: Land and Water in Low-Down California* (Albuquerque: University of New Mexico Press, 1999); Ward, *Border Oasis*; Norris Hundley, *Water and the West: The Colorado River Compact and the Politics of Water in the American*

West (Berkeley: University of California Press, 1975); Fradkin, *A River No More;* Martin, *A Story That Stands Like a Dam;* Marc Reisner, *Cadillac Desert: The American West and Its Disappearing Water* (New York: Viking, 1986).

13. Ward, *Border Oasis*, 5–6.

14. Hundley, *Water and the West*, 94, 162–163, 191, 233; Ward, *Border Oasis*, 4–5.

15. 21 Opinions of the Attorney General 274 (1898); Meyers, "The Colorado River: The Treaty with Mexico," 370; Fradkin, *A River No More*, 298–299; Stephen C. McCaffery, "Water, Politics, and International Law," in *Water in Crisis: A Guide to the World's Fresh Water Resources*, ed. Peter H. Gleick (New York: Oxford University Press, 1993), 96, citing J. Moore, *A Digest of International Law* (Washington, D.C.: U.S. Government, 1906), 653–654 (quote).

16. Ward, *Border Oasis*, 16, 23–25; McCaffery, "Water, Politics, and International Law."

17. Hundley, *Water and the West*, 175–176, 204–205; Meyers, "The Colorado River: The Treaty with Mexico," 367–368.

18. Meyers, "The Colorado River: The Treaty with Mexico," 368, 379.

19. Ibid., 370.

20. Environmental Defense Fund, *A Delta Once More*, 31.

21. Robert W. Adler and Michele Straube, "Lessons From Large Watershed Programs: A Comparison of the Colorado River Basin Salinity Control Program with the San Francisco Bay-Delta Program, Central and South Florida, and the Chesapeake Bay Program," National Academy of Public Administration, Learning From Innovations in Environmental Protection, Research Paper Number 10 (2000).

22. Bureau of Reclamation, Quality of Water, *Colorado River Basin, Progress Report 22* (2005), 9; Donald Worster, *Rivers of Empire: Water, Aridity and the Growth of the American West* (New York: Oxford University Press, 1985), 34; Godfrey Sykes, *The Colorado Delta* (American Geographical Society, 1937), 68.

23. U.S. Environmental Protection Agency, *The Mineral Quality Problem in the Colorado River* (1971).

24. Loretta C. Lohman et al., U.S. Department of the Interior, Bureau of Reclamation, *Estimating Economic Impacts of Salinity of the Colorado River* (1988).

25. Dale Pontius, *Colorado River Basin Study* (Denver: Western Water Policy Review Advisory Commission, 1997); Joseph F. Freidkin, "The International Problem with Mexico Over Salinity of the Lower Colorado River," in *Water and the American West*, ed. David Getches (University of Colorado School of Law, 1988); Jose Trava, "Sharing Water with the Colossus of the North," in *Western Water Made Simple* (Washington, D.C.: Island Press, 1987); Ward, *Border Oasis*, 44; David H. Getches, "From Askhabad, to Wellton-Mohaw, to Los Angeles: The Drought in Water Policy," *University of Colorado Law Review* 64 (1993): 523.

26. International Boundary and Water Commission, Minutes No. 241 and 242 (July 14, 1972, and August 30, 1973); *U.S. Code* 3, §§ 1571 et seq.; Ward, *Border Oasis.*

27. Bureau of Reclamation, *Colorado River Basin, Progress Report 22*, 9; Adler and Straube, "Lessons From Large Watershed Programs"; U.S. General Accounting Office, Water Quality, *Information on Salinity Control Projects in the Colorado River Basin*, GAO/RCED-95-98 (1995); Richard L. Gardner and Robert A. Young, "Assessing Salinity-Control Programs on the Colorado River," *Resources* 80 (1985): 10; U.S. Department of the Interior, Bureau of Land Management, *Salinity Control on BLM-Administered Public Lands in the Colorado River Basin, A Report to Congress* (1987); U.S. Department of the Interior, Bureau of Land Management, *BLM Salinity Reduction Efforts 1996–1998, Audit Report 93-I-810, Implementation of the Colorado River Basin Salinity Control Program* (March 1993).

28. *Environmental Defense Fund v. Costle*, 657 F.2d 275 (D.C. Cir. 1981).

29. Bureau of Reclamation, Quality of Water, *Colorado River Basin, Progress Report 20* (2005), 29–32.

30. U.S. Bureau of Reclamation, *Report on Public and Agency Review of the Salinity Control Program and Suggested Revisions to the Program* (1994); U.S. Department of the Interior, *Audit Report 93-I-258, Operation and Maintenance Contracts, Colorado River Basin Salinity Control Program* (1992); U.S. Department of the Interior, *Audit Report 93-I-810, Implementation of the Colorado River Basin Salinity Control Program* (1993); U.S. Department of the Interior, Quality of Water, *Colorado River Basin Progress Report 22* (2005); U.S. Bureau of Reclamation, *Colorado River Basin Salinity Control Program, Report to Congress on the Bureau of Reclamation Basinwide Program* (1996), 14.

31. Colin Fletcher, *River: One Man's Journey Down the Colorado, Source to Sea* (New York: Knopf, 1997).

32. Leopold, *A Sand County Almanac*, 150–151; Sykes, *The Colorado Delta*, 48–49, 65.

33. Environmental Defense Fund, *A Delta Once More*, 15–16, 30; Glenn et al., "Ecology and Conservation Biology of the Colorado River Delta," 10.

34. National Environmental Policy Act, *U.S. Code* 42, §§ 4331(a), (b)(2), 4332(2)(C), (F) (emphasis added).

35. *Lujan v. Defenders of Wildlife*, 504 U.S. 555 (1992).

36. *Defenders of Wildlife et al. v. Norton*, 257 F.Supp.2d 53 (D.D.C. 2003).

37. Salton Sea Authority, *Salton Sea Restoration, Final Preferred Project Report* (July 2004); U.S. Department of the Interior, Bureau of Reclamation, *Salton Sea Study Status Report* (January 2003); Pacific Institute, *A Proposal to Preserve and Enhance Habitat at the Salton Sea* (October 2001); EIS/EIR Draft, *Salton Sea Restoration;* Pacific Institute, *Haven or Hazard: The Ecology and Future of the Salton Sea* (February 1999); DeBuys, *Salt Dreams;* Fradkin, *A River No More*.

38. Milton Friend, "Avian Disease at the Salton Sea," *Developments in Hydrobiology: The Salton Sea* (2002); Kristin M. Reifel et al., "Possible Importance of Algal Toxins in the Salton Sea, California," *Developments in Hydrobiology: The Salton Sea* (2002); Joseph R. Jehl, Jr., and Robert L. McKernan, "Biology and Migration of Eared Grebes at the Salton Sea," in *Developments in Hydrobiology: The Salton Sea* (2002).

39. Brent M. Hadad, *Rivers of Gold: Designing Markets to Allocate Water in California* (Washington, D.C.: Island Press, 2000); Salton Sea Authority, *Salton Sea Restoration, Final Preferred Project Report*.

40. Salton Sea Restoration Act, § 101(b)(1)(A).

41. Salton Sea Authority, *Salton Sea Restoration, Final Preferred Project Report;* Robert W. Adler, "Toward Comprehensive Watershed-Based Restoration and Protection for Great Salt Lake," *Utah Law Review* (1999): 121–125.

42. David H. Getches, "Colorado River Governance: Sharing Federal Authority as an Incentive to Create a New Institution," *University of Colorado Law Review* 68 (1997): 537.

43. Glen Canyon Action Network et al., "Utah Environmental Leaders Launch Four-State, Ten Day Roadshow for River Restoration," Press Release (March 2, 2001).

44. Robert Jerome Glennon and Peter W. Culp, "The Last Green Lagoon: How and Why the Bush Administration Should Save the Colorado River Delta," *Ecology Law Quarterly* 28 (2002): 903.

45. Glenn et al., "Ecology and Conservation Biology of the Colorado River Delta," 1176; Environmental Defense Fund, *A Delta Once More*, 20, 42; Pitt et al., "Two Nations, One River."

46. Environmental Defense Fund, *A Delta Once More*, 24.

47. Lower Colorado River Multi-Species Conservation Program, *Lower Colorado River Multi-Species Conservation Program, Volume II: Habitat Conservation Plan: Final* (Sacramento, CA: J&S 00450.00., 2004).

48. Glenn et al., "Ecology and Conservation Biology of the Colorado River Delta," 11.

49. Ibid., 13.

50. Pitt et al., "Two Nations, One River," 848–849; Karl Flessa, personal communication.

51. Robert W. Adler et al., *The Clean Water Act 20 Years Later* (Washington, D.C.: Island Press, 1993), 221–222.

52. Rachel Carson, in *The Norton Book of Nature Writing*, ed. Robert Finch and John Elder (New York: W.W. Norton & Co., 2002), 480 (emphasis added).

Chapter 9

1. Glen Canyon Institute, *Citizens' Environmental Assessment (CEA) on the Decommissioning of Glen Canyon Dam, Report on Initial Studies* (December 2000).
2. S. Rosenkrans, *The Effect of Draining Lake Powell on Water Supply and Electricity Production* (Environmental Defense Fund, 1997); Thomas Myers, *Water Balance of Lake Powell: An Assessment of Groundwater Seepage and Evaporation* (prepared for Glen Canyon Institute, 1999); Thomas Myers, *Sediment Hydrology on the Colorado River: The Impacts of Draining Lake Powell* (prepared for Glen Canyon Institute, 1998); Anders Beck, *Salinity in the Colorado River Basin: Past, Present and Future* (prepared for Glen Canyon Institute, 1999); Kathleen Webb, *Bioinventory of Glen Canyon Prior to Inundation by the Lake Powell Reservoir* (prepared for Glen Canyon Institute, 2000); Ross Mulholland, *Endangered and Threatened Species in the Lower Colorado River Basin, Delta and Sea of Cortez* (prepared for Glen Canyon Institute, 1999); Christina Rinderle and David Wegner, *Glen Canyon Species Status Report: Amphibians and Reptiles, Historic Summary* (prepared for Glen Canyon Institute, 2000).
3. Henry J. Pollack, *Uncertain Science–Uncertain World* (New York: Cambridge University Press, 2003).
4. Colin Fletcher, *The Man Who Walked Through Time* (New York: Vintage Books, 1989), 109 (quote); Colin Fletcher, *River: One Man's Journey Down the Colorado, Source to Sea* (New York: Knopf, distributed by Random House Inc., 1997).
5. Peter B. Moyle and Georgina M. Sato, "On the Design of Preserves to Protect Native Fishes," in *Battle Against Extinction: Native Fish Management in the American West*, ed. W.L. Minckley and James E. Deacon (Tucson: University of Arizona Press, 1991), 161; R.D. Ohmart et al., *The Ecology of the Lower Colorado River from Davis Dam to the Mexico-United States International Boundary: A Community Profile*, U.S. Fish and Wildlife Service Report 85 (7.10) (1988), 210–217; Society for Ecological Restoration International, Science and Policy Working Group, *The SER International Primer on Ecological Restoration* (Tucson: Society for Ecological Restoration International, 2004), 3, 6–7; Brian Walker and David Salt, *Resilience Thinking: Sustaining Ecosystems and People in a Changing World* (Washington, D.C.: Island Press, 2006).
6. Society for Ecological Restoration International, *Primer on Ecological Restoration*, 2.
7. Marc Reisner, *Cadillac Desert: The American West and Its Disappearing Water* (New York: Viking, 1986); Philip L. Fradkin, *A River No More: The Colorado River and the West* (Berkeley: University of California Press, 1995).
8. Margaret Bushman LaBianca, "The Arizona Water Bank and the Law of the River," *Arizona Law Review* 40 (1998): 659, 673–674.
9. Norris Hundley, *Water and the West: The Colorado River Compact and the Politics of Water in the American West* (Berkeley: University of California Press, 1975); Charles J. Meyers, "The Colorado River: The Treaty with Mexico," *Stanford Law Review* 19 (1967).
10. Kenneth Strzepek and Davis N. Yates, "Assessing the Effects of Climate Change on the Water Resources of the Western United States," *Water and Climate in the Western United States*, ed. William M. Lewis, Jr. (Boulder: University Press of Colorado, 2003); Gregory Hobbs, "The Role of Climate Change in Shaping Western Water Institutions," *University of Denver Water Law Review* 7 (2003); Niklas S. Christensen et al., "The Effects of Climate Change on the Hydrology and Water Resources of the Colorado River Basin," *Climatic Change* 62 (2004): 337–363.
11. David Lavender, *River Runners of the Grand Canyon* (Grand Canyon, AZ: Grand Canyon Natural History Association, 1986), 14, 35–36.

12. Stephen F. Williams, "The Requirement of Beneficial Use as a Cause of Waste in Water Resource Development," *Natural Resources Journal* 23 (1983): 7; Hundley, *Water and the West;* Peter H. Gleick, "Global Freshwater Resources: Soft-Path Solutions for the 21st Century," *Science* 3002 (2003): 1524–1528.

13. Gregory A. Thomas, "Conserving Aquatic Biodiversity: A Critical Comparison of Legal Tools for Augmenting Streamflows in California," *Stanford Environmental Law Journal* 15 (1996): 3; Harrison C. Dunning, "Revolution (and Counter-Revolution) in Western Water Law: Reclaiming the Public Character of Water Resources," *Fordham Environmental Law Journal* 8 (1997): 439.

14. O.K. Buros, 2 ABC's of Desalting (Int'l Desalination Association, undated); James E. Miller, *Review of Water Resources and Desalination Technologies* (Sandia National Laboratories, 2003); *Desalination and Water Technology Roadmap: Desalination & Water Purification Research & Development Program Report #95,* Sandia National Laboratory and Bureau of Reclamation (2003).

15. Comprehensive Everglades Restoration Plan, Aquifer Storage and Recovery Program (2001); Doug McChesny, Washington State Department of Ecology, *2001 Report to the Legislature: Artificial Storage and Recovery of Ground Water* (2001); R. David Pyne, *Groundwater Recharge and Wells: A Guide to Aquifer Storage Recovery* (Boca Raton, FL: Lewis Publishers, 1994).

16. Bureau of Reclamation, *1996–2000 Consumptive Uses Report, Provision Consumptive Uses and Losses Report 2001–2005;* U.S. Department of the Interior, Bureau of Reclamation, *The Colorado River: A Natural Menace Becomes a Natural Resource,* Project Planning Report No. 34-8-1, The Colorado River (1945); E.C. LaRue, *Water Power and Flood Control of Colorado River Below Green River, Utah,* U.S. Geological Survey, Water-Supply Paper 556 (1925); Hundley, *Water and the West,* 15.

17. U.S. Bureau of Reclamation, *Colorado River System Consumptive Uses and Losses Report 1996–2000* (2004); Jason I. Morrison et al., *The Sustainable Use of Water in the Lower Colorado River Basin* (Pacific Institute and Global Water Policy Project, 1996); Terry R. Anderson and Donald R. Leal, *Free Market Environmentalism* (New York: Palgrave, 2001), 91–94; Reisner, *Cadillac Desert.*

18. Jason I. Morrison, *The Sustainable Use of Water in the Lower Colorado River Basin* (A Joint Report of the Pacific Institute and the Global Water Policy Project, November 1996); Gleick et al., *Waste Not, Want Not: The Potential for Urban Water Conservation in California* (Pacific Institute for Studies in Development, Environment, and Security, 2003); U.S. Bureau of Reclamation, Lower Colorado Region, *Water Conservation Report 1996–2001; Western Resource Advocates, Smart Water, A Comparative Study of Urban Water Use Efficiency across the Southwest* (2003).

19. Philip L. Fradkin, *A River No More,* 177; Edward Abbey, *The Monkey Wrench Gang* (Philadelphia: Lippincott, 1975), 174 (quote).

20. Zane Grey, *The Rainbow Trail* (New York: Grosset and Dunlap, 1915); Katie Lee, *All My Rivers Are Gone: A Journey of Discovery Through Glen Canyon* (Boulder: Johnson Books, 1998); Russell Martin, *A Story That Stands Like a Dam: Glen Canyon and the Struggle for the Soul of the West* (New York: Henry Holt, 1989); Fradkin, *A River No More.*

21. National Park Service, *2005 Glen Canyon National Recreation Area Profile;* Barry Wirth, "Glen Canyon Dam," *Lake Powell Magazine* (Summer 2001): 8; *Lake Powell Vacations,* http://www.lakepowellvacations.com/houseboats/rental.shtml (last accessed December 28, 2006).

22. National Park Service, *Final Environmental Impact Statement, Personal Watercraft Rulemaking, Glen Canyon National Recreation Area* (2003); National Park Service, Glen Canyon National Recreation Area, *Lake Powell Pure Now & Forever,* http://nps.gov/archive/glca/lpp.htm (accessed December 28, 2006).

23. *U.S. Code* 43, §§ 620, 620(g); Martin, *A Story That Stands Like a Dam,* 136–137.

24. Lee, *All My Rivers Are Gone*, 36, 86–89, 102, 128.
25. Western Area Power Administration, Salt Lake City Area/Integrated Projects (2005).
26. *U.S. Code* 43, § 602 (emphasis added).
27. U.S. Department of Energy, Energy Information Administration, *Electric Power Annual 2004* (2005).
28. American Wind Energy Association, *Wind Power Outlook 2005*; Western Area Power Administration, *Annual Report 2004* (2005); U.S. Department of Energy, *Energy Efficiency and Renewable Energy, Solar Energy Technologies Program*, http://www1.eere.energy.gov/solar (accessed May 5, 2006).
29. Joel B. Stronberg, *Common Sense: Making the Transition to a Sustainable Energy Economy* (American Solar Energy Society, 2005).
30. Matthew Daly, "Study Backs Breaching Washington State Snake River Dams," *Salt Lake Tribune*, September 5, 2002; The Aspen Institute, *Dam Removal: A New Option for a New Century* (Queenstown, MD: The Aspen Institute, 2002); World Commission on Dams, *Dams and Development, A New Framework for Decisionmaking* (London: Earthscan Publications, 2000); Emily H. Stanley and Martin W. Doyle, "Trading Off: The Ecological Effects of Dam Removal," *Frontiers in Ecology and the Environment* 1, no.1 (2003): 15–22.
31. John C. Schmidt, personal communication; Steven W. Carothers and Dorothy A. House, "Decommissioning Glen Canyon Dam: The Key to Colorado River Ecosystem and Recovery of Endangered Species?" *Arizona Law Review* 42 (2000): 215, 237–238.
32. Henry Beston, *The Outermost House: A Year on the Great Beach of Cape Cod* (Doubleday & Co., Inc., 1929).
33. Daniel Glick (photographs by Michael Melford), "A Dry Red Season, Drought Drains Lake Powell: Uncovering the Glory of Glen Canyon," *National Geographic*, April 2006, 64–81; Elliot Porter and David Brower, ed., *The Place No One Knew: Glen Canyon on the Colorado* (San Francisco: Sierra Club, 1963), 14.
34. Lee, *All My Rivers Are Gone*, 111; Martin, *A Story That Stands Like a Dam*, 176–178.
35. Wallace Stegner, *Mormon Country* (New York: Hawthorn Books, 1970); Wallace Stegner, *Beyond the 100th Meridian* (Lincoln: University of Nebraska Press, 1982); Wallace Stegner, "Glen Canyon Submersus," in *The Norton Book of Nature Writing*, ed. Robert Finch and John Elder (New York: W.W. Norton & Co., 2002), 509 (emphasis added).
36. Quoted in *The Place No One Knew*, ed. Porter and Brower, 120.
37. Stegner, "Glen Canyon Submersus," in *The Norton Book of Nature Writing*, ed. Finch and Elder, 506; Lee, *All My Rivers Are Gone*, 208–209.
38. Steven Carothers and Bryan T. Brown, *The Colorado River Through Grand Canyon* (Tucson: University of Arizona Press, 1991).

Coda

1. Clarence E. Dutton, *Tertiary History of the Grand Cañon District* (University of Arizona Press, 2001), 149–50 (emphasis added), first published in 1882 by the U.S. Geological Survey.
2. Cid Rickets Sumner, quoted in *The Place No One Knew: Glen Canyon on the Colorado*, ed. Elliot Porter and David Brower (San Francisco: Sierra Club, 1963), 80.
3. Ann Zwinger, *Down Canyon* (Tucson: University of Arizona Press, 1995), 102 (emphasis added).

Index

About the Author

Robert W. (Bob) Adler is the Associate Dean for Academic Affairs and the James I. Farr Chair and Professor of Law at the University of Utah, S. J. Quinney College of Law, where he has been teaching since 1994. Before becoming a law professor, Adler worked as an attorney for the Pennsylvania Department of Environmental Resources, the Trustees for Alaska, and the Natural Resources Defense Council. He has a J.D. from the Georgetown University Law Center, and a B.A. in ecology from Johns Hopkins University. He is the coauthor of a previous book by Island Press, *The Clean Water Act: Twenty Years Later*, and numerous other books, book chapters, and articles about environmental law and policy.